Structure and Dynamics of Molecular Systems

Structure and Dynamics of Molecular Systems

Edited by

R. Daudel

J.-P. Korb

J.-P. Lemaistre

and

J. Maruani

Centre de Mécanique Ondulatoire Appliquée,
CNRS, Paris, France

D. REIDEL PUBLISHING COMPANY

A MEMBER OF THE KLUWER ACADEMIC PUBLISHERS GROUP

DORDRECHT / BOSTON / LANCASTER

Library of Congress Cataloging in Publication Data

CIP

Main entry under title:

Structure and dynamics of molecular systems.

"Invited lectures given at the international seminar of the same title held at the Centre de Mécanique Ondulatoire Appliquée du Centre National de la Recherche Scientifique in Paris (France) from October 1983 to May 1985,"–Pref.

Includes index.

1. Molecular structure–Congresses. 2. Molecular dynamics–Congresses. I. Daudel, Raymond.

QD461.S924 1985 541.2'2 85–2010

ISBN 90–277–1977–2

Published by D. Reidel Publishing Company,
P.O. Box 17, 3300 AA Dordrecht, Holland.

Sold and distributed in the U.S.A. and Canada
by Kluwer Academic Publishers,
190 Old Derby Street, Hingham, MA 02043, U.S.A.

In all other countries, sold and distributed
by Kluwer Academic Publishers Group,
P.O. Box 322, 3300 AH Dordrecht, Holland.

Printed in The Netherlands

TABLE OF CONTENTS

Preface vii

M. SANA / From Theoretical Chemistry to Molecular Shape and Reactive Transformation 1

Y. G. SMEYERS and A. HERNÁNDEZ-LAGUNA / Theoretical Determination of Nuclear Probability Density Distributions: Two-Rotor Molecules 23

P. G. MEZEY / Topological Theory of Molecular Conformations 41

P. G. MEZEY / Topological Model of Reaction Mechanisms 57

A. BLUMEN, G. ZUMOFEN and J. KLAFTER / Reactions in Disordered Media 71

G. ZUMOFEN, A. BLUMEN and J. KLAFTER / Random Walks on Fractals 87

J.-P. LEMAISTRE and A. BLUMEN / Localization of Excitonic Eigenstates in Molecular Crystals 101

PH. PÉE, R. BROWN, F. DUPUY, Y. REBIÈRE, PH. KOTTIS and J.-P. LEMAISTRE / Fusion and Transfer of Triplet Excitons in Isotopically Mixed Naphthalene Crystals 119

J.-P. LEMAISTRE and J.-P. KORB / Concentration Dependences of Homogeneous Broadenings and Cross-Relaxation Rates in Condensed Media: The Role of Pair Distributions 137

R. BOSCAINO, F. M. GELARDI and R. N. MANTEGNA / The Effect of Spectral Diffusion on the Saturation Transient Regime 149

M. SEIGNEURET, J. DAVOUST and P. F. DEVAUX / Simulation of the Electron Spin Resonance Spectra of Spin-Labeled Fatty Acids in Protein-Containing Model Membranes 171

C. CHACHATY, J.-P. QUAEGEBEUR and B. PERLY / Nuclear Magnetic Resonance and Molecular Motions in Lyotropic Liquid Crystals 187

L. TRAHMS / NMR Studies of the Gel Phase of Lecithins and Cephalins 203

J. TABONY / Interpretation of Proton N.M.R. Spin-Lattice Relaxation Time Minima in Heterogeneous Systems by the Effects of Bounded Two-Dimensional Diffusion. Comparison with Neutron Scattering Measurements 225

J.-P. KORB / Theory of Spin Relaxation by Translational Diffusion in Two-Dimensional Systems 245

INDEX OF SUBJECTS 269

INDEX OF NAMES 275

PREFACE

This volume is the first of a set of two which contain the invited lectures given at the international seminar of the same title held at the Centre de Mécanique Ondulatoire Appliquée du Centre National de la Recherche Scientifique in Paris (France) from October 1983 to May 1985. They are intended to provide a survey of topics of current interest relative to the structure and the dynamics of molecular systems. The papers have been selected on the basis of their relevance to the following four topics: i) molecular conformations and transformations; ii) molecular relaxation and motion; iii) charge, spin and momentum distributions in molecular solids; iv) collective phenomena in condensed matter. The first volume deals mostly with the first two topics, the second volume mostly with the last two. Each volume consists of about fifteen self-contained, reference contributions covering recent achievements in active branches of molecular physics and physical chemistry.

The first four papers of the present volume deal with theoretical aspects of structure and reactivity problems, with particular attention being paid to topology considerations, which have joined symmetry considerations as an important tool in approaching chemistry problems. The treatment of nuclear probability density distributions is performed on a model basis for a simple system, even though it has come to the attention of theoreticians through experimental results for complex systems. The following six papers are relevant to particular phenomena occurring in disordered solids as well as in mixed crystals: spatial or spectral localization of electronic or magnetic excitation, fractal structures, random walks, cross relaxation, triplet annihilation as well as chemical reactions. The last five papers deal with the structural and dynamical information one can derive from magnetic resonance and relaxation techniques for such systems as model membranes, liquid crystals, gel phases, micelles and interfaces. Many of these papers contain an abundant bibliography which should help the interested reader go deeper into the subject.

The editors wish to express their gratitude to the Direction de la Chimie of the C.N.R.S. for lending financial support to the international seminar.

THE EDITORS

FROM THEORETICAL CHEMISTRY TO MOLECULAR SHAPE AND REACTIVE TRANSFORMATION

M. Sana*
Laboratoire de Chimie Quantique
Bâtiment Lavoisier
Place Louis Pasteur, 1
B-1348 Louvain-la-Neuve (Belgium)

ABSTRACT. In the present work we will try to show how theoretical chemistry can be a tool for analysing some chemical events such as the recognition and the classification of the molecular forms, or such as the reactive transformation of the molecules.

Starting with the Born-Oppenheimer approximation, we can separate the electronic motion from the nuclear one. This approximation gives rise to a useful concept: the potential energy (hyper)surface, on which the nuclei are moving. Then, with the help of Thom's concepts, we can try to analyse the potential surface and partition it into molecular regions. After having considered the eigenvalues of the electronic equation, we can go further and, with the help of the theory of graphs, define an electronic picture for any molecular structures, including those which appear along the reaction pathway: we thus obtain the so-called electronic reaction mechanism.

1. FROM THE POTENTIAL ENERGY SURFACE (PES) TO THE MOLECULAR SHAPE

1.1. PES and Born Oppenheimer approximation

The time independent Schrödinger equation for a molecular system which contains N nuclei and n electrons is :

$$\mathcal{H}(r,R)\,Y(r,R) = WY(r,R) \tag{1}$$

where the total hamiltonian $\mathcal{H}$ stands for :

$$\mathcal{H} = -\frac{1}{2}\sum_a^N \frac{\Delta_a}{m_a} - \frac{1}{2}\sum_i^n \Delta_i - \sum_{a,i}^{N,n} \frac{Z_a}{|r_i - R_a|} + \frac{1}{2}\sum_{i\neq j}^{n,n} \frac{1}{|r_i - r_j|} + \frac{1}{2}\sum_{a\neq b}^{N,N} \frac{Z_a Z_b}{|R_a - R_b|}$$

Then, it contains kinetic (T) and potential (V) operators :

R. Daudel et al. (eds.), Structure and Dynamics of Molecular Systems, 1–22.

$$\mathcal{H} = \underbrace{T_N(R)}_{\mathcal{H}^{(1)}} + \underbrace{T_E(r) + V(r,R)}_{\mathcal{H}^{(0)}}$$

As the kinetic operator relative to the nuclei (T_N) is proportional to the inverse of the nucleus mass (m_a, $\forall$ $a = 1, N$) which is larger than the electron mass (m_e), we can consider T_N as a perturbation with respect to the remaining part of $\mathcal{H}$. Using the Rayleigh-Schrödinger perturbation theory, we find a zeroth order wave-equation which does not contain any more the T_N term(1) :

$$\{T_E(r) + V(r|R)\} \Psi(r|R) = E(|R) \Psi(r|R) \tag{2}$$

This is the so-called electronic equation; it describes the electron motion in the fixed nucleus field. Their eigenvectors ($\Psi(r|R)$) as well as their eigenvalues ($E(|R)$) are defined parametrically to the nuclear coordinates (R). At the first order of the perturbational treatment, we find the total wavefunction ($Y(r,R)$). Let us expand Y in the basis of the eigenvectors of (2). Then, if the following approximations may be assumed (the Born-Oppenheimer approximations) :

$$Y(r,R) = \sum_\mu \chi_\mu(R) \Psi_\mu(r,R) \simeq \chi_\mu(R) \Psi_\mu(r|R) \tag{3a}$$

$$< \Psi_\mu | \nabla_R | \Psi_\nu >_r \simeq 0 \tag{3b}$$

the first order wave-equation becomes :

$$\{T_N(R) + E_\mu(R)\} \chi_\mu(R) = W_\mu \chi_\mu(R) \tag{4}$$

This is the so-called nuclear equation, which describes the motion of the nuclei in the potential energy field E(R). We conclude that, as long as we can put the relations (3) into practice, we can describe separately the motion of the electrons (eq.(2)) from this of the nuclei (eq. (4)); moreover, the eigenvalues of (2) must be considered as a potential energy hypersurface for the nuclear motion; this last concept comes from the B.O. approximation.

The potential energy surface is a function of the 3N cartesian coordinates for the N nuclei :

$$E = E(\vec{R}) = E(\vec{R}_1,\ldots,\vec{R}_N) = E(X_1Y_1Z_1,\ldots,X_NY_NZ_N) \tag{5}$$

Nevertheless, it remains invariant under global rotation and translation for the entire system. We further define a new coordinate frame (s), which only contains 3N-6 independent variables :

$$E = E(\vec{s}) = E(s_1,\ldots,s_k) \qquad \text{with} \quad k = 3N-6 \tag{6}$$

s is the so-called internal coordinate system. We can also define an other frame, the internal displacement system S, related to the previous one by :

$$\vec{S} = \vec{s} - \vec{s}_o = \{ S_1, \ldots, S_k \} \tag{7}$$

This last frame is of any use to write a local analytical form for the potential surface. So, let us expand the potential in a Taylor series:

$$E(\vec{S}) = E_o + \vec{g}_o' \vec{S} + \frac{1}{2} \vec{S}' \tilde{H}_o \vec{S} + \ldots \tag{8}$$

where $E_o = E(S = o) = E(s_o)$

$\vec{g}_o = (\vec{\nabla}_S E(S))_{S=o}$: the potential energy gradient

$\tilde{H}_o = (\vec{\nabla}_S' \vec{\nabla}_S E(S))_{S=o}$: the force constant matrix

In solving the equation (2) for s_o, we obtain E_o :

$$E_o = \frac{\langle \Psi(r|R_o) \mid \mathcal{H}(r|R_o)^{(o)} \mid \Psi(r|R_o) \rangle_r}{\langle \Psi(r|R_o) \mid \Psi(r|R_o) \rangle_r} \tag{9}$$

By differentiating this last relation with respect to s (at $s = s_o$), we find analytical estimates of g_o and H_o(1-6). However, this is not always feasible, specially beyond the SCF level; in such cases g_o and H_o are only available from numerical approach(7-9).

1.2. PES and stationary points

The relation (8) is of great interest because it enables one to locate stationary points and to know the local surface curvature. A stationary point is a point of the potential energy surface where the gradient vector $g_\blacksquare$ vanishes :

If $\vec{g}_\blacksquare = \vec{o}$

$$\text{then } \vec{S}_\blacksquare = - \tilde{H}_o^{-1} \vec{g}_o \tag{10a}$$

$$\text{and } \vec{s}_\blacksquare = \vec{s}_o - \tilde{H}_o^{-1} \vec{g}_o \tag{10b}$$

If we translate the origin of the Taylor expansion at the stationary point and rotate the internal coordinate frame in such a way as to diagonalise the force constant matrix, the potential function takes the simple form :

if $\tilde{U}' \tilde{H}_o \tilde{U} = \tilde{h}$ (diagonal) with $\tilde{U}' \tilde{U} = \tilde{1}$

and $\vec{V} = \tilde{U}' (\vec{S} - \vec{S}_\blacksquare)$

$$E(\vec{V}) = E_\blacksquare + \frac{1}{2} \vec{V}' \tilde{h} \vec{V} \tag{11a}$$

or $$\Delta\vec{E}(V)_{\blacksquare} = E(\vec{V}) - E_{\blacksquare} = \frac{1}{2}\sum_{i=1}^{k} V_i^2\, h_{ii} \qquad (11b)$$

That means the energy variation, which occurs when we are leaving the stationary point, depends on the diagonal h_{ii} terms. So the current stationary point corresponds to an absolute or an asymptotic maximum ($\Delta E(\vec{V}) \geqslant 0$, $\forall\vec{V}$) if h is a positive or semi-positive defined matrix[10] :

$$g_i = 0 \quad \text{and} \quad h_{ii} \geqslant 0 \qquad \forall i = 1, k \qquad (12)$$

The corresponding nuclear structure is a stable equilibrium structure (see Figure 1).

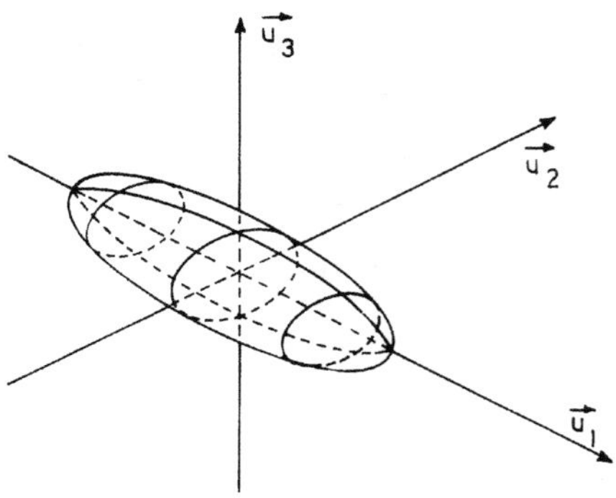

Figure 1. Potential isoenergy surface around a stationary point corresponding to an equilibrium structure in a 3-dimension internal coordinate space.

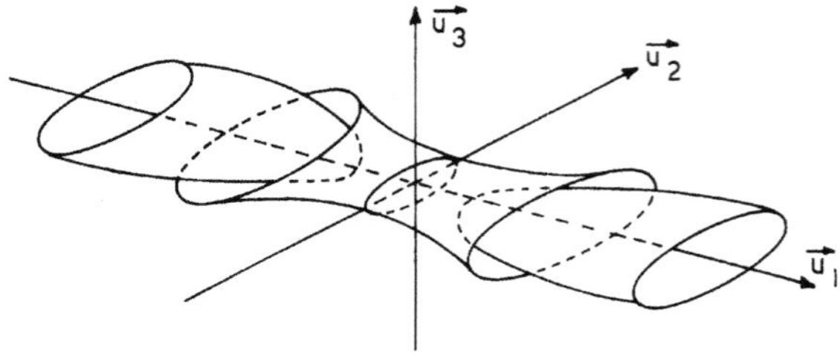

Figure 2. Potential isoenergy surfaces around a stationary point of first order minimax type in a 3-dimension internal coordinate space.

Another critical point is of particular chemical interest: the first order minimax (see Figure 2), which can be defined in the following way :

$$g_1 = o,\ h_{11} < o,\ g_i = o \quad \text{and} \quad h_{ii} > o \qquad \forall i = 2, k \tag{13}$$

Murrell and Laidler[11] showed that, as long as a potential barrier separates the reactants from the products, a first order minimax (also called saddle point) is the highest energy point lying on the lowest energy path between both reactants and products. As shown on Figure 3, if this highest energy point is a minimax of order greater then one, it exists a path "A" of lower energy than the path "B", path "A" which does not cross such a minimax.

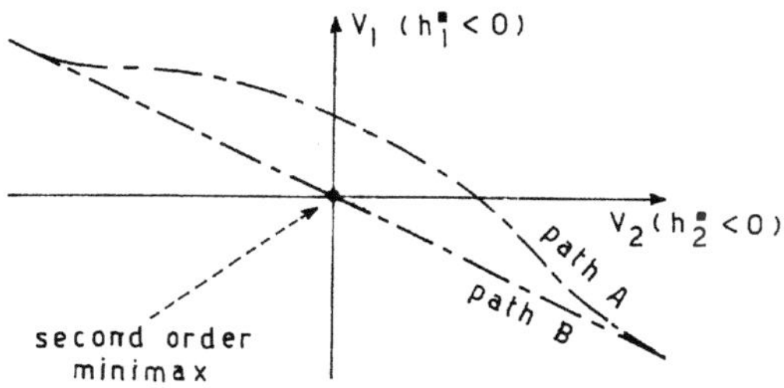

Figure 3. The first order minimax is the highest energy point on the lowest energy path which connects reactants and products as long as a potential barrier separates both.

1.3. PES and reaction pathway

Some define the reaction path as the lowest energy path connecting reactants and products on the potential surface[12] or alternately as the path of highest slope starting from the highest energy point of the previous lowest energy path[13]. Those definitions are incomplete[14-17]. We can prove this statement as follow: let us suppose a potential energy surface (E) described in two coordinate frames (S and V) related to each other by the local relation :

$$\vec{S} = \tilde{U}\vec{V} \tag{14}$$

On the one hand, the first order derivatives of E in both frames are related using the chain rule by :

$$\vec{g}(V) = \tilde{U}'\,\vec{g}(S) \tag{15}$$

On the other hand, as $\vec{g}(S)$ is a vector, its direction on the E(V) surface will be given by :

$$\vec{d}(V) = \tilde{U}^{-1}\,\vec{g}(S) \tag{16}$$

From (15) and (16), we find that g(V) and d(V) are only identical if :

$$\vec{g}(S) = \vec{o} \qquad \text{for any U} \tag{17a}$$

or $\tilde{U}' = \tilde{U}^{-1}$ for any g(S) (17b)

Equation (17a) tells that the stationary points location does not depend on the coordinate frame in use, this is an intrinsic property. In other respects, (17b) shows the non-intrinsic character of the paths as long a U is not a unitary transformation matrix. The Figure 4 illustrates this point. The steepest descent paths which connect the reactants and the products on E(S) and E(V) surfaces are only confused at the stationary points.

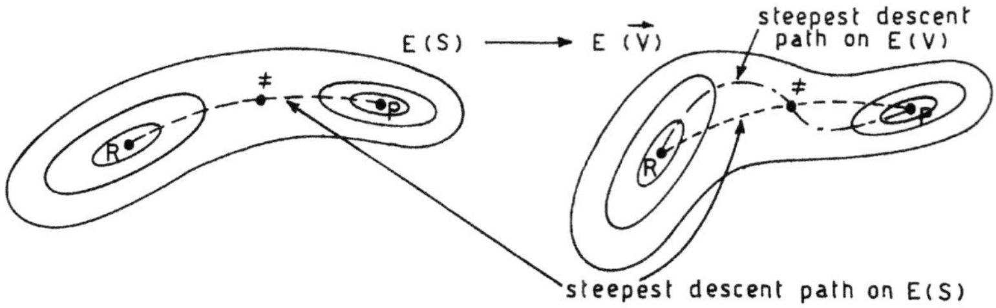

Figure 4. Steepest descent reaction pathway on a potential energy surface expressed in S or V coordinate frames.

Some years ago Fukui et al.[15] have introduced the concept of intrinsic reaction pathway. This concept has been expressed more recently in terms of internal coordinate frame[17]. We know that the classical equations of Lagrange describe the motion independently with respect to the coordinate frame in use. So we have :

$$\frac{d}{dt}\frac{\partial T}{\partial \dot{S}} + \frac{\partial E}{\partial S} = 0 \qquad (18)$$

with $E = E_o + \vec{g}_o{}'\vec{S} + \frac{1}{2}\vec{S}'\tilde{H}_o\vec{S}$ (19a)

and $T = \frac{1}{2}\dot{\vec{\xi}}'\tilde{M}\dot{\vec{\xi}} = \frac{1}{2}\dot{\vec{S}}'\tilde{G}^{-1}\dot{\vec{S}}$ (19b)

if $\tilde{G} = \tilde{B}\tilde{M}^{-1}\tilde{B}'$ (the G matrix of Wilson[18])

and $\dot{\vec{S}} = \tilde{B}\dot{\vec{\xi}} = \tilde{B}(\dot{\vec{R}} - \vec{R}_o)$

Then, the motion equations become :

$$\ddot{\vec{S}} + \tilde{G}\vec{g}_o + \tilde{G}\tilde{H}_o\vec{S} = \vec{o} \qquad (20)$$

Those second order differential equations can be uncoupled by using a coordinate frame which diagonalizes the product GH; the relation (20) becomes now :

$$\ddot{\vec{Q}} + \vec{g}(Q) + \tilde{\Lambda}\vec{Q} = \vec{o} \tag{21}$$

where $\vec{S} = \tilde{L}\vec{Q}$

if $\tilde{L}^{-1}\tilde{G}\tilde{H}\tilde{L} = \tilde{\Lambda}$ (diagonal)

The Q coordinates are nothing else but the normal local coordinates for the chemical system of interest. So the relation (21) suggests a new general definition for the intrinsic reaction pathway: The intrinsic reaction path is the steepest descent path everywhere tangent to the acceleration vector. Nevertheless, this definition does not give to the concept of reaction path any dynamical meaning. It only remains a reference curve on the potential surface which corresponds to the least change of energy to go from reactants to products. This is a trajectory with zero speed; we only hoop that real trajectories (of low energy) remain close to this particular curve.

1.4. PES and molecular shape

Having considered particular points and curves on the potential energy surface, we further look for the partition of the internal coordinate space into some regions. We request that any of those regions may be regarded as the existing space for a chemical entity to which one can give a particular name. We perform this work by applying the Thom's technics[19]. Let us define the potential surface in terms of normal local coordinate frame Q. Moreover, we choose a vector field on this surface, anywhere defined by :

$$E(Q) = E_o + \vec{G}'\vec{Q} + \frac{1}{2}\vec{Q}'\tilde{\Lambda}\vec{Q}$$

$$\vec{F} = -(\vec{\nabla}_Q E)_{Q=o} = -\vec{g} \tag{22}$$

They are nothing else but the internal force vectors. They generate force paths which converge to the absolute minima of the potential energy surface. Those stationary points are ω-limits or attractors for the force field. So, any points in the internal coordinate space having a particular minimum as ω-limit, through the field F, can be considered as lying in the basin associated to this minimum. On Figure 5, we illustrate the partition of the internal space into basins; so any nuclear structures of DHF chemical supersystem can be associated to one of the DH + F or D + HF attractors. That means that, in a classical experiment with kinetic energy dissipation, the nuclear structures will evolute to one of those stationary points after an infinite time. The concept of reaction pathway becomes now the curve formed by two Force pathes having both one minimax of first order as α-limit and as ω-limit, two different absolute minima lying in two basins next to each other.

The present partition[20] is not able to recognise unstable structures such as maxima on the potential surface (i.e. some particular dissociated structures). To be more discriminant in the internal space partition, we could replace the force field F by the gradient field of the force norm :

$$\vec{\Gamma} = - \{\vec{\nabla}_Q(\vec{F}'(Q)\,\vec{F}(Q))^{1/2}\}_{Q=o} = - \mathcal{H}\vec{g}/(\vec{g}'\,\vec{g})^{1/2} \tag{23}$$

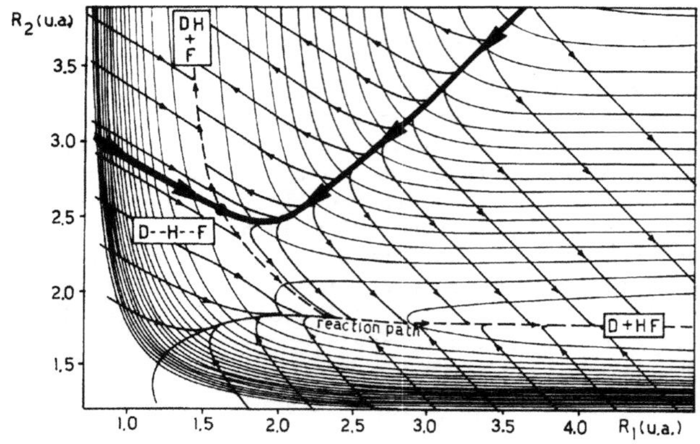

Figure 5. Space partition into basins through the force field.

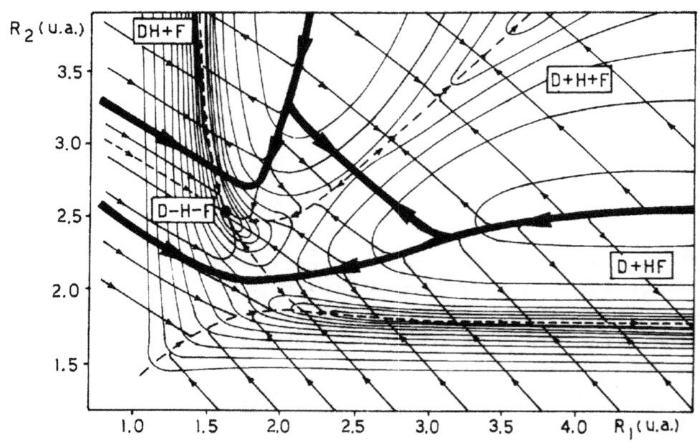

Figure 6. Space partition into basins through the gradient of the force norme.

Any stationary point on E becomes a minimum on $|F|$ surface. The number of attractors increases from the first description. The basin number increases in the same way but their size are reduced. As shown on Figure 6, we find new basins which are associated to minimax (transition basin) or maximum (dissociation basin). We note also that minimax appear on the Force norm surface, in particular along the reaction paths and on the basins boarders; their correspond to inflexion points lying on the potential surface. So, we further consider the basins of the Figure 6 as the regions in the internal coordinate space surrounding a stationary point where the eigenvalues of the force constant matrix keep the same signs.

The internal coordinate space partition enables us to go from a continuous representation of the chemical systems to a discrete set of names and existence regions. Moreover, we dispose now of a mean to classify the chemical forms. If m_0 and m_- stand respectively for the number of zero and negative eigenvalues associated to a stationary point (having a basin in charge), then we can find the following classification:

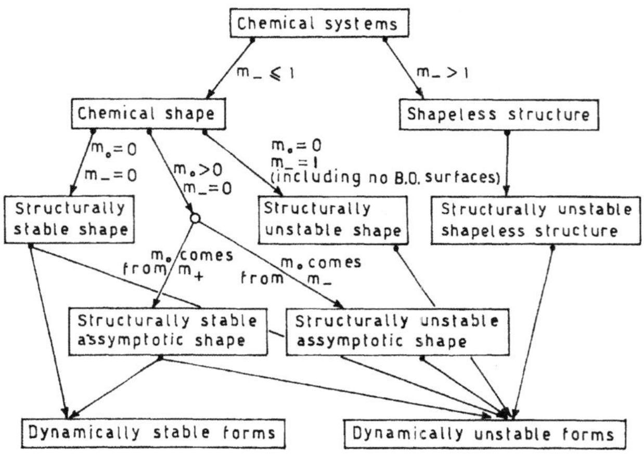

The local potential energy surface curvature is responsible for the structural stability. Any structurally unstable chemical forms can never be expected as having a long live time. On another hand the structurally stable forms are only persistent if their are dynamically stable. Indeed, the width of the nuclear motion is closely related to the translational-, rotational- and vibrational state of the system of interest.

1.5. PES : a useful tool

The potential energy surface, which exists in the Born-Oppenheimer approximation frame, is an help in the molecular analysis. As previously shown, it enables to recognize the molecular forms and to classify them. The capabilities of this tool can be summarized as follow :

-1- Recognition of the chemical forms: any stationary point of the PES being their first representation moment;

-2- Classification of the chemical forms in terms of their structural stabilities: the local curvature of the PES around the stationary points;

-3- Recognition of the connexion paths from a chemical form to an other one: the intrinsic reaction pathway on the PES which is the first representation moment of the chemical transformation;

-4- Recognition of the existence domain for the chemical forms: the basin associated to any stationary point in the gradient field of the force norm;

-5- Classification of the chemical forms in terms of their dynamical stabilities: the accessible region on the PES at a given temperature; this is the second moment of the chemical representation.

So, the PES leads to an explicit description in terms of nuclear structure. Moreover, the PES contains the energy description of the molecular forms :

-1- The first moment of the energetical description of a chemical form is given by the total energy of the associated stationary point;

-2- The first moment of the description of the chemical transformation from the energetical view point is given by the energy differences between stationary points lying along the reaction pathway;

-3- The second moment of the previous representations will be obtained by correcting the related energy values for the temperature dependence.

To fulfil the previous molecular representation, we have still to complete it by the electronic structure description. In fact, this last is implicitely contained in the potential energy surface. We have only considered in this section the eigenvalues of the equation (2), while the electronic description comes from the eigenvectors of the same equation (2).

2. FROM THE MOLECULAR SHAPE TO THE ELECTRONIC STRUCTURE AND CHEMICAL REACTIVITY

2.1. The electronic structure from the wavefunction

The problem is now how to obtain a simple picture of the electronic structure for a given chemical form and how to describe its evolution along a reaction pathway. As molecule can be regarded as atoms at finite distances from each other, bridged by an electronic cloud, a picture in terms of vertices and edges seems suitable in order to represent the molecular cohesion; that is nothing else that a graph. Untill now, the quantum chemistry has estabished two different approaches for obtaining such electronic graphs.

The Bader's approach[21-22] is first of all a topological analysis of the electron density. As shown on Figure 7, the map of the gradient density path enables to partition the R^3 space into different atomic volumes. Such volumes can be classify in two groups (depending on the fact that the atom have or no lone pairs) and three subgroups (depending on the number of surrounding atoms), as shown on Figure 8.

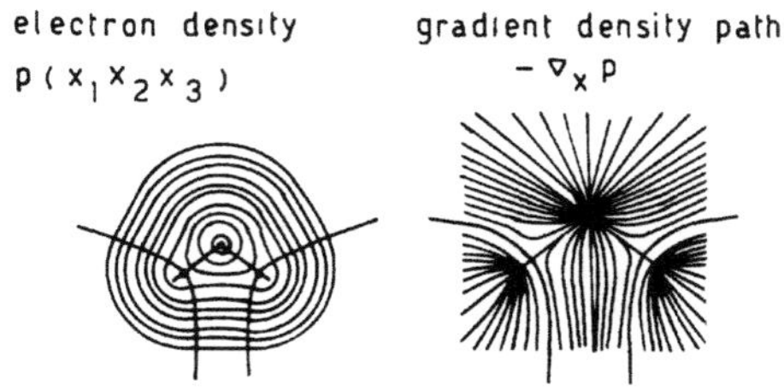

Figure 7. Space partition into atomic regions using Bader's topological approach.

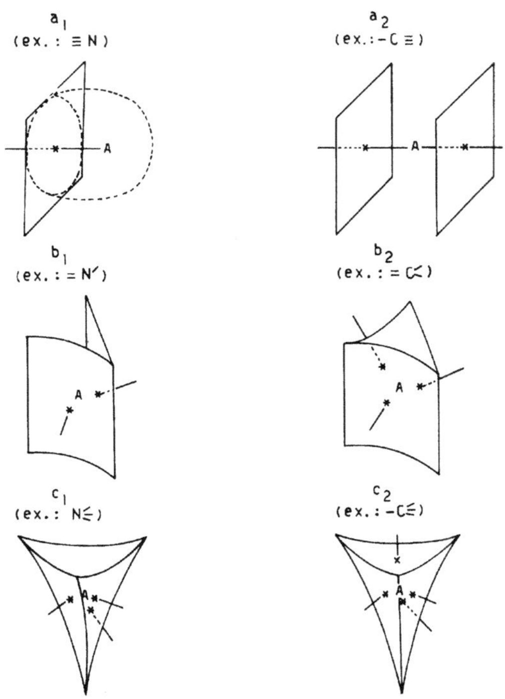

Figure 8. Different types of atomic regions using the Bader's topological approach.

It exists also on the electron density map some particular paths, perpendicular to the basin boarder at a critical point (minimax of second order), which connects two nuclei attractors of neighbouring basins); such paths form molecular graphs in the sence previously defined (see Figure 9).

C_2H_4 $C-C_3H_4$

Figure 9. Molecular structure graphs in the Bader's topological frame.

From an other point of view the loge approach of Daudel[23] enables the space partition into core, lone pair and bond region by analysing the wavefunction itself and searching for regions where appear dominant electronic events (i.e. to find one, two, three,... electrons). We give on Figure 10 an illustration. Nevertheless, for large molecules, the loge approach in its original formulation becomes an hard and expansive work. Then, other technics have been proposed.

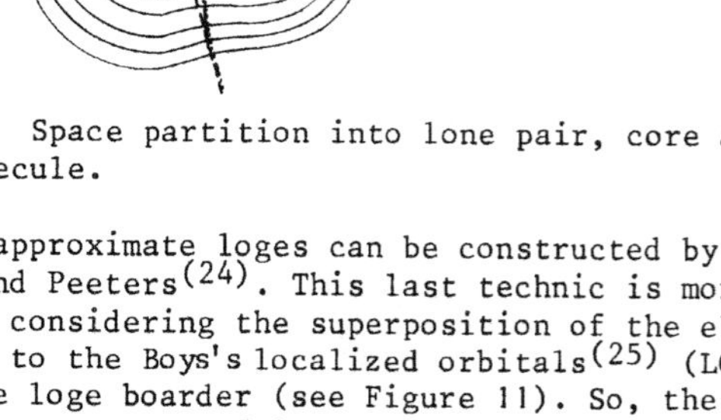

Figure 10. Space partition into lone pair, core and bond regions for the BH molecule.

Specially approximate loges can be constructed by using the pseudo-loge of Leroy and Peeters[24]. This last technic is more simple than the others; by considering the superposition of the electron density map associated to the Boys's localized orbitals[25] (LO), one can define the approximate loge boarder (see Figure 11). So, the space partition through LO picture seems an useful tool. Moreover, it has been shown[26] that the charge centroid associated to loge and to localized orbitals are closed to each other.
According to those results, we further employ the Boys's orbitals as a cheap representation for the electronic structure; their first moment of

representation are called centroid and defined as follow :
Localized orbitals :

$$\rho = \sum_i \varphi_i(CO)^2 = \sum_i \varphi_i(LO)^2$$

with $\vec{\varphi}(LO) = \tilde{T}\vec{\varphi}(CO)$

and $\tilde{T}'\tilde{T} = \tilde{I}$

Charge centroid :

$$\vec{C} = \langle \varphi(LO) \mid \vec{r} \mid \varphi(LO) \rangle$$

where $\vec{r}$ is the position vector of any point in the R^3 space.

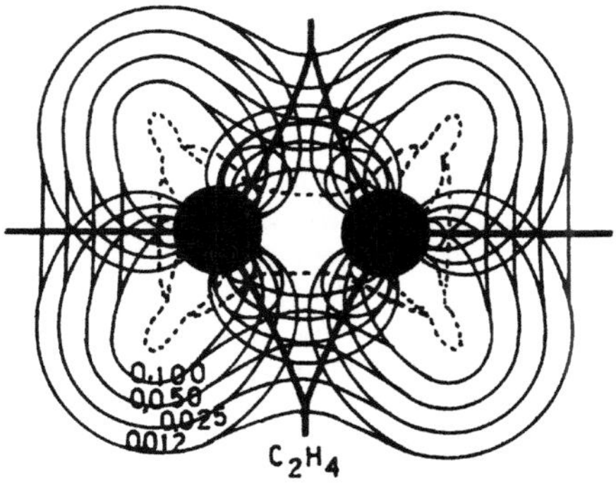

Figure 11. Graphical construction of pseudo-loge boarder for the molecule of ethylene.

The charge centroids form in R^3 space a particular design (for example, see Figure 12). By using the graph terminology, we can tell that such a picture is nothing else than a line-graph of the molecule. That means a graph where edges become vertices and reciprocally. It appears then that the Boy's procedure gives a graph of the electronic structure. Such graphs correspond in most of cases to the usual Lewis or Linnett structure for closed or open-shell structure respectively.
Let us now see what happens when we follow the charge centroid along a reaction pathway. Due to continuity reasons in the wavefunction evolution, the charge centroids move continuously from their initial position in the reactants to their final position in the products. This movement is illustrated by Figure 13. Many other examples are available from (9,27-29). On Figure 13, we see that the B_3H_4 bond breaks to form a new C_1H_4 bond. Simultaneously one electron doublet from C_1C_2 triple bond gives the new C_2B_3 bond. The electron motion shows that the hydrogen H_4 is transfered into C_1 as an hydrid (H^-).

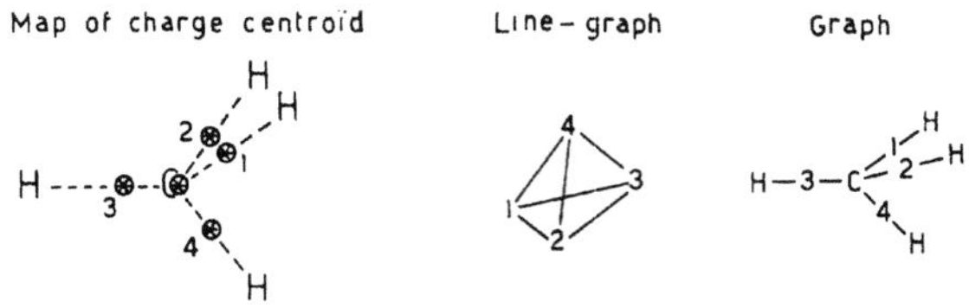

Figure 12. From Boys's centroids to graph picture in the methane molecule.

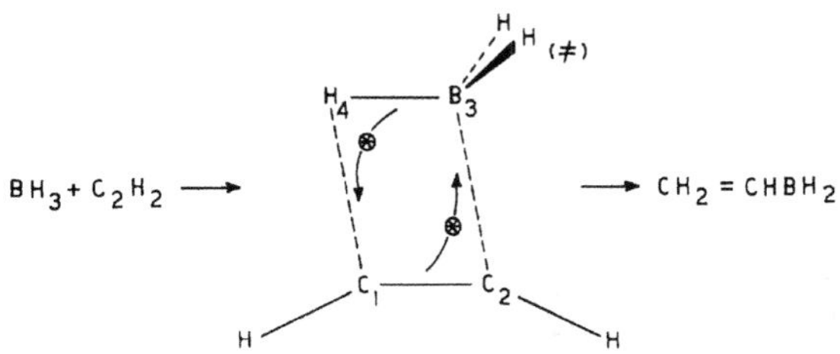

Figure 13. Centroid (*) evolution along the reaction pathway in case of C_2H_2 hydroboration.

We conclude that the charge centroids can be considered as the first moment of the electronic structure representation ; they enable us to represent the electronic mechanism of a reaction by describing the centroid displacement along the reaction pathway.

2.2. The graphs of the electronic structure

In the previous section we have shown that LO centroids can generate a graph. From a lot of results, we find that some general rules can be established to obtain those graphs without any computations. Let us resume those rules in terms of graph :

Suppose a chemical compound described by a graph M :

with $M = (A, U)$

where $A = \{a_1, \ldots, a_N\}$ is the set of the atomes (the vertices);

and $U = \{u_1, \ldots, u_n\}$ is the set of the electrons (the edges).

A *and* U *are nothing else then the vertices and the edges of the graph* M. *In terms of oriented edges (or arrows), the edges can be incoming (superscript*$^-$ *in further) or outcoming (superscript*$^+$*) relatively to an atom* $a_i \in A$. *If we note* $d(a_i)$ *(degree of vertex* a_i*), the number of vertices* connected to a_i, we can write :

$d^+(a_i)$ = *number of valence electrons lying on* a_i ;

$d^-(a_i)$ = *number of vacancies in valence shell of* a_i.

RULE : $d(a_i) = d^+(a_i) + d^-(a_i) \leqslant v$ *(where* v *stands for 2 or 8 depending on the nature of the atom* a_i*); this rule is nothing else then the Octet rule when* $d(a_i)=v$*; for compound which are deficient in electron (such us* BH_3*), we have the possibility to keep some vacancies* $(v - d)$ *on a particular atom. The quantity* $d^+ - d^-$*, when it is greater than zero, counts the number of unshared electrons.*
It is clear that connexions between atoms must be made with respect to the following requirements :

RULE :
- *no crossing of edges;*
- *minimum length for any edges;*
- *edges having a spacial distribution in agreement with the nuclear symmetry;*
- *any pair of edges lying in the same region forms an electron doublet unrespectively with their orientation.*

Further, the set $\Gamma(a_i)$ *of vertices or atoms directly connected to* a_i *by incomming and outcoming edges must constain a limited number of terms :*

RULE :

$$\Gamma^+(a_i) \leqslant g$$

$$\Gamma^-(a_i) \leqslant g$$

$$\Gamma^-(a_i) = \Gamma^+(a_i) + \Gamma^-(a_i) \leqslant g + 1$$

where g *equals 1 for the H atom and 4 for the atoms of the second or third period in the Mendeleev's table.*
We see that the surrounding of a_i *can be larger than* g *by one unit; this facility enables to code the electronic structure of borane type molecules or transition structure arising along a reaction pathway.*

Applying these rules, we obtain a description of the electronic structure in agreement with those from the LO pictures as well as the usual description coming from Lewis's (for closed shell molecules) or Linnett's (for open-shell structure) representation. Figure 14 gives some illustrations of the molecular graph codification for the electron structure. We see that the graph picture we propose is adequate to code the electronic structure of radicals, ions and molecules in their ground state. Moreover, Figure 15 shows that borane compounds are easely described and that special events of borane chemistry occur also in the carbon or nitrogen chemistry[9,29]. Particularly, the X→H→Y two electrons / three centers bonds could be considered as protonated bond:

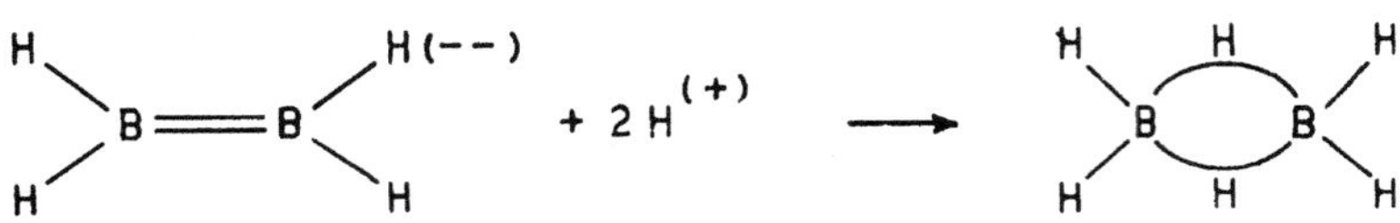

Atome centered radical : 1 electron / 1 center

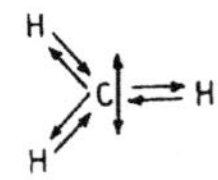

Lone pair : 2 electrons / 1 center

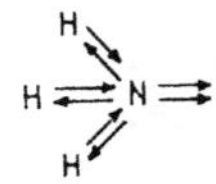

Single bond : 2 electrons / 2 centers

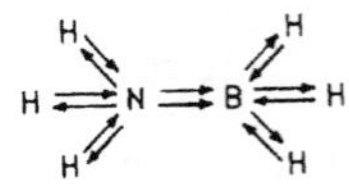

Aromatic bond : 3 electrons / 2 centers

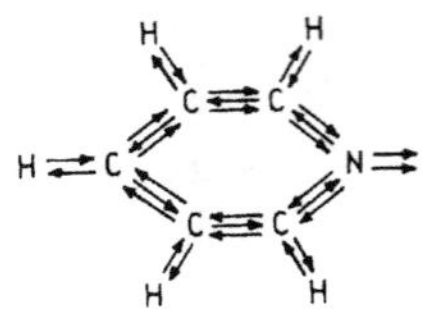

Double bond : 4 electrons / 2 centers

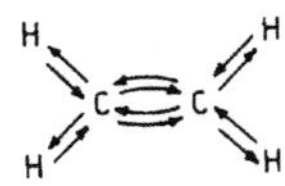

Triple bond : 6 electrons / 2 centers

Protonated bond : 2 electrons / 3 centers

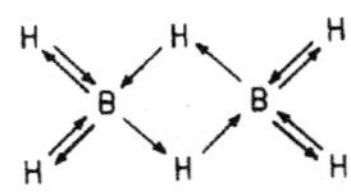

Hydrogen bridge : 4 electrons / 3 centers

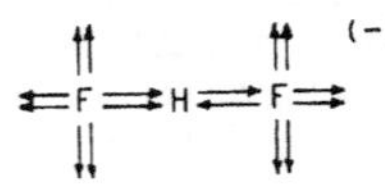

1,3-dipole bond : 8 electrons / 3 centers

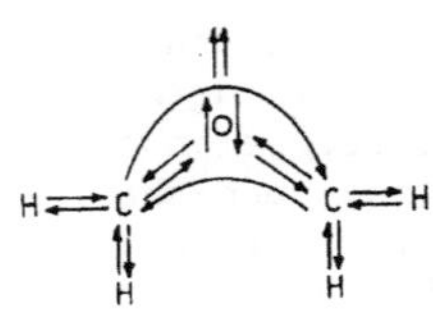

:10 electrons / 3 centers

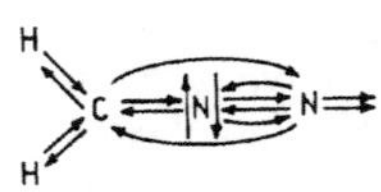

Figure 14. Electron structure graphs for some small compounds.

2.3. The graphs of electronic structure and the reaction mechanisms

The graph rules previously detailed are appropriate for the transition points. Illustrations are given on Figure 16. Those molecular pictures agree which the ones obtained from LO centroids. However, in the cyclic transition states, the global orientations of the edges is taken from the centroid migration along the reaction pathway any times a symmetry plane does not cut the cycle. Nevertheless, we have shown in other work that correlation(31) appears between atomic net charges in reactants and bond withdrawing or donating of an atom in the reactants.

Electron number	Graph of structure	Compounds with similar structure
10	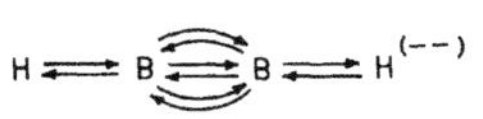	C_2H_2; $N_2H_2^{++}$; BCH_2^-; BNH_2; CNH_2^+...
10	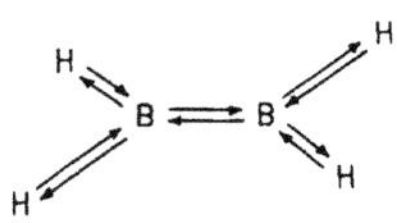	$C_2H_4^{++}$; BCH_4^+
10	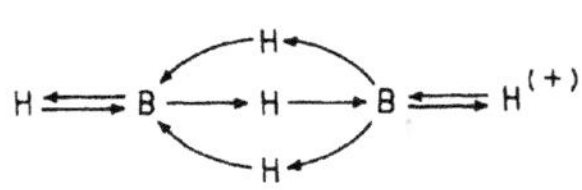	BCH_5^{++}
12	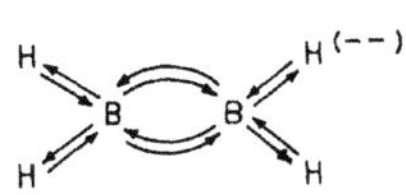	C_2H_4; $N_2H_4^{++}$; BCH_4^-; BNH_4; CNH_4^+
12	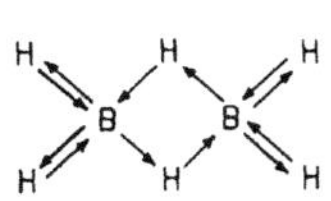	$C_2H_6^{++}$; BCH_6^+
14	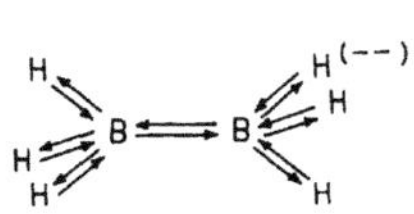	C_2H_6; $N_2H_6^{++}$; BCH_6^-; BNH_6; CNH_6^+

Figure 15. Some typical events of borane chemistry occur in carbon and nitrogen chemistry too and reciprocally(32).

As example, let us consider the acetylene hydroboration. In the borane, the unsymmetrical reactant, the boron atom is deficient in electrons relatively to the hydrogen atoms. The boron must be expected then as bond withdrawing. The new C...B bond comes from the triple bond of the C_2H_2 and the H-B bond gives the H...C new bond in the product (see Figures 12 and 16).

$CH_4+CH_3 \rightarrow CH_3+CH_4$ (29) : 3 electrons / 3 centers

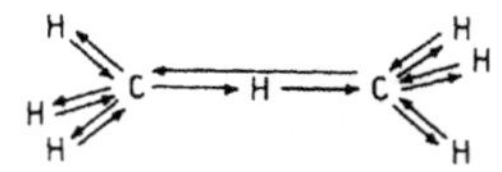

$CH_4+H^- \rightarrow H^-+CH_4$: 4 electrons/ 3 centers

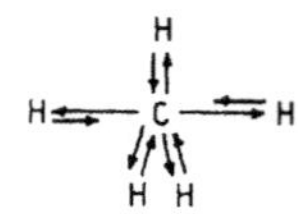

NH_2-CHO→NH=CHOH : 6 electrons / 4 centers

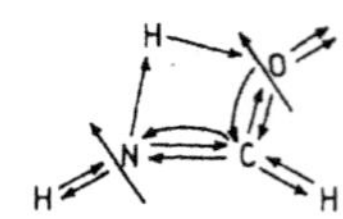

$C_2H_2+BH_3 \rightarrow CH_2$=$CHBH_2$ (30) : 4 electrons / 4 centers

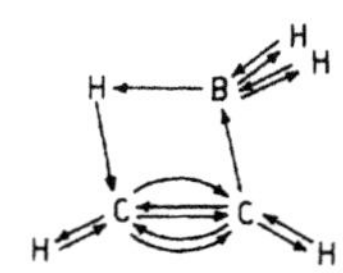

$C_2H_2+N_2O \rightarrow$ c-$C_2H_2N_2O$ (9) : 6 electrons / 5 centers

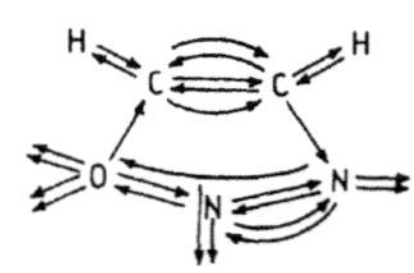

$C_2H_3N_3 \rightarrow$ c-$C_2H_3N_3$ (9,27) : 6 electrons / 5 centers

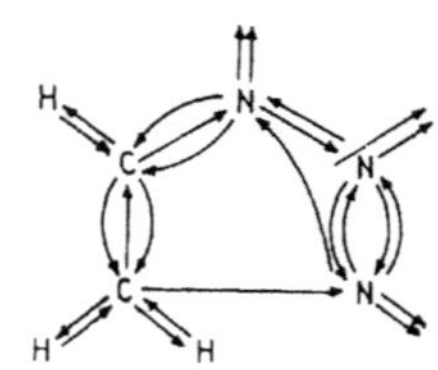

$C_2H_2+N_2H_2 \rightarrow C_2H_4+N_2$ (30) : 6 electrons / 6 centers

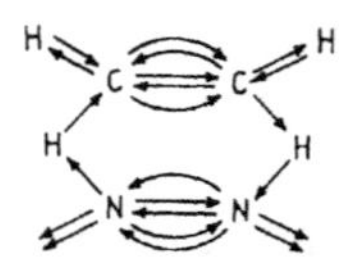

Figure 16. Electronic picture of some transition chemical form.

Similarly, in 1,3-dipolar cycloadditions(31,33-34), we can divide the transition graph into parts respectively corresponding to the dipolarophile (X-Y) and the dipole (a-b-c) :

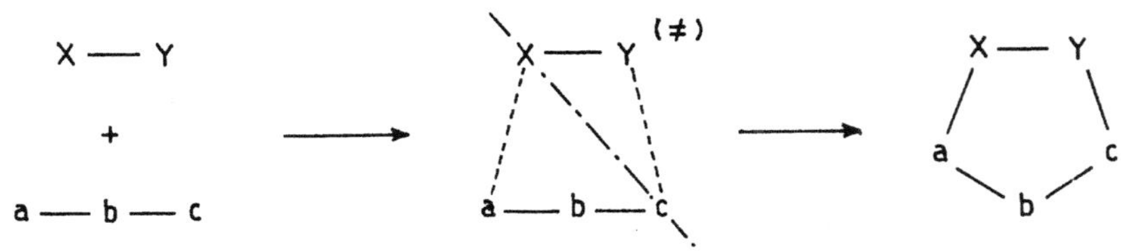

where "X" and "c" are deficient in electron relatively to "Y" and "a" respectively. The normal addition (the experimentaly observed), is defined by placing in a complementary way the partial graphs of the dipole and the dipolarophile :

If 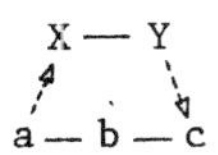 (examples: CH_2NH, CH_2O, NHO, CHN, ...)

and (examples: CH_2N_2, NHN_2, ON_2, ONCH, ...)

then the normal addition is(31) :

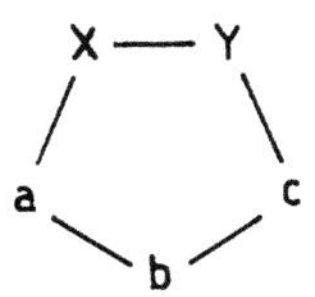

instead of the reverse one which is not observed in general :

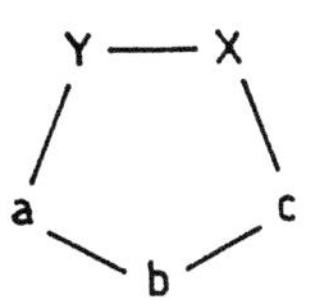

From the examples of Figure 16, it appears that the transition structures are intermediate between the structures of the reactants and of the products. We can give a more quantitative view of this by writing the graphs in terms of connectivity matrices. For the hydroboration of acetylene we find (see Figures 12 and 15) :

for reactants :

$$M_o = \begin{pmatrix} H & & & & & & & \\ 0 & H & & & & & & \\ 2 & 0 & C_1 & & \text{Sym.} & & & \\ 0 & 2 & 6 & C_2 & & & & \\ 0 & 0 & 0 & 0 & B_3 & & & \\ 0 & 0 & 0 & 0 & 2 & H_4 & & \\ 0 & 0 & 0 & 0 & 2 & 0 & H & \\ 0 & 0 & 0 & 0 & 2 & 0 & 0 & H \end{pmatrix}$$

for products :

$$M_1 = \begin{pmatrix} H & & & & & & & \\ 0 & H & & & & & & \\ 2 & 0 & C_1 & & \text{Sym.} & & & \\ 0 & 2 & 4 & C_2 & & & & \\ 0 & 0 & 0 & 2 & B_3 & & & \\ 0 & 0 & 2 & 0 & 0 & H_4 & & \\ 0 & 0 & 0 & 0 & 2 & 0 & H & \\ 0 & 0 & 0 & 0 & 2 & 0 & 0 & H \end{pmatrix}$$

and for transition structure :

$$M_{\neq} = \begin{pmatrix} H & & & & & & & \\ 0 & H & & & & & & \\ 2 & 0 & C_1 & & \text{Sym.} & & & \\ 0 & 2 & 5 & C_2 & & & & \\ 0 & 0 & 0 & 1 & B_3 & & & \\ 0 & 0 & 1 & 0 & 1 & H_4 & & \\ 0 & 0 & 0 & 0 & 2 & 0 & H & \\ 0 & 0 & 0 & 0 & 2 & 0 & 0 & H \end{pmatrix}$$

We see that the last connectivity matrix is related to the previous by :

$$M_{\neq} = \frac{1}{2}(M_o + M_1)$$

That means that as long a we follow the reaction pathway we find a continuous evolution in the electronic description which passes through a transition structure closed to the mean structure between reactants and products. This can be regarded as the less change of electronic structure during the reaction.

CONCLUSION

Here, we have summarized some features which can be deduced from the potential energy hypersurface concept. Mainly, we have carefully discussed the following features :

- Some stationary points are related to the stable or unstable structures for the chemical forms. One can associate to them basins which correspond to particular existing region and names for the molecules ;
- It exists some low energy paths which connect different stable chemical structures;
- Electronic structures can be described in terms of graphs. The electron reorganization during chemical reactions can be analysed with the same tool, leading to the concept of electronic mechanism.

So this paper tries precise the exact meaning of molecular designs from the structural viewpoint as well as from the electronical one. It also shows which approximations have to be done to go from a continuous representation of the molecules, as those coming from the Schrödinger equation, to a discrete picture, as the one handled by chemists.

REFERENCES

1. P.Pulay: Mol. Phys. 17, 197 (1969).
2. H.B.Schlegel in *Computational Theoretical Organic Chemistry*, Ed. I.G.Csizmadia, R.Daudel, Reidel Dordrecht, C67, 129 (1980).
3. J.D.Goddard in *Computational Theoretical Organic Chemistry*, Ed. I.G.Csizmadia, R.Daudel, Reidel Dordrecht, C67, 161 (1980).
4. S.Bratoz: *Coll. Int. CNRS*, 82, 287 (1958).
5. A.Gerratt, M.Mills: *J. Chem. Phys.*, 49, 1917 (1968).
6. K.T.Thomsen, P.Swanström: *Mol. Phys.*, 26, 735 (1973).
7. M.Sana: *Theoret. Chim. Acta*, 60, 543 (1982).
8. M.Sana: *Internat. J. Quantum Chem.*, 19, 139 (1981).
9. M.Sana: Surface de Potentiel et Réactivité Chimique, Agrégation à l'Enseignement Supérieur, UCL (1983).
10. R.E.Stanton, J.W.McIver Jr., *J. Amer. Chem. Soc.*, 97, 3632 (1975).
11. J.N.Murrel, K.J.Laidler: *Trans. Farad. Soc.*, 64, 371 (1968).
12. J.W.McIver Jr., D.M.Silver: *J. Amer. Chem. Soc.*, 94, 2625 (1972).
13. E.A.McCullough Jr., D.M.Silver: *J. Chem. Phys.*, 62, 4050 (1975), and D.M.Silver, Ibid, 57, 586 (1966).
14. A.Tachiba, K.Fukui: *Theoret. Chim. Acta*, 42, 311 (1978); 51, 175 (1979); 51, 189 (1979).
15. K.Fukui, S.Kato, H.Fujimoto: *J. Amer. Chem. Soc.*, 97, 1 (1975).
16. Z.Ishida, K.Morokuma, A.Komornicki: *J. Chem. Phys.*, 66, 2153 (1979).
17. M.Sana, G.Reckinger, G.Leroy: *Theoret. Chim. Acta*, 58, 145 (1981).
18. E.B.Wilson, J.C.Decius, P.C.Cross: *Molecular Vibrations*, Ed. McGraw-Hill, New York (1955).
19. R.Thom: *Structural Stability and Morphologenesis*, Ed. Benjamin Reading Massachussets (1975).
20. M.Sana: *C.R. Acad. Sc. Paris*, 297(II), 569 (1983).
21. R.F.W.Bader, S.G.Anderson, A.J.Duke: *J. Amer. Chem. Soc.*, 101, 1389 (1979).
22. R.F.W.Bader, T.T.Nguyen-Dang, Y.Tol: *J. Chem. Phys.*, 70, 4316 (1979).
23. R.Daudel: *Compte Rendu*, 240,90 (1955). M.Roux, R.Daudel: *C.R. Acad. Sc. Paris*, 240, 90 (1955).
24. G.Leroy, D.Peeters in *Localization and Delocalization in Quantum Chemistry*, Ed. O.Chalvet et al., Reidel, Dordrecht, I, 210 (1975).
25. S.J.Boys: *Rev. Mod. Phys.*, 32, 296 (1960).
26. R.Daudel, M.E.Stephens, L.A.Burke, G.Leroy: *Chem. Phys. Lett.*, 52, 426 (1977).
27. G.Leroy, M.Sana, L.A.Burke, M.T.Nguyen in *Quantum Theory of Chemical Reactions*, Eds. R.Daudel, A.Pullman, L.Salem, A.Veillard, Reidel, Dordrecht, 1, 91 (1979).
28. M.T.Nguyen, M.Sana, G.Leroy, K.J.Dignam, A.F.Hegarty: *J. Amer. Chem. Soc.*, 102, 573 (1980) and 101, 1988 (1978).
29. G.Leroy, J.L.Villaveces, M.Sana: *Theoret. Chimica Acta*,65,109 (1984).
30. M.Sana, G.Leroy: *Theochem*, in press (1984).

31. M.Sana, G.Leroy, G.Dive, M.T.Nguyen:
J. Mol. Struct., 89, 147 (1982).
32. Geometrical parameters of all those structures have been determined in 6-31G· basis set at SCF level by using the numerical computation of force constant matrices. Their are available partly from Ref. (30) and from Ref. (9).
33. G.Leroy, M.Sana: *Tetrahedron*, 31, 2091 (1975).
34. G.Leroy, M.Sana: *Tetrahedron*, 32, 709 (1976).

* Chercheur Qualifié au Fonds National de la Recherche Scientifique Belge

THEORETICAL DETERMINATION OF NUCLEAR PROBABILITY DENSITY DISTRIBUTIONS. TWO-ROTOR MOLECULES

Yves G. Smeyers and A. Hernández-Laguna
Instituto de Estructura de la Materia, C. S. I. C.
c/ Serrano, 119, 28006-Madrid, Spain

ABSTRACT. In the present work, we attempt to give a mathematical description for the non-rigidity of some deformable molecules, as a function of temperature. For this purpose, we propose the calculation of probability density maps of finding a two-rotor molecule in a given conformation and at a given temperature, as a function of the rotation angles. The theory is limited to relatively light rotors so that the calculation must be performed into the quantum mechanics framework. From the probability density maps, we consider the calculation of some observable parameters. The theory is applied to the double internal rotation in the series of the standard and deutered xylene molecules. Nuclear probability density maps are given for various temperatures. The main conclusion is that the deuteration does not affect, as least in a large extent, the conformation distributions.

1. INTRODUCTION

Most of the Physical Chemistry properties of a deformable molecules, such as IR, NMR, EPR spectra, thermodynamic parameters, dipolar moment, and even chemical reactivity,...., depend on the "dynamic structure" of this deformable molecule. Therefore, the determination of distribution of probabilities of finding a deformable molecule in a given structure and at a given temperature is a very important problem. Some authors have already considered to determine this distribution of probabilities, both theoretically and experimentally [1-2], but generally these calculations have been carried out in a very crude way.

In the following, we pretend to determine rigorously these probability distributions in the case of a two-rotor molecule, as a first step for studying more complex systems in a future. Furthermore,

R. Daudel et al. (eds.), Structure and Dynamics of Molecular Systems 23 40.

we shall restrict our considerations to relatively light rotors, in such, a way that the calculations must be performed into the framework of the quantum mechanics. Finally, we shall suppose that the non-rigid system is in thermodynamic equilibrium, without interaction with the environment.

In the calculation of the nuclear probability distributions, we shall distinguish three stages:

1) The determination of the potential energy function, in which the rotors are moving;

2) The resolution of the Schrödinger equation corresponding to the double internal hindered rotation;

3) The calculation of the probability densities using the solutions of the Schrödinger equation, and the state populations obtained at a given temperature.

From the probability density distributions, probability density maps versus the rotation angles may be drawn. Examples of such maps may be found in [3]. These resemble completely to the potential energy maps, but now their contours depend on the temperature. In addition, for very deformable molecules, there exists a threshold surface below of which , all the conformations are equiprobable. From the nuclear probability density distributions, observable parameters may be evaluated as a function of temperature.

From this point of view, the determination of the nuclear probability density distribution of a non-rigid molecule seems to be a very simple problem. The calculations, however, are so complicated that the introduction of some simplifications, as well as the use of some very sophisticated technics, are necessary. Group theory for non-rigid systems will be employed for shortening not only the potential energy function determination, but especially for simplifying the Schrödinger equation resolution and reducing the probability distribution calculations to reasonable dimensions.

2. POTENTIAL ENERGY FUNCTION

Let us now consider the hamiltonian operator of a two-rotor molecule, without any symmetry. The Schrödinger equation corresponding to the double hindered rotation of such a system can be written as [4-5] :

$$-\left[\sum_{j=}^{2}\sum_{i=1}^{2} \frac{\partial}{\partial\theta_i} B_{ij}(\theta_1, \theta_2) \frac{\partial}{\partial\theta_j} + V(\theta_1, \theta_2)\right]\Psi = E\Psi \quad (1)$$

where B_{ii} is the so called rotation "constant" of the rotor i; B_{ij} the kinetic interaction term, between the rotors i and j; and $V(\theta_1, \theta_2)$ the potential energy of the two coupled rotors rotating in the field of the frame. The rotation constants and kinetic interaction terms depend, although partially, on the rotation angles. However, they may be considered really as constants, when the rotors are "symmetric". The Schrödinger equation may be then written as:

$$\left[- B_{11} \frac{\partial^2}{\partial \theta_1^2} - 2 B_{12} \frac{\partial^2}{\partial \theta_1 \partial \theta_2} - B_{22} \frac{\partial^2}{\partial \theta_2^2} + V(\theta_1, \theta_2) \right] \Psi = E \Psi \qquad (2)$$

where the rotation constants and the interaction terms have the form:

$$B_{11} = \frac{h^2}{2} \frac{I_2}{I_1 I_2 - \Lambda_{12}^2} \qquad \text{and} \qquad B_{12} = \frac{h^2}{2} \frac{\Lambda_{12}}{I_1 I_2 - \Lambda_{12}^2}$$

In these expression I_1 is the reduced moment of inertia of one of the rotors, and Λ_{12} the coupling term [5] :

$$I_1 = A_1 \left(1 - \sum_{i=x,y,z} A_1 \lambda_{1i}^2 / M_{ii} \right)$$
$$\Lambda_{12} = \sum_{i=x,y,z} A_1 A_2 \lambda_{1i} \lambda_{2i} / M_{ii} \qquad (3)$$

where A_i is the moment of inertia of the rotor i, M_{ii} the moments of inertia of the whole molecule, and λ_{1i} and λ_{2i} the director cosines.

The heavier is the frame, the better is this approximation. Furthermore, it is seen that, in that case, the kinetic interaction terms may be neclected.

A special case is that of two equivalent symmetric rotors, in which the Schrödinger equation takes the simple form:

$$\left[- B \frac{\partial^2}{\partial \theta_1^2} - 2B' \frac{\partial^2}{\partial \theta_1 \partial \theta_2} - B \frac{\partial^2}{\partial \theta_2^2} + V(\theta_1, \theta_2) \right] \Psi = E \Psi \qquad (4)$$

We shall limit our calculations to this restricted case in the following.

2. 1. Expansion of the Potential Energy

As it is well known, the potential energy operator $V(\theta_1,\theta_2)$ may be developed in terms of a double Fourier expansion as a function of the rotation angles [6-7]:

$$V(\theta_1,\theta_2)=\sum_K^{\infty}\sum_L^{\infty}\Big[A_{KL}^{CC}\cos K\theta_1\cos L\theta_2+A_{KL}^{CS}\cos K\theta_1\sin L\theta_2+ \\ +A_{KL}^{SC}\sin K\theta_1\cos L\theta_2+A_{KL}^{SS}\sin K\theta_1\sin L\theta_2\Big] \quad (5)$$

This expansion runs up to infinity. However, the experience shows that it converges well, and only few terms are necessary for reproducing the principal characteristics of the potential.

In order to delimit the expansion, let us underline that the cosine-cosine terms, which do not depend on the sense of the rotations, describe the global interactions of the rotors between themselves and with the frame. The sine-sine terms, which depend on the rotation sense, describe the "cog-wheel" effects between the rotors. Finally, the sine-cosine mixed terms describe the interactions of the rotors with the particularities of the frame [8-9].

The "cog-wheel" effects are relatively weak, but if they should be taken into account, it would be necessary to conserve at least one sine-sine term in the potential energy expansion. Furthermore, because of the limit conditions, where the sine functions are zero, the sum on the cosine functions must run on two terms more than that on the sines. As a result, the minimal expansion for reproducing the principal features of the potential energy function will be limited, for the cosine functions to the values of K and L from zero to two, and for the sine functions to the value of one. In these conditions, the minimal expansion for a two-rotor molecule will contain only 16 terms [9].

2. 2. Introduction of the Symmetry

As it is also well known, the hamiltonian operator commute with all the operations of the "symmetry" group of the molecule. When the two rotors are symmetric, it is easily seen that the kinetic operator commute with these symmetry operations, the potential energy operator must thus commute too.

If the molecule possesses some induced symmetries, the number of terms of the minimal expansion could be reduced significantly. Let us consider a molecule with two equivalent rotors. This equivalence may be accomplished in different ways [7] : 1) The two rotors are superimposable by rotation, i. e. , the two rotors are identical and the frame possesses a two-fold symmetry axis; 2) The two rotors are superimposable by reflexion, i. e. , the two rotors are the mirror images of each others, and the frame possesses a perpendicular symmetry plane between them; 3) The two rotors are superimposable by rotation and reflexion.

These three different ways for the equivalence may be easily expressed into the Altmann's formalism by the so called "isodynamic" operations [10] (see also [3]). The potential energy function must be then invariant under these operations.

When the two rotors are equivalent by rotation, the potential must be invariant under the angle exchange operation:

$$\hat{W}\, f(\theta_1, \theta_2) \equiv f(\theta_2, \theta_1) = f(\theta_1, \theta_2) \tag{6}$$

This requirement on the potential implies that the coefficients of the Fourier expansion (5) fulfil the restrictive conditions:

$$\begin{aligned} A^{CC}_{KL} &= A^{CC}_{LK} \qquad & A^{CS}_{KL} &= A^{SC}_{LK} \\ A^{SC}_{KL} &= A^{CS}_{LK} \qquad & A^{SS}_{KL} &= A^{SS}_{LK} \end{aligned} \tag{7}$$

When the two rotors are equivalent by reflexion, the potential must be invariant under the double switch-angle exchange operation:

$$\hat{V}\, \hat{W}\, f(\theta_1, \theta_2) \equiv f(-\theta_2, -\theta_1) = f(\theta_1, \theta_2) \tag{8}$$

This requeriment implies now that the coefficients of the Fourier expansion (5) fulfil the alternative conditions:

$$\begin{aligned} A^{CC}_{KL} &= A^{CC}_{LK} \qquad & A^{CS}_{KL} &= - A^{SC}_{LK} \\ A^{SC}_{KL} &= -A^{CS}_{LK} \qquad & A^{SS}_{KL} &= A^{SS}_{LK} \end{aligned} \tag{9}$$

When the two rotors are equivalent by rotation and reflexion both conditions (7) and (9) must be fulfil. Thus, we have necessarily:

$$A^{CC}_{KL} = A^{CC}_{LK} \qquad A^{SS}_{KL} = A^{SS}_{LK}$$
$$A^{CS}_{KL} = A^{SC}_{LK} = A^{SC}_{KL} = A^{CS}_{LK} = 0 \tag{10}$$

If in addition, the rotors have n-fold symmetry axis, the potential energy function must be also invariant under the corresponding n-fold isodynamic rotations [9].

As a result, it we consider two-rotor molecules with full equivalence and n-fold rotors, the minimal expansion takes the simple form:

$$V(\theta_1, \theta_2) = \sum_{L \geqslant K}^{2} \sum^{2} A^{CC}_{KL}(\cos nK\theta_1 \cos nL\theta_2 + \cos nL\theta_1 \cos nK\theta_2) + A^{SS}_{11} \sin n\theta_1 \sin n\theta_2 \tag{11}$$

which contains only seven terms.

In the following, we shall consider the series of the xylene molecules, which are two-rotor molecules with full equivalence and rotors of order three. Intermediate cases are considered in [7] and [9].

2. 3. Numerical Determination of the Potential Energy Function

In order to determine the potential energy function of the xylene molecules, it is sufficient to calculate the formation energies of so many conformers as variables we have in the expansion (11). Thus using the minimal expansion seven conformations will be enough. This number is reduced to five in the case of the para-xylene of higher symmetry [9].

For calculating the formation energies, we shall use the quantum semi-empirical method CNDO/2, since this procedure appears to yield relatively good results in the case of the two-rotor molecules, even when compared with the "ab initio" methods [11-12].

The calculation of the formation energies of the different conformers should be done into the normal coordinate space. Unfortunately, these coordinates are generally not known. Two procedures can be then used: 1) the first would be to calculate the energies in the space of the internal coordinates without minimizing the geometry; 2) to perform the same calculation with minimization of the geometry. Since there exit programs for geometry optimization, the second procedure is often used. We think that this second procedure is not necessarily the best one. If the interactions between the different vibration modes are weak, then the internal coordinates are good normal coordinates. On the contrary, if the interactions are large, the general problem must be solved, taking into account explicitely these interactions.

3. RESOLUTION OF THE SCHROEDINGER EQUATION

In order to calculate the probability density distribution, the next step to be done is to solve the Schrödinger equation in an analytical way, i. e. , to determine the eigenfunctions in an analytical form. The Schrödinger equation to be solved for the double hindered rotation is given by (4) in the case of a two equivalent rotor molecules.

3. 1. Expansion of the Rotational Wave-Functions

In order to obtain analytical expressions for the eigenfunctions, let us develop these onto the basis of the solutions of the double free rotor equation [13]:

$$\Psi = \sum_K \sum_L \Big[C_{KL}^{CC} \cos K\theta_1 \cos L\theta_2 + C_{KL}^{CS} \cos K\theta_1 \sin L\theta_2 + C_{KL}^{SC} \sin K\theta_1 \cos L\theta_2 + C_{KL}^{SS} \sin K\theta_1 \sin L\theta_2 \Big] \quad (12)$$

The eigenfunctions take then the form of a rotational configuration interaction function, the coefficients of which may be determined variationally. The problem is then reduced to the diagonalization of an interaction hamiltonian matrix between rotational configurations.

The expansion (12), however, must be in practice limited to a reasonable number of terms. If the expansion is large enough, the eigenfunctions and eigenvalues will be good approximations for the rotational wave-functions and energies of the lowest levels, respectively. Concerning the solutions of the highst levels, these may be approached by the solutions of the free rotor equation. The problem subsists, however, for the intermediate levels.

For the calculation of the probability densities, all the significantly populated levels have to be taken into account. This number may be important, especially at high temperatures. For this reason, a large configuration interaction calculations would be necessary. For example, up to K and L = 15, in order to cover the intermediate levels. In this case, the hamiltonian matrix to be diagonalized would be of 961 order, and it would be necessary to calculate about one million of matrix elements!

3. 2. Introduction of the symmetry

The calculation becomes still more tedious when the probability density has to be computed, because a summation has to be performed on all the mixed products of the expansion coefficients, weighed by two trigonometric functions, for all the populated levels (see equation (17)). Therefore, it is convenient to develop the configuration interaction wave-function on the basis of the symmetry eigenvectors, which transform them-selves according to the irreducible representations of the non-rigid operation group of the molecule:

$$\Phi_i^G = \sum_j c_{ij}^G \varphi_j (\theta_1, \theta_2) \tag{13}$$

where the superindex G denotes the component of the symmetry species to which the vector belongs. $\varphi_j(\theta_1, \theta_2)$ is a product of two trigonometric functions. As it is well known the symmetry eigenvectors factorize in boxes the hamiltonian matrix. Thus, it would be only necessary to diagonalize a few smaller submatrices.

When the character table of the non-rigid operation group is known, the symmetry eigenvectors can be easily deduced by projection using the classical projector:

$$\hat{P}_j = \frac{l_j}{N} \sum_R^N \chi_j(R) \hat{R} \tag{14}$$

where N is the group order, l_j the dimension of the representation j, and $\chi_j(R)$ its character corresponding to the operation $\hat{R}$.

Unfortunately the character tables corresponding to the non-rigid molecule groups are relatively little studied. The character table of a two-equivalent - C_3 - rotor molecule is, however, well known [13-15] . Furthermore, the symmetry vectors have been deduced for the study of the rotation levels of acetone [13] . For these reasons, we limite the present work to this kind of induced symmetry.

3. 3. Application of the symmetry

The non-rigid molecule group for an acetone-like two-rotor molecule may be written into the Altmann's formalism as follow [13-15] :

$$(\mathcal{C}_3^I \times \mathcal{C}_3'^I) \quad (W^I \times V^I) = G_{36} \tag{15}$$

It is the matter of a group of order 36, in which appear the $\hat{V}$, $\hat{W}$ and $\hat{W}\hat{V}$ operations defined in (6) and (8), together with the $\hat{\mathcal{C}}_3$, $\hat{\mathcal{C}}_3^2$, $\hat{\mathcal{C}}'_3$ and $\mathcal{C}'^2_3$ describing the periodicity of the rotors.

The corresponding character table possesses 4 one-dimensional irreducible representations, A_1, A_2, A_3, and A_4, 4 two-dimensional irreducible representations, E_1, E_2, E_3, and E_4, and 1 four-dimensional irreducible representation G.

By using the symmetry eigenvectors for developing the rotational wave-functions, in a basis of 961 configurations, the hamiltonian matrix is factorized in 16 submatrices: Four boxes of order 36,30,30 and 25, corresponding to the representations A_1, A_2, A_3, and A_4, respectively; eight boxes of order 55, 45, 55 and 45, corresponding to the representations E_1, E_2, E_3, and E_4, respectively, and four boxes of order 110, corresponding to the representation G.

4. DETERMINATION OF THE PROBABILITY DENSITY DISTRIBUTION

When the eigenfunctions and eigenvalues of the hamiltonian matrix have been deduced, the probability density of finding the rotors in a given conformation, as a function of the rotation angles and the temperature, may be calculated.

4.1. The methodology

This calculation will be carried out in several steps. First, the level populations are determined resorting to a Boltzman distribution:

$$P_K^G(T) = \frac{\exp(-E_K/RT)}{\sum_j \exp(-E_j/RT)} \tag{16}$$

where the temperature dependence is introduced. In this expression, the superindex G denotes again the component of the symmetry species to which the K level belongs. The summation must run on all the significantly populated levels, i. e. , the hamiltonian matrix eigen values must be completed by solutions of the double free rotor equation up to convergence of the partition function (the denominator of (16)).

In a second step, the probability densities of finding the two-rotor molecule, in a K state of symmetry G, are computed as a function of the rotation angles θ_1 and θ_2. In this aim, the squares of the wave-function modulus of the K state are determined for pairs of values of θ_1 and θ_2:

$$d_K^G(\theta_1, \theta_2) = \sum_i \sum_j c_{Ki}^G c_{Kj}^G \varphi_i(\theta_1,\theta_2) \varphi_j(\theta_1,\theta_2) \qquad (17)$$

These results are then weighed by the population of the K state, and the summation on all the K states is performed in order to obtain partial probability densities for each component of symmetry species:

$$D^G(\theta_1,\theta_2) = \sum_K d_K^G(\theta_1,\theta_2) P_K^G(T) \qquad (18)$$

Finally, the summation on all the partial probability densities is achieved:

$$D(\theta_1,\theta_2) = \sum_G D^G(\theta_1, \theta_2) \qquad (19)$$

These calculations may be carried out for angle values included between 0° and 360°. However, a last summation on the isoenergetic conformations, which are equiprobable, must be performed in order to taken into account the three-fold indistinguishability of the rotors:

$$\mathbb{D}(3\theta_1, 3\theta_2) = \sum_{n,m=0}^{2} \sum^{2} D(\theta_1 + \frac{2n}{3}, \theta_2 + \frac{2m}{3}) \qquad (20)$$

The probability density distribution of finding a two-equivalent - C_3 - rotor molecule at a given temperature is given then by the "density matrix" (20), as a function of the rotation angles.

4. 2. Checking of the results

If the rotational eigenfunctions are normalized, and all the populated levels are taken into account in the summation (18), the density matrix (20) should fulfil the following normalization relation:

$$\int_{0}^{120^\circ}\int_{0}^{120^\circ} \mathbb{D}\,(3\theta_1, 3\theta_2)\, d\theta_1\, d\theta_2 = 1 \tag{21}$$

The summation on the states (18) is limited to the solutions of the hamiltonian matrix, whereas the population calculation is extended to all the significantly populated levels up to convergence of the partition function. One may thus expect that equation (21) will not hold exactly. This equation permits, however, to check the quality of the probability distribution, by comparing the density matrix value with unity.

As it is well known, the solutions of the hamiltonian matrix may be classified into three cathegories [13] : 1) the pure vibration states in which the rotors vibrate around their equilibrium positions; 2) The vibration-rotation states in which the rotors are rotating in a coupled manner, and vibrating one with respect to each other; 3) The pure rotation states in which the rotors are rotating freely. For establishing this classification, the saddle points and the maxima of the potential energy surface are taken as references. So, the states situated above the potential energy maxima are considered as pure rotation states. It is noticeable that these states will contribute in a constant way to the probability density distribution.

In order to obtain thus a reliable probability density distribution, all the populated vibration and vibration-rotation states must be taken into account. One way to check this requirement is to verify the existence of free rotation states among the solutions of the hamiltonian matrix. These can be obtained using a sufficiently large configuration interaction expansion.

5. APPLICATIONS

As examples of two-C_3-rotor molecules, equivalent by rotation and reflexion, the ortho- meta-, and para-xylene molecules are chosen. In the calculations, the potential energy surfaces and momenta of inertia reported in [16] have been used. These data had been computed in the rigid rotor approximation, without relaxation of the geometry, and using standard bond lenght and bond angle values.

5.1. Probability Density Map Calculations

From the potential energy surfaces, the minimal expansions for the potential energy function of the three xylenes have been computed. The expansion coefficients are given in Table I. The accuracy of the data did not allow a larger expansion. As to be expected, seven coefficients are found for the minimal expansions of ortho- and meta-xylenes, whereas five were enough for that of para-xylene of higher symmetry.

TABLE I. Coefficients of the minimal expansions for the potential energy functions, in cal/mol, corresponding to the double hindered rotation in xylenes. Saddle points and maxima of energy potential

coefficients	K	L	ortho.	meta	para
A_{KL}^{CC}	0	0	1,360.75	25.85	14.00
	1	0	-961.75	-8.42	0
	1	1	576.75	2.50	-2.62
	2	0	43.25	-5.81	-5.69
	2	1	-59.25	0.07	0
	2	2	18.25	-0.04	0.00
A_{KL}^{SS}	1	1	-247.50	-1.36	-1.24
Saddle points			889	17.40	12.64
Maxima			4,064	38.80	26.64

As to be expected too, the coefficients found for ortho-xylene were much larger than those encountered for meta- or para-xylenes, because of the vecinity of the two rotors. It is noticeable that the coefficients of order three (K or L = 1), due to the interactions between the rotors are larger, in ortho-xylene, than those of order six (K or L = 2) due to the interactions with the frame. They are of the same magnitude in meta-xylene, and do not exist in para-xylene, except for a small cog-wheel effect between the rotors.

The Schrödinger equation (4) corresponding to the double hindered rotation was solved for these three molecules using different expansion lenghts. In order to verify the quality of the configuration interaction expansions for describing the internal dynamics of these molecules, the percentages of pure vibration, vibration-rotation and pure rotation states were calculated, at different temperature, together with the percentages of states yielded by the diagonalization of the hamiltonian matrix.

In the cases of the meta- and para-xylenes relatively short expansions were sufficient (169 terms) in order to obtain all the vibration and vibration-rotation states, owing the smallness of coefficients, inclusive at very low temperature. On the contrary, in the case of the ortho-xylene, 361 terms (19 x 19) were at least necessary for describing all the vibration and vibration-rotation states. In table II, the percentages of vibration, vibration-rotation and pure rotation states are given, together with that of states yielded by the diagonalization of the hamiltonian matrix, as a function of temperature. It is seen that this last percentage is reasonably close to 100, in accordance with expression (21). To extend the configuration interaction expansion could appear to be advisable, but a larger expansion makes the probability density calculations too heavy.

TABLE II. Percentages of pure vibrational (V), vibration-rotation (VR) and pure internal rotation states (R), together with that of states yielded by the diagonalization of the hamiltonian matrix (CI), as a function of temperature for the double hintered rotation in the non-deutered ortho-xylene

Temperature in °C	V	VR	R	CI
-25°	48.55	50.33	1.12	98.91
0°	44.26	53.96	1.78	98.26
25°	40.53	56.84	2.63	97.45
50°	37.27	59.10	3.63	96.49
75°	34.40	60.81	4.79	95.39
100°	31.87	62.06	6.07	94.17
125°	29.63	62.93	7.45	92.86

From the eigenfunctions and eigenvalues of the hamiltonian matrices, the probability density distributions of the three xylenes were computed for the angle values: 0°, 0°; 0°, 30°; 30°, 30°; -30°, 30°; 30°, 60°; and 60°, 60°; and at the temperature of 0°, 25°, 50°, 75° and 100°C.

In the cases of meta- and para-xylene, completely plane probability density maps are encountered, at any temperatures. This result confirms that the rotors are rotating completely freely in these two molecules.

On the contrary, in the case of ortho-xylene undulating surfaces, the profile of which depends on the temperature, were encountered. In figures 1,2 and 3, the probability density maps calculated at 0°, 50° and 100° are given. It is seen that the profile of the probability density map of ortho-xylene is qualitatively the same as that of the potential energy map, i. e. , the symmetry is the same, and the probability density map may be described by the minimal expansion (11).

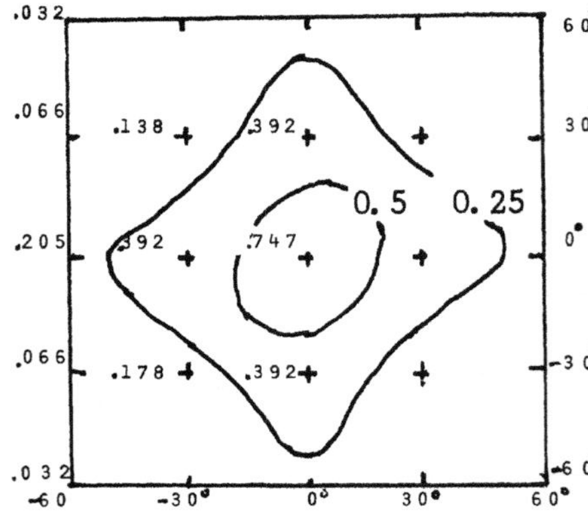

Fig. 1. Probability density map for standard o-xylene calculated at 0°C

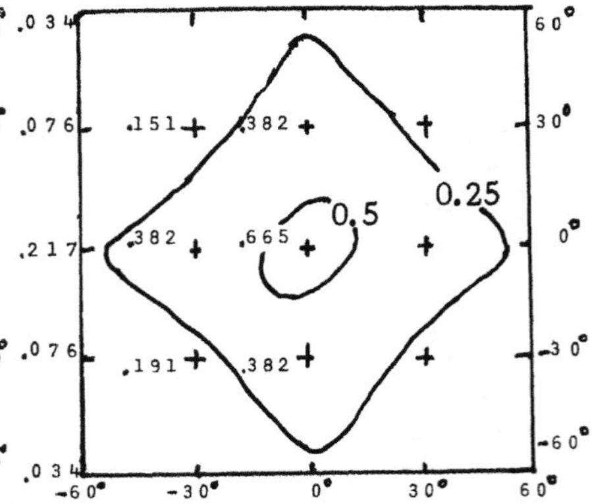

Fig. 2. Probability density map for standard o-xylene calculated at 50°C

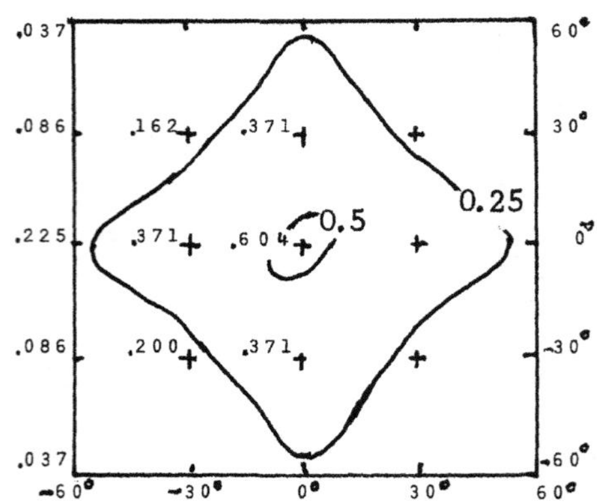

Fig. 3. Probability density map for standard o-xylene calculated at 100°C

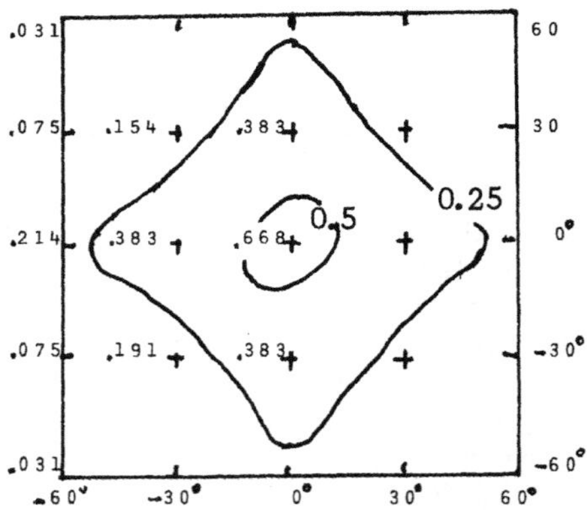

Fig. 4. Probability density map for deutered o-xylene calculated at 50°C

As to be expected, the most probable conformation is the doubly staggered one, which corresponds to the potential energy well. But, it is seen that, in addition, other conformations are also widely allowed. In particular, the less probable conformation, the doubly eclipsed, which corresponds to the maximum of potential, is also permited, although in a less extent.

Another important factor which can influence the probability density map profile is the inertia momentum of the rotors. In order to study the influence of this factor, we have repeated all the calculations for the ortho-xylene with two deutered methyl groups. As to be expected, we found a very different rotation level distribution. We have surprisingly encountered, however, very similar probability density distributions. In figure 4, we give the probability density map for the deutered ortho-xylene calculated at 50°C. There is practically no difference with the map calculated for the standard ortho-xylene at the same temperature. Anyway, and as pointed already in [3] , the very small difference disappears completely when the temperature increases.

5. 2. Mean Parameter Determinations

The probability density distribution of finding a deformable molecule in a given conformation describes the internal dynamics of this deformable molecule. Notice, however, that we can take advantage of this density distribution for calculating the density distribution of other molecular parameters which depend on the conformation. In the same way, we can determine mean values for these parameters as a function of temperature.

For this purpose, we need only to know the conformational dependency of the parameter under study in form of a matricial distribution as a function of the rotation angles. For obtaining the density distribution, it is sufficient to weigh the matricial representation of the parameter by the probability density matrix calculated at a given temperature:

$$\mathbb{M}\,(3\theta_1,\, 3\theta_2) = M(3\theta_1,\, 3\theta_2)\;\mathbb{D}\,(3\theta_1,\, 3\theta_2) \tag{22}$$

Finally integrating on the angular variables, we obtain a mean value for the parameter at that temperature:

$$\overline{M} = \iint \mathbb{M}\,(3\theta_1,\, 3\theta_2)\, d\theta_1\, d\theta_2 \tag{23}$$

A molecular parameter, which seemed of interest to be determined, was the NMR coupling constant of a hydrogen atom of one of the rotors with a carbon atom of the benzene ring in the ortho-xylene molecule. The theoretical values obtained for this constant show indeed large variations with the conformation. Unfortunately, a tunelling effect, due to the three-fold degeneracy of the potential energy curve of a methyl group, damps completely these variations. The determination was lacking completely in interest.

Another parameter measurable, the value of which depends on the conformation, is the dipole moment. The internal symmetry of the ortho-xylene permits to limit the calculations to the component of the dipole moment, which coincides with the two-fold symmetry axis of this molecule. By using the INDO method, we calculated the components of the dipole moments for different conformations, and we expressed them in spherical coordinates, taking as z axis the two-fold symmetry axis, and as x axis the perpendicular axis lying in the benzene plane. Then, we have determined the probability density distributions of the dipole moment according to (22), at 0°, 25°, 50°, 75°, and 100°. Finally, we calculated the mean values for the dipole moment at these temperatures.

TABLE III. Dipole moment distribution for different conformations of ortho-xylene as a function of temperature

Dipole moment modulus in D.					Dipole orientation in degrees	
θ_1	θ_2	0°C	50°C	100°C	θ	φ
0	0	0.14790	0.13163	0.11956	0.0	0.0
0	30	0.06813	0.06631	0.06445	22.53	45.77
0	60	0.03020	0.03199	0.03331	34.80	0.0
0	-30	0.06813	0.06631	0.06445	22.53	-45.77
30	0	0.06813	0.06631	0.06445	22.53	-134.23
30	30	0.02627	0.02814	0.02951	0.0	0.0
30	60	0.00894	0.01031	0.01164	28.65	-49.06
30	-30	0.02237	0.02444	0.02616	33.67	-90.0
60	0	0.03020	0.03144	0.03331	34.80	180.00
60	30	0.00894	0.01031	0.01164	28.65	130.94
60	60	0.00393	0.00423	0.00464	0.0	0.0
60	-30	0.00894	0.01031	0.01164	28.65	-130.94
-30	0	0.06813	0.06631	0.06445	22.53	134.23
-30	30	0.02237	0.02444	0.02616	33.67	90.0
-30	60	0.00894	0.01031	0.01164	28.65	49.06
-30	-30	0.02627	0.02814	0.02951	0.0	0.0

In table III we give the dipole moment distributions, in spherical coordinates, as a function of the rotation angles and at 0°, 50°, and 100°C. In this table, it is seen how the different contributions are assembled in a bundle, and how they are more probable in the center of this bundle.

In table IV we give the mean values for the dipole moment calculated according to (23) at different temperatures. A significant variation of this parameter with the temperature is observed. In the same table, and for comparison purpose, we have included the theoretical value of dipole moment obtained for the most stable conformation, supposing a frozen molecule (0°K), and the averaged value, supposing all the conformations equiprobable (very high temperature). Here again the relative non-rigidity of the molecule can be estimated.

TABLE IV. Mean dipole-moment values in D. calculated at various temperatures in °C

Temperature	Mean value
Frozen molecule	0.19791
0°	0.16937
25°	0.16845
50°	0.16764
75°	0.16690
100°	0.16487
Free rotation	0.15432

6. CONCLUSION

In the present work, we attempted to give a mathematical description for the non-rigidity of some deformable molecules, in order to open a new way in the study of internal molecular dynamics. In this first work, we limited our-selves to symmetrical two-rotor molecules for simplicity. This limitation prevented more illustrative applications. But, new calculations are now in progress with less symmetrical molecules.

The application to the determination of the mean dipole moment of ortho-xylene, as a function of temperature, is however a good illustration of the possibility or our method. We hope to be able to offer more useful applications, in a next future, especially in the quantum pharmacology and quantum biology fields.

The principal result of this work, however, is to verify that deuteration of the methyl groups in ortho-xylene does not have any influence on the conformation distribution. This fact was known long time ago by the experimentalists, who used this property for determining conformations [2] !

BIBLIOGRAPHY

[1] L. Farnell, W.G. Richards and C.R. Ganellin, J. Theor. Biol., 43, 339 (1974); W.G. Richard in "Quantum Pharmacology", Butterworths, London, 1977.

[2] C. Chachaty, B. Perly and T. Zemb, in "Studies in Physical and Theoretical Chemistry", J. Maruani et al. Eds., (Elsevier, Amsterdam, 1983), 23, 273-290.

[3] Y.G. Smeyers, in "Current Aspect of Quantum Chemistry", R. Carbó Ed., (Elsevier, Amsterdam, 1981), 21, 293-312.

[4] J.E. Kilpatrick and K.S. Pitzer, J. Chem. Phys., 17, 1064 (1949).

[5] C.R. Quade and C.C. Lin, J. Chem. Phys., 38, 540 (1963); C.R. Quade, ibid., 47, 1073 (1967); J.V. Knopp and C.R. Quade, ibid., 48, 3317 (1968).

[6] J. Maruani, A. Hernández Laguna and Y.G. Smeyers, J. Chem. Phys., 63, 4515 (1975).

[7] J. Maruani, Y.G. Smeyers and A. Hernández-Laguna, J. Chem. Phys., 76, 3123 (1982).

[8] Y.G. Smeyers and A. Hernández-Laguna, J. Chim. Phys., 75, 83 (1978).

[9] Y.G. Smeyers, Theochem, 107, 3 (1984).

[10] S.L. Altmann, Proc. R. Soc. Lonson, A 298, 184 (1967).

[11] Y.G. Smeyers and M.N. Bellido, Int. J. Quantum Chem., 23, 507 (1983).

[12] Y.G. Smeyers, A. Huertas-Cabrera and S. Suhai, Theoret. Chim. Acta (Berl.)., 64, 97 (1983).

[13] Y.G. Smeyers and M.N. Bellido, Int. J. Quantum Chem., 19, 553 (1981).

[14] P.R. Bunker, in "Vibrational Spectra and Structure", J.R. Durig Ed., (Dekker, New York, 1975), 3, 1-129.

[15] Y.G. Smeyers, Fol. Chim. Theoret. Lat., 6, 139 (1978).

[16] Y.G. Smeyers and A. Hernández-Laguna, in "Chemical and Biochemical Reactivity", E.D. Bergman et al. Eds., (Israel Academy of Sciences, Jerusalem, 1974), 6, 449-456.

TOPOLOGICAL THEORY OF MOLECULAR CONFORMATIONS

Paul G. Mezey
Department of Chemistry
University of Saskatchewan
Saskatoon, Canada, S7N OWO

ABSTRACT. Some of the recent advances in the topological theory of molecular conformational analysis are reviewed. A formal proof is given for the continuity of energy functionals over a special "reduced" nuclear configuration space, obtained by reducing entire families of equivalent configurations to individual points of a space. The possibility for the existence of "potential defying" conformers is pointed out. Global topological transformations of the energetically important conformational domains of potential surfaces are discussed, which transformations simplify the enumeration of various conformers.

1. INTRODUCTION

Various molecular conformations may play very different roles in chemical processes, which roles are determined in part by the relative stabilities of conformers and by the facility of interconversion processes among them. The stabilities of conformers, as well as their interconversion processes, can be analysed within the framework of the topological theory of molecular conformations.

In the last decade the organic chemistry applications of theoretical conformation analysis and the variety of computational studies on the energy surfaces of conformational rearrangements have proved that quantum chemistry is a powerful complement of experimental conformational research.[1,2] In principle, very similar theoretical methods can be used for the study of chemical reactions and conformational changes, since in both types of processes there is a change in the nuclear geometry. This change is usually much smaller in conformational processes, nevertheless, in some conformational changes the nuclear geometry may undergo very large variations. Yet, the original interpretation of conformational changes as processes in which no bonds are broken or formed,[3] is the usually accepted definition that distinguishes the reaction processes from the simpler conformational rearrangements, even if in the most common cases the internal motions of the latter type do cause some minor changes in the experimentally observed bond length values.

R. Daudel et al. (eds.), Structure and Dynamics of Molecular Systems, 41–56.

Differential and algebraic topology have been proposed[4-10] as exceptionally suitable quantum chemical tools for the analysis of molecular structure. The chemically most important properties of all chemical species composed from a given set of N nuclei can be deduced from a topological analysis of multidimensional potential surfaces. Development and the eventual implementation of the analytic-topological methods may lead to a global description of all of the rearrangements possible for a given set of nuclei. A global model accounts for both "conformational" and "reactive" types of nuclear motions. Within this model conformational changes can also be regarded as special monomolecular reactions. Formally, a global topological model describes all the reaction mechanisms possible on the actual potential energy hypersurface of the N nuclei, satisfying the constraints of the electronic state, in particular, that of the spin multiplicity.

Topological invariants of the space over which the energy functional is defined lead to constraints on the number of possible chemical species and interconversion processes, as well as to bounds on the electronic energy[11-13]. Continuity of the actual energy functional is a condition for the application of these topological invariants. Whereas the hypersurface of the energy functional over the nuclear configuration space, and similarly, over the w-dimensional ($w \geq N$) nuclear charge space[11] the electronic energy functional, both may show pathological behaviour at certain points, nevertheless, in the simplest and the conformationally most important domains, e.g. in the "reactive domains"[4] D_μ of small μ indices, the energy functional can be regarded as a continuous hypersurface (see Appendix). The application of some topological relations is possible only if the energy functional is differentiable. Critical point relations, e.g. the Morse inequalities[14-16] apply only if the second derivatives exist. We assume that the potential surface is twice differentiable in the important domains which are the topological open sets of low energy, given in the

$$Y = \bigcup_i C(\lambda,i) \qquad (1)$$

form, where the Y sets are unions of the open "catchment regions", that is, the unions of the $C(\lambda,i)$ base sets from the actual base defining the n-dimensional topological space (M,T_C) of the T_C reaction topology[6,7].

The topological model may considerably simplify the analysis of nuclear rearrangements. Conformational rearrangements can be represented by sequences of catchment regions. Simple relations in the M space (reduced nuclear configuration space), describing the internal rearrangements confined to the lower regions of the energy surface, can also be represented by the digraph relations of the open sets within this topology. Such digraphs ("directed graphs") are the simplest representations of the conformational rearrangements. Directions on these graphs are defined by energy differences between critical points. Intersection properties of open sets, that is, the neighbour relations of these sets define edge sets for the graphs (digraphs) of conformational mechanisms[17].

The energy of activation of a conformational change is usually much smaller than that of a chemical reaction. This fact may be utilized in our model. One may consider the rearrangements of nuclei corresponding exclusively to conformational changes, by the analysis of the lower regions of hypersurface E(K), that is, by analysing the simply connected level sets F of the energy surface[18,19], where conformational mechanisms are restricted to open sets of the

$$X = \bigcup_i F \cap C(\lambda, i) \qquad (2)$$

type. Here $F \cap C(\lambda, i)$ represents a conformer within level set F.

2. REPRESENTATION OF CONFORMERS IN THE REDUCED NUCLEAR CONFIGURATION SPACE M.

2.1. Potential energy functional E(K) over the reduced nuclear configuration space M.

Consider a general nuclear configuration space ${}^{3N}R$ of dimension 3N, defined in terms of the mass-weighted Cartesian coordinates of the N nuclei in a fixed laboratory frame, and the Born-Oppenheimer potential energy hypersurface $E_{BO}(x)$, of a fixed electronic state Ψ, where $x \in {}^{3N}R$. $E_{BO}(x)$ is not necessarily differentiable for every x (e.g., if x is a point of a conical intersection) and it is convenient to replace $E_{BO}(x)$ with a functional E(x) with the required differentiability properties (e.g., twice continuously differentiable). One may require that $E_{BO}(x) = E(x)$ in all of ${}^{3N}R$, except in some D_{excl} excluded domains containing these special x points of non-analyticities.

An explicit construction for a reduced nuclear configuration space M has been proposed[9] that is based on equivalence classes introduced into the 3N dimensional nuclear configuration space ${}^{3N}R$. Consider the set $\{TY\}$ of transformations in ${}^{3N}R$

$$T(\Theta_1, \Theta_2, \Theta_3)Y(\beta_1, \beta_2, \beta_3), \qquad (3)$$

$$0 \le \Theta_i \le 2\pi, \quad -\infty < \beta_i < \infty, \qquad (4)$$

where $T(\Theta_1, \Theta_2, \Theta_3)$ and $Y(\beta_1, \beta_2, \beta_3)$ represent the ${}^{3N}R$ space effects of rotations and translations, respectively, of an entire nuclear configuration within the laboratory frame: Θ_i and β_i refer to the ith laboratory frame axis. Explicit forms of these transformations have been given earlier[9]. An equivalence relation is defined for points of ${}^{3N}R$: x and $x' \in {}^{3N}R$ are regarded equivalent if and only if there exists a TY transformation for some parameters Θ_i and β_i, i=1,2,3 which interrelate x and x' by

$$x' = T(\Theta_1, \Theta_2, \Theta_3)Y(\beta_1, \beta_2, \beta_3)x. \qquad (5)$$

Then, set $\{TY\}$ generates an equivalence class partitioning of ${}^{3N}R$ into various subsets $K \subset {}^{3N}R$, where these K equivalence classes are defined in terms of nuclear configurations in ${}^{3N}R$ which are related to each other by rigid translation and rotation of the molecule within a laboratory frame. These equivalence classes give a partitioning of space ${}^{3N}R$,

$$\bigcup_{\alpha} K_{\alpha} = {}^{3N}R \tag{6}$$

and a new space M (the reduced nuclear configuration space) can be defined as the quotient space, that is, by taking the K sets as elements:

$$M = \{K\} \tag{7}$$

A metric d has been proposed for M turning it into a metric space[20], that metric inherits some of the important properties of the euclidean metric ϱ of ${}^{3N}R$:

$$d(K,K') = \min\{\varrho(x,x') : x \in K,\ x' \in K'\}. \tag{8}$$

It is easy to prove[20] that this minimum exists for certain x, x' pairs of points in every pair of K and K' sets, the minimum is positive whenever $K \neq K'$ and d(K,K') is a proper metric in M. In the actual proof an additional feature has also been demonstrated: for any fixed $x \in K$ there exists an $x' \in K'$ for each $K' \in M$ such that the minimum distance d(K,K') is realized:

$$d(K,K') = \varrho(x,x') . \tag{9}$$

For a general polyatomic molecule each equivalence class $K \subset {}^{3N}R$ is a six dimensional subset of ${}^{3N}R$. Attempts to visualize features of such multidimensional objects, embedded in higher dimensional spaces, are of little help and are often dangerously misleading, when analysing their properties. Nevertheless, a cautious reader may find some comfort in visualizing a drastically oversimplified three dimensional analogue of a system of K sets as annual growth rings of a hypothetical, infinitely tall and infinitely thick tree.

Since the reduced nuclear configuration space M is a metric space (M,d) with the metric d(K,K'), on this space M one may re-define the energy functional, based on E(x) of ${}^{3N}R$. The definition of the K equivalence classes implies that in the absence of external fields, disregarding rigid rotation and rigid translation of molecules,

$$E(x_1) = E(x_2) \tag{10}$$

whenever

$$x_1,\ x_2 \in K \tag{11}$$

Consequently, one may take, by definition

$$E(K) = E(x) \tag{12}$$

with any

$$x \in K. \tag{13}$$

This relation (12) defines E(K) over the entire M.

Evidently, continuity properties of E(K) are of fundamental importance in the applications of the M space model to chemical problems. Fortunately, if E(x) is a continuous functional over ${}^{3N}R$ in the metric ρ then E(K) is also a continuous functional over the reduced nuclear configuration space M, in the metric d. Whereas for the actual functionals this conclusion can be deduced from certain general properties of quotient spaces, instead of relying on abstract theorems, we shall give a direct proof for the continuity of E(K) in the Appendix. Parts of this proof will be also used in a subsequent study on dynamic problems.

2.2. Conformers as catchment regions in subsets of M.

There is a pointwise correspondence between any steepest descent path $P_{x'} \subset {}^{3N}R$ of E(x) with origin $x' \in {}^{3N}R$ and a path $P_{K'} \subset M$ of E(K) with origin $K' \in M$ where $x' \in K'$. We shall refer to path $P_{K'}$ as a steepest descent path of E(K) on M. This implies that a <u>catchment region topology</u> T_C can be introduced directly into M, as follows:

Let us denote by D, $D \subset M$, the collection of all K equivalence classes for which the $x \in K$ points of ${}^{3N}R$ are in $D_{excl} \subset {}^{3N}R$:

$$D = \left\{K:\ K \in M,\ x \in K \Rightarrow x \in D_{excl}\right\}, \tag{14}$$

which set D is the union of its maximum connected components D(i)

$$D = \bigcup_i D(i)\ . \tag{15}$$

Within each D(i) we choose arbitrarily a point that we shall denote by K(-1,i). The formal index -1 serves to distinguish these elements from the critical points $K(\lambda,i)$ of the energy hypersurface E(K), where $\lambda \geq 0$ is the number of negative eigenvalues of the Hessian matrix $\underline{\underline{H}}(K(\lambda,i))$, and index i is an index of ordering.

We define a function

$$\eta(K): M \rightarrow M, \tag{16}$$

that assigns to each point $K \in M$ the extremity $K_e = K(\lambda,i)$ of steepest descent path P_K of origin K, if $P_K \cap D = \emptyset$, or $K_e = K(-1,i)$ if P_K enters D(i):

$$\eta(K) = \begin{cases} K(\lambda,i) & \text{if } K_e(P_K) = K(\lambda,i),\ P_K \cap D = \emptyset, \\ K(-1,i) & \text{if } P_K \cap D(i) \neq \emptyset. \end{cases} \tag{17}$$

In order to make the above definition applicable for the entire set M, we take,

$$P_K = K = K_e(P_K), \tag{18}$$

whenever K itself is a critical point,

$$K = K(\lambda, i) \tag{19}$$

or whenever it falls within D,

$$K \in D. \tag{20}$$

The catchment regions $C(\lambda,i)$ in M are defined as

$$C(\lambda,i) = \left\{K:\ K \in M,\ \eta(K) = K(\lambda,i)\right\}, \tag{21}$$

that is, as the collection of all those points K in M from where a P_K steepest descent path leads to (is assigned to) critical point (distinguished point) $K(\lambda,i)$, for $\lambda \geq 0$ ($\lambda = -1$, respectively). In order to define a catchment region topology with respect to the given energy functional E(K) of a specified electronic state, we generate a defining sub-base from these catchment regions[9,20]. Take the family C' of catchment regions on M:

$$C' = \left\{C(\lambda,i)\right\} \tag{22}$$

and the family C" of their closures in the metric d of M:

$$C'' = \left\{\overline{C}(\lambda,i)\right\} . \tag{23}$$

Set C, given as

$$C = C' \cup C'', \tag{24}$$

is a defining sub-base for a unique topology T_C on M, called catchment region topology (or reaction topology) of M. Topology T_C is a family of subsets of M. These and only these subsets are considered open sets (T_C-open sets) of the reduced nuclear configuration space M in the given topologization. Each such open set can be obtained as a union of finite intersections of sets from sub-base C. The above construction ensures that these open sets indeed fulfill the conditions required for a topology on M (for some more mathematical details see ref [9]).

The set M, together with its actual topology T_C is called the topological space (M, T_C). In the above model we have assumed that no degenerate critical point exists on E(K), that always can be achieved by an arbitrarily small distortion of E(K). This topological model of the reduced nuclear configuration space M, with respect to energy hypersurface E(K), leads, in a natural way, to a topological representation of chemical species, that reflects the fundamental non-geometrical (in fact topological) properties of quantal particles. The topological concept of chemical structure also has some practical implications with regard to quantum chemical synthesis design. In the anticipated applications of potential energy hypersurfaces in theoretical synthesis planning, it is convenient to define chemical structure and reaction mechanism in terms of properties of energy hypersurfaces[9].

In topological space (M, T_C) chemical structures are represented by catchment regions $C(\lambda,i)$, that are T_C-open sets by definition. A catchment region $C(\lambda,i)$ with index $\lambda = 0$, i.e. the n-dimensional basin associated with a critical point $K(\lambda,i)$ that is a minimum point of E(K), represents a stable molecule . A catchment region $C(\lambda,i)$ with index $\lambda = 1$, i.e. the (typically but not necessarily) n-1 dimensional basin associated with a critical point $K(\lambda,i)$ that is a saddle point of index 1 of E(K), represents a transition structure ("transition state"). Catchment regions $C(\lambda,i)$ with higher λ indices ($\lambda > 1$) represent unstable formal species in the topological model. Although their direct chemical importance is minimal, nevertheless, these formal species are also needed for a complete partitioning of the potential energy hypersurface E(K) and that of M:

$$M = \bigcup_{\lambda,i} C(\lambda,i) \ . \tag{25}$$

Evidently, as equivalence classes generated by function $\eta(K)$, any two different catchment regions are disjoint:

$$C(\lambda,i) \cap C(\lambda',i') = \emptyset \tag{26}$$

unless $\lambda=\lambda'$ and $i=i'$ (where $\emptyset$ is the empty set).

The above T_C reaction topology (catchment region topology) is suitable for the global analysis of relations among all chemical species formed from the given collection of N nuclei, subject to the electronic constraints imposed by the given energy hypersurface. Yet, in a conformational problem only a very limited number of these chemical species are relevant. Any conformational problem is subject to two (somewhat related) constraints:

(i) conformational changes in general require much less activation energy than reaction processes. Hence, a conformational analysis can be restricted to the lower energy regions of a hypersurface.

(ii) in conformational processes there are limits on how far various molecular moieties can be displaced from one another. By contrast, in chemical reactions very large (formally, infinitely large) separations of molecular fragments or of newly formed molecules may routinely occur, e.g. in a dissociation reaction or elimination process. In a conformational process there is a limit on the "crowdedness" (or, on the density of matter) within a molecular neighbourhood in three space.

Hence it is natural to restrict a conformational analysis to subsets of M, e.g. to those domains where the energy expectation value functional is less than some fixed value E'. A level set F of M, can then be defined with respect to E(K) and E' as

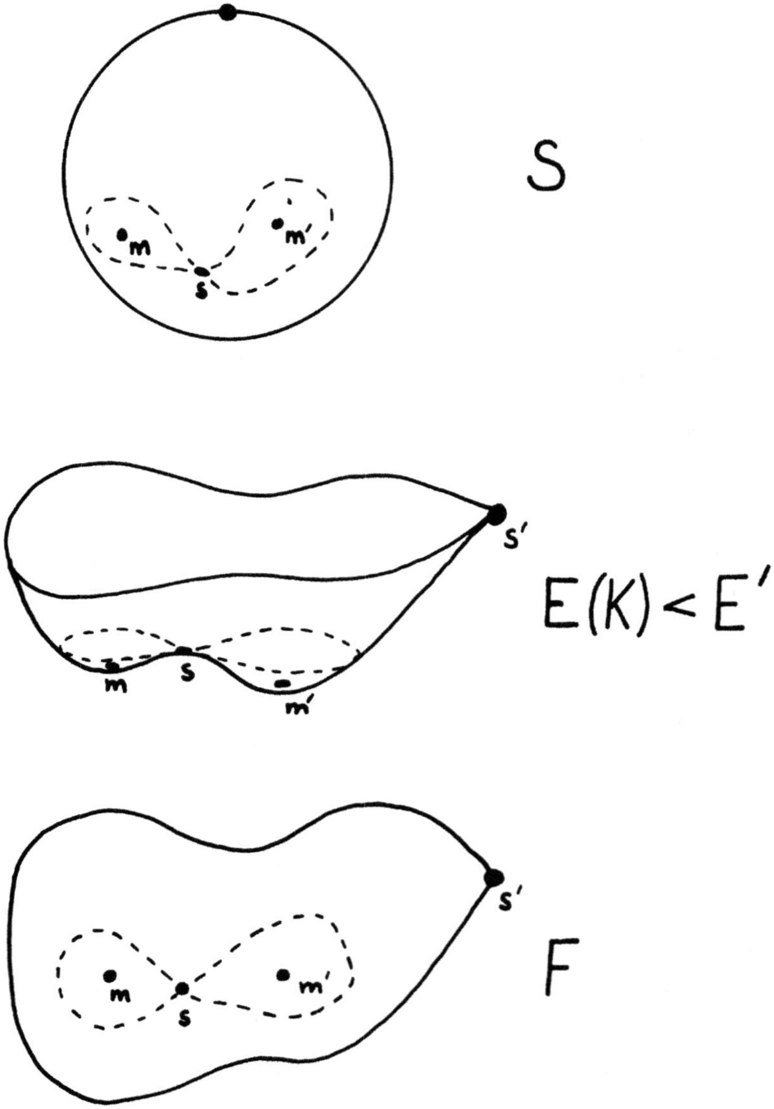

Figure 1. Level set F of a two dimensional conformational domain $E(K)<E'$ and its image on compact manifold (sphere) S.

$$F = \left\{ K: E(K) < E' \right\}. \tag{27}$$

The <u>relative catchment region topology</u> $T_{C/F}$ on set F is defined as the following family of open sets:

$$T_{C/F} = \left\{ G: G = F \cap G_C,\ G_C \in T_C \right\}. \tag{28}$$

The resulting topological space (F, $T_{C/F}$) is simpler to analyse than (M, T_C), and it can be applied in conformation analysis, e.g. if the chosen E' value is less than the minimum dissociation energy required to break a bond in the given molecule[18].

In order to exploit the spatial constraints of type (ii), a technique, based upon the hyperspherical coordinate representation of nuclear geometries, has been proposed[19]. In such a representation there is one distance-like coordinate, the hyperradius r, and all other coordinates are angles. There are many possible choices for the definition of the n-1 angle parameters. The hyperradius r, can be obtained as

$$r^2 = \sum_{i=2}^{N} d_i^2 \tag{29}$$

where d_i is the mass-weighted distance of the i-th nucleus from the center of mass of nuclei 1,...i-1. Hyperradius r serves then as a natural measure of the "crowdedness" of nuclear configurations, and a conformational analysis can be restricted to a multidimensional hyperspherical shell, constrained by suitably chosen r_{min} and r_{max} values. An integration of the hyperspherical constraints with the level set constraints of type (27) has been discussed earlier[19].

Constraints of type (i) and (ii), when applied to all possible conformations accessible by a given molecule, define a subset of M. We denote this subset, the <u>conformational domain</u>, by F. This conformational domain is necessarily a <u>connected</u> subset of M, and the relative catchment region topology can be introduced on F by applying relation (28). Evidently, a constraint of type (i) implies that within F all catchment regions $C(\lambda,i)$ with $\lambda = 0$, are <u>accessible conformers</u> of one another. Similarly, within F all catchment regions $C(\lambda,i)$ with $\lambda = 1$ are transition structures of conformational interconversion processes.

A topological characterization of such conformational domains F of a given catchment region partitioning is obtained by exploiting their topological equivalence with abstract, multidimensional polyhedra. Catchment regions $C(\lambda,i)$ of F correspond the n-λ dimensional faces of such a polyhedron, as it can be easily demonstrated by mapping F to a compact manifold[18]. For example, a simply connected two dimensional conformational domain F can be mapped onto a punctured sphere, and the catchment regions of F induce a partitioning of this sphere, equivalent to various faces, edges and vertices of a polyhedron. Although such analogies are of no help in higher dimensions, this sphere is analogous to the globe, with geographical basins corresponding to C(0,i) type

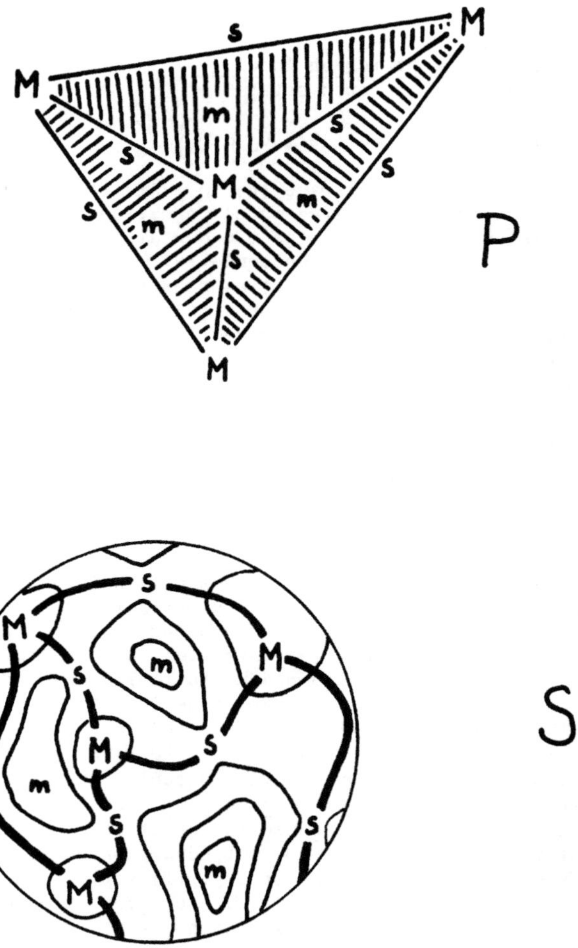

Figure 2. Two dimensional potential energy function defined over manifold (sphere) S. Symbols **m**, s and M stand for minimum, saddle point and maximum, respectively. Heavy lines show the boundaries of catchment regions, thin lines show equipotential contours. One additional minimum is on the far side of the sphere. P is the topologically equivalent reaction polyhedron (conformational polyhedron).

catchment regions of stable conformers, (i.e. to faces of the polyhedron), mountain ridges corresponding to C(1,i) transition structures (i.e. to edges of the polyhedron) and mountain peaks analogous to the C(2,i) type zero dimensional catchment regions (to vertices of the polyhedron).

Some of the topological invariants of such reaction polyhedra, in our case, of conformational polyhedra, (since within F only conformational changes occur), have been already used as a characterization tool[9,18,19]. Homology groups and cohomology groups of these reaction polyhedra are such invariants, in particular, the rank B_λ of the λth homology group ($\lambda = 0,1,...n$) is an important topological invariant, which has been used as a constraint in the enumeration problem of conformers[18,19].

2.3. Comments on the role of vibrational energy.

A recent finding[21,22,23] on the role of vibrational stabilization on a reactive energy surface suggests the possibility that "potential defying conformers" may exist. As it has been shown, the existence of the IHI molecule is profoundly affected by vibrational stabilization and destabilization in various domains of the corresponding nuclear configuration space. This example is very special, since no true non-degenerate minimum exists on the Born-Oppenheimer ground electronic state potential surface, only degenerate minima at infinite nuclear separations. Nevertheless, the formal decrease of zero point energy in the neighbourhood of a saddle point of the hypersurface leads to a bound state in its neighbourhood.

Upon inclusion of a minimum vibrational energy component similar "potential defying" chemical species, stable molecules or transition structures, may emerge in domains where the qualitative features of potential energy hypersurfaces do not indicate them. The existence of such potential defying species can be investigated using a topological technique[9]. A continuous energy functional $E^o(K)$ can be defined over M (hence, over F) which incorporates the zero point vibrational energy at potential minima and also at various other critical points of E(K). At intermediate points $K \in M$, $E^o(K)$ smoothly interpolates between energy values at critical points. A semiclassical interpretation of $E^o(K)$ can be given by a formal separation of nuclear geometry changes at critical points of E(K) into two classes: to those along positive curvature directions, which are treated quantum-mechanically and give rise to some zero point energy contribution, and to those along the negative curvature directions, which are usually treated semi-classically. A catchment region analysis of the new hypersurface $E^o(K)$, in terms of its critical points, reflects the relative stabilities of various species when the minimum, intrinsic vibrational energy is also taken into account, in addition to the potential energy.

Assuming that in the topological model the vibrational energy contribution to the total energy can be "switched on" in a continuous manner, all fundamental changes in the catchment region structure of the nuclear configuration space can be detected by monitoring the occurrence of degenerate critical points. The potential energy hypersurface E(K), defined on a differentiable manifold $\underset{\sim}{F}$, obtained from F, can be continuously deformed into a hypersurface $E^{o}(K)$ which latter hypersurface incorporates the minimum vibrational energy contribution. This continuous deformation can be accomplished by a (generally non-unique) homotopy H(K, t) defined as

$$H(K, t): \underset{\sim}{F} \otimes I \rightarrow {}^{1}R, \; K \in \underset{\sim}{F}, \; t \in I \tag{30}$$

$$H(K, 0) = E(K) \tag{31}$$

$$H(K, 1) = E^{o}(K) \tag{32}$$

$$H(K, t) = E_{(t)}(K): \underset{\sim}{F} \rightarrow {}^{1}R \tag{33}$$

where functionals $E_{(t)}$ are twice continuously differentiable hypersurfaces defined on $\underset{\sim}{F}$ for every $t \in I$, where I = [0, 1], the unit interval. Choosing <u>any</u> homotopy H(K, t) satisfying the above criteria, a necessary condition for the existence of potential defying conformers on $E^{o}(K)$ is the existence of at least one degenerate critical point of hypersurface $H(K, t) = E_{(t)}(K)$ for some value t',

$$t' \in (0,1). \tag{34}$$

Note, that the topological invariants and critical point relations must also apply to both hypersurfaces E(K) and $E^{o}(K)$, since they both are defined on the same manifold $\underset{\sim}{F}$. Consequently, the number and type of potential defying chemical species, in our case, the number and type of potential defying conformers, are constrained by these topological relations[9]. In particular, if a potential-defying critical point appears on $E^{o}(K)$, then it must be counterbalanced by some other potential defying feature, e.g. by the appearance or disappearance of another critical point, in order to keep the topological invariants unchanged. Hence, potential defying features, if they occur at all, can never occur alone.

APPENDIX

We present here a direct proof (involving only elementary "ε,δ" considerations and no sophisticated theorems) that continuity of the energy functional E(x) over the mass-weighted 3N dimensional nuclear configuration space ${}^{3N}R$ of metric ρ implies the continuity of the E(K) energy functional over the reduced nuclear configuration space M of metric d.

In particular, we want to show that for each $K \in M$ and for each positive number $\varepsilon_K > 0$ there exists a positive number $\delta_K > 0$, such, that for any point $K' \in M$ fulfilling condition

$$d(K,K') < \delta_K, \tag{A1}$$

the relation

$$|E(K) - E(K')| < \varepsilon_K \tag{A2}$$

follows, subject to the analogous continuity condition for E(x) in ^{3N}R.

Choose any point $K \in M$ and any number $\varepsilon_K > 0$. Define a subset $\underset{\sim}{K}$ of M as

$$\underset{\sim}{K} = \left\{ K'' : |E(K) - E(K'')| \geq \varepsilon_K \right\} \tag{A3}$$

that is, as the collection of all those K'' points for which the energy difference is larger than or equal to ε_K. If there is no such $\underset{\sim}{K}$ set, then (A2) is valid for any $\delta_K > 0$ and we are ready. Otherwise, define a quantity $\underset{\sim}{\delta}$ as the greatest lower bound for distances between K and the elements of $\underset{\sim}{K}$:

$$\underset{\sim}{\delta} = \inf\left\{ d(K,K'') : K'' \in \underset{\sim}{K} \right\} . \tag{A4}$$

From property (9) it follows that for any fixed

$$x \in K \tag{A5}$$

there must exist certain

$$x'' \in K'' \tag{A6}$$

within each set $K'' \in \underset{\sim}{K}$ such that

$$d(K, K'') = \rho(x, x''). \tag{A7}$$

Hence,

$$\underset{\sim}{\delta} = \inf\left\{ \rho(x,x'') \right\} \tag{A8}$$

is also true.

We show now that $\underset{\sim}{\delta}$ cannot be zero, hence it must be positive, $\underset{\sim}{\delta} > 0$. In order to show this, consider the continuity condition with the same ε_K value at point x for functional E(x) over ^{3N}R, which functional is continuous according to our assumption: at the point

$$x \in K \subset {}^{3N}R \tag{A9}$$

there must exist some

$$\delta > 0 \tag{A10}$$

such that for each

$$x''' \in {}^{3N}R \tag{A11}$$

satisfying

$$\rho(x,x''') < \delta \tag{A12}$$

the relation

$$\left| E(x) - E(x''') \right| < \varepsilon_K \tag{A13}$$

follows. We show now that for such a positive $\delta > 0$,

$$\delta \leq \underset{\sim}{\delta} . \tag{A14}$$

If, contrary to (A14), $\delta > \underset{\sim}{\delta}$ were true, i.e. δ were greater than the infimum in (A4), then there would exist some

$$x''' \in K''' \in \underset{\sim}{K} \tag{A15}$$

for which

$$\underset{\sim}{\delta} < \delta(x,x''') < \delta . \tag{A16}$$

But then (A15) would also imply that

$$\left| E(K) - E(K''') \right| \geq \varepsilon_K \tag{A17}$$

and substitutions E(x) = E(K) and E(x''') = E(K''') would give

$$\left| E(x) - E(x''') \right| \geq \varepsilon_K \tag{A18}$$

that contradicts continuity condition (A13) of E(x). Hence $\delta > \underset{\sim}{\delta}$ is impossible and (A14) is true, implying that

$$\underset{\sim}{\delta} > 0 . \tag{A19}$$

Since according to (A4) no $K' \in M$ of distance

$$d(K, K') < \underset{\sim}{\delta} \tag{A20}$$

can be an element of $\underset{\sim}{K}$;

$$K' \notin \underset{\sim}{K}, \tag{A21}$$

that implies, by virtue of (A3), that

$$\left| E(K) - E(K') \right| < \varepsilon_K \tag{A22}$$

We know already that $\underset{\sim}{\delta} > 0$, hence this positive $\underset{\sim}{\delta}$ can be chosen as δ_K,

$$\delta_K = \underset{\sim}{\delta} > 0 \tag{A23}$$

Consequently, (A2) is valid for any $K' \in M$ fulfilling (A1), that is, E(K) is a continuous functional at any point $K \in M$ if E(x) is a continuous functional over ^{3N}R at every point $x \in K$.

Q.E.D.

Note that differentiability of E(x) over ^{3N}R implies that at the given point x E(x) can be replaced locally by a linear functional E'(x). Differentiability properties of E(K) can be deduced from a similar, although slightly more involved derivation, applied to E(x) and to these linear functionals E'(x).

ACKNOWLEDGEMENTS

This work was supported by a grant from the Natural Sciences and Engineering Research Council of Canada. It is a pleasure to acknowledge the assistance of Frances Chow and H.I. Minty.

REFERENCES

1. T. Oie, G.M. Maggiora, R.E. Christoffersen and D.J. Duchamp, 'Development of a flexible intra- and intermolecular empirical potential function for large molecular systems', Int. J. Quantum Chem. Quant. Biol. Symp., **8**, 1-47(1981).
2. P.G. Mezey, 'Analysis of conformational energy hypersurfaces', Progr. Theor. Org. Chem., **2**, 127-161(1977).
3. E.L. Eliel, N.L. Allinger, S.L. Angyal and G.A. Morrison, Conformational Analysis, Wiley, New York, 1965.
4. P.G. Mezey, 'Reactive domains of energy hypersurfaces and the stability of minimum energy reaction paths', Theor. Chim. Acta, **54**, 95-111(1980).
5. P.G. Mezey, 'Critical level topology of energy hypersurfaces', Theor. Chim. Acta, **60**, 97-110(1981).
6. P.G. Mezey, 'Catchment region partitioning of energy hypersurfaces', Theor. Chim. Acta, **58**, 309-330(1981).
7. P.G. Mezey, 'Topology of energy hypersurfaces', Theor. Chim. Acta, **62**, 133-161(1982).
8. P.G. Mezey, 'The topological model of non-rigid molecules and reaction mechanisms', p. 335-353 in Symmetries and properties of non-rigid molecules: a comprehensive survey, Eds.: J. Maruani and J. Serre, Elsevier, Amsterdam, 1983.
9. P.G. Mezey, 'Reaction topology: manifold theory of potential surfaces and quantum chemical synthesis design', p. 75-98 in Chemical applications of topology and graph theory, Ed.: R.B. King, Elsevier, Amsterdam, 1983.
10. P.G. Mezey, 'Classification schemes of nuclear geometries and the concept of chemical structure', J. Chem. Phys., **78**, 6182-6186(1983).
11. P.G. Mezey, 'Level set topology of the nucleai charge space and the electronic energy functional', Int. J. Quantum Chem., **22**, 101-114(1982).
12. P.G. Mezey, 'The symmetry of electronic energy level sets and total energy relations in the abstract nuclear charge space', Mol. Phys., **47**, 121-126(1982).

13. P.G. Mezey, 'Constraints on electronic energy hypersurfaces of higher multiplicities', J. Chem. Phys., **80**, 5055-5057(1984).
14. M. Morse, Calculus of Variations in the large, Am. Math. Soc. Colloq. Publ., Vol. **18**, 1934.
15. M. Morse and S.S. Cairns, Critical point theory in global analysis and differential topology, Academic, New York, 1969.
16. J. Milnor, Morse theory, Ann. Math. Studies, Vol. **51**, Princeton Univ. Press, Princeton, 1973.
17. P.G. Mezey, 'Quantum Chemical reaction networks, reaction graphs and the structure of potential energy hypersurfaces', Theor. Chim. Acta, **60**, 409-428(1982).
18. P.G. Mezey, 'An approach to conformation analysis on multidimensional potential surfaces', Int. J. Quantum Chem. Quant. Biol. Symp., **10**, 153-160(1983).
19. P.G. Mezey, 'The hyperspherical coordinate representation of potential surfaces of large molecules', Int. J. Quantum Chem. Quant. Biol. Symp., **11**, 000(1984).
20. P.G. Mezey, 'The differentiable manifold model of quantum-chemical reaction networks', Int. J. Quantum Chem. Quant. Chem. Symp., **17**, 137-152(1983).
21. J. Manz, R. Meyer, E. Pollak and J. Romelt, 'A New Possibility of chemical bonding: vibrational stabilization of IHI', Chem. Phys. Letters, **93**, 184-187(1982).
22. D.C. Clary and J.N.L. Connor, 'Existence of a bound state for the three-dimensional IHI Molecule on a purely repulsive potential energy surface', Chem. Phys. Letters, **94**, 81-84(1983).
23. E. Pollak, 'Do vibrationally adiabatic molecules exist in three dimensions?', Chem. Phys. Letters, **94**, 85-89(1983).

TOPOLOGICAL MODEL OF REACTION MECHANISMS

Paul G. Mezey
Department of Chemistry
University of Saskatchewan
Saskatoon, Canada, S7N OWO

ABSTRACT. The family of all possible reaction mechanisms among molecules involving a fixed set of atoms and constrained by a given electronic state form an algebraic group, the fundamental group of reaction mechanisms. Some of the recent developments leading to the topological model and to this group of reaction mechanisms are reviewed. A proof is given for an equivalence relation, used in establishing the link between two topological models of quantum chemical reaction mechanisms, one based on homotopy equivalence classes of reaction paths, the other based on topological open sets of a nuclear configuration space. The algebraic structures of reaction mechanisms are of some importance in computer-based quantum chemical synthesis planning and molecular design.

1. INTRODUCTION

The concept of reaction mechanism plays a fundamental role in the interpretation of molecular processes and in the classical picture how the chemist usually visualizes a chemical reaction. Usually, a chemical reaction mechanism is thought of as a geometrical rearrangement of molecules and molecular fragments, that may be represented by an assumed reaction path, most commonly, by a minimum energy path[1]. Such a path describes a specific geometrical rearrangement of the nuclei within a classical mechanical model. Implicit in the correct interpretation of the above model, however, is the understanding that any given reaction path may serve, at most, as a rough guideline to the interpretation of the actual chemical process. Molecules are quantum mechanical entities, exhibiting wave-packet properties and their rearrangements do not follow any specific path in the classical sense. This realization has led to the concept of reaction channels. Such a reaction channel can be represented as a family (more precisely, as an open set) of reaction paths which paths are equivalent in certain sense. All formal, classical reaction paths following roughly the same general route from reactant to product, are regarded to belong to the same reaction channel.

R. Daudel et al. (eds.), Structure and Dynamics of Molecular Systems, 57–70.

There is an extensive literature on the mathematical-geometrical properties of reaction paths[1-3], yet relatively little attention has been paid to the global mathematical properties of reaction mechanisms represented by families of equivalent reaction paths. It may have been thought that the properties of reaction mechanisms are already described in terms of properties of individual reaction paths, and there would be little to gain by analysing the collective properties of families of reaction paths. This turns out not to be the case.

Using the concept of families of equivalent reaction paths instead of individual paths, brings about a remarkable simplification in the analysis of potential energy hypersurfaces. The most fundamental relations among reaction paths and among reaction mechanisms can be formulated in terms of an algebraic structure, much like molecular symmetry properties can be formulated in terms of an algebraic structure: group theory. The very large set of all individual reaction paths on a hypersurface has only very few algebraic properties, none of which properties is directly applicable to a significant simplification of the quantum chemical reaction problem. On the other hand, reaction mechanisms, defined in terms of a suitable equivalence relation of reaction paths, have a rich algebraic structure, containing an important group: the fundamental group of reaction mechanisms[4-6].

The natural way to a precise definition of quantum mechanical reaction mechanisms and to their algebraic structure is through differential and algebraic topology[7-9]. Applied topology in physics and chemistry shows some common features with applied group theory, especially in its potential to simplify complex problems and to provide proofs for the existence of relations between seemingly very different problems, e.g. between different reaction mechanisms. However, the special facility of topology to deal with both continuous and discrete problems within a unified framework, suggests it as an ideal mathematical tool for quantum mechanical applications, perhaps even more powerful than group theory. Topology appears to be the natural mathematical language for quantum mechanical models exhibiting a very special blend of discrete features (e.g. quantized states) and continuous ones (eg.wavefunctions). Whereas the underlying analogies with group theory are generally strong, in certain sense the fundamental group of reaction mechanisms and the related algebraic structures discussed in this study formally short-circuit this analogy between topology and group theory. There is more than a mere analogy: it is, indeed, the topological analysis that leads to the very definition of these groups.

Various branches of group theory have been applied extensively in studies on individual reaction paths: on the symmetry of minimum energy reaction paths[10,11], narcissistic reactions[12], transition "state" symmetries[13,14], the role of permutation groups, symmetry groups of non-rigid molecules[15-18], and many other related problems. The topological approach discussed in this study has a different goal and uses different tools: the analysis is centered on the relations between different quantum chemical reaction mechanisms and the primary tools are the one dimensional homotopy group and various groupoids. The approach taken also differs from earlier ones in its scope.

Instead of the analysis of individual nuclear configurations or certain type of large amplitude motions it has the potential for a global analysis of a complete set of reaction mechanisms. Hence, in view of the spectacular and still accelerating development of computer hardware and with super-computers of 10-100 "gigaflop" capabilities on the horizon, such global topological techniques are likely to find application in quantum chemical molecular design and synthesis planning[3].

2. GROUP THEORY OF REACTION MECHANISMS ON POTENTIAL SURFACES

2.1. The global problem of reaction mechanisms

The collective mathematical (algebraic) properties of the family of all reaction mechanisms associated with a given potential energy hypersurface, are of great theoretical interest and also of potential importance in using calculated energy surfaces for the analysis of chemical reactions as possible aids in quantum chemical synthesis planning. The techniques used for the study of reaction mechanisms differ from those used for investigating individual reaction paths. The essential difference between the two types of problems is the fact that individual reaction paths can be analysed locally, using well known techniques, whereas the relations between reaction mechanisms depend on some global properties of the potential energy hypersurface, hence they require some form of global analysis. A local analysis of an individual reaction path may appear simpler than any global analysis, nevertheless, the problem of a direct, exhaustive analysis of all possible reaction paths on a given hypersurface is an intractable one. On the other hand, a global approach leads to a well defined algebraic problem of a discrete set of all reaction mechanisms, it retains the most important chemical information and brings about a remarkable simplification.

A simultaneous treatment of entire molecular families within a comprehensive theoretical-mathematical model is particularly useful for the study of interrelations among molecules and chemical reactions. One approach that has been suggested recently is the global analysis of all reaction mechanisms of all chemical species formed from a fixed set of nuclei, which species are associated with a given potential energy hypersurface[4-6]. This analysis is based on homotopy theory: quantum chemical reaction mechanisms are represented by homotopy equivalence classes[4-6] of reaction paths on the hypersurface, that is, by homotopy classes where within each equivalence class the reaction paths are continuously deformable into one another (homotopic to one another). Within this model the results of algebraic topology on the homotopy classes of paths in a general topological space[8,9] are directly applicable to reaction paths and to reaction mechanisms[4]. Before describing the application of the general results, we shall consider some chemical concepts which justify the mathematical model.

This topological model has a simple chemical interpretation. The key to the application of the general results of homotopy theory[8,9] is the representation of reaction mechanisms by homotopy equivalence classes of reaction paths. This representation, however, is based on the actual chemical equivalence of all those reaction paths which lead from some fixed reactant to some fixed product, and which paths are "not too different" from one another. The above "chemical" condition can be formulated in rigorous topological language: two reaction paths are regarded "not too different" if they can be continuously deformed into each other below some fixed energy value A. Energy bound A can be chosen such that it restricts the detailed analysis to the chemically most important low energy regions of the energy hypersurface. Note, however, that the analysis is valid for any choice of A. Such an energy condition is sufficient for an equivalence class (homotopy class) classification of all reaction paths on the entire energy hypersurface. Hence, the same conditions also lead to a significant reduction of the very large set of all reaction paths to a much smaller, hence more managable set: to the family of equivalence classes. A straightforward application of homotopy theory shows that the family of all reaction mechanisms has an algebraic structure. Further simplification is possible by considering only a subset of the reaction mechanisms, which subset is still sufficient to describe any chemical process on the surface. One finds that the family of all fundamental reaction mechanisms on the potential energy hypersurface is a group, the fundamental group of reaction mechanisms. This terminology follows from the topological concept of the fundamental group of a topological space. However, this terminology also reflects the chemical significance of this group, since the group generators ("generator mechanisms") of this group are sufficient to generate any reaction mechanism on the hypersurface, hence these groups have, indeed, some fundamental chemical role[4-6]. As it has been pointed out[6], homotopy equivalence classes on a chemical potential surface may serve as natural constraints for energy-constrained Feynman path integrations. Essentially the same homotopy class analysis is applicable for more general energy functionals, for example, for energy functionals defined over the space of parameters of the generator coordinate (G.C.) model of molecular processes[19].

2.2. Reaction mechanisms as homotopy classes of reaction paths

Our first step is to restrict the analysis to the chemically important lower regions of the potential energy hypersurface. We choose an energy bound A (that bound may in fact be a rather high value). A level set F=F(A) of the potential energy hypersurface E(K) over the nuclear configuration space M is defined as the collection of all those points of M where the energy is less than this fixed value A:

$$F = \{K : K \in M, E(K) < A\} \tag{1}$$

Here we shall regard a maximum connected component F that is typically a multiply connected set. (For example, a surface with some neighborhoods of hilltops "removed" above energy A, is multiply connected. A simple two dimensional example is shown in Figure 1).

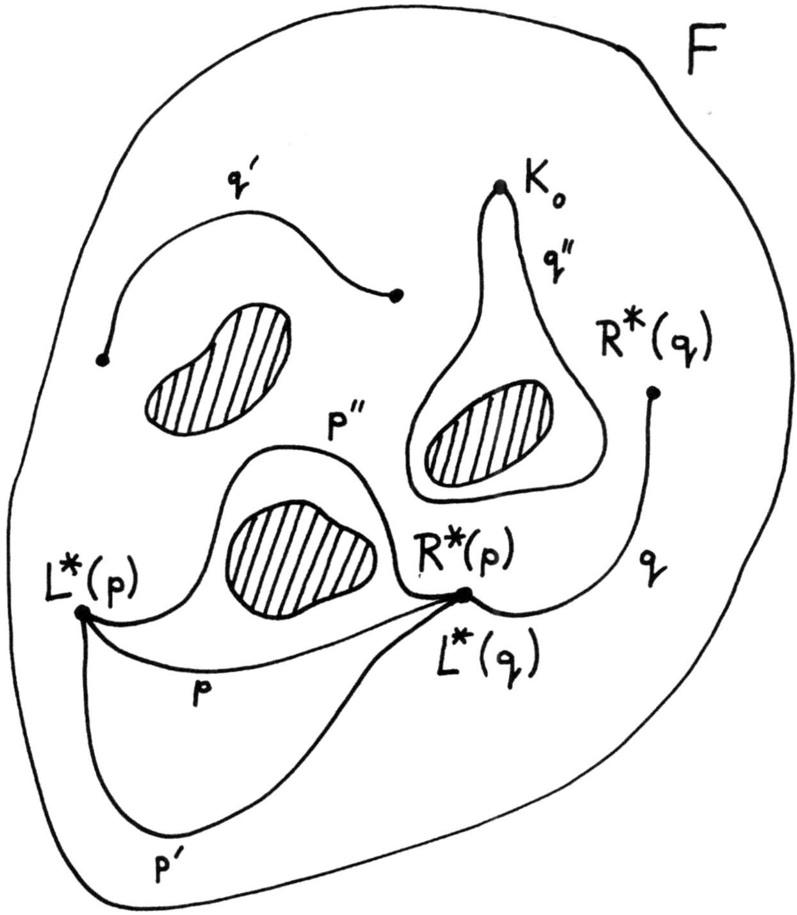

Figure 1. A simple two dimensional example for a level set F. The shaded areas do not belong to F, they correspond to neighborhoods of hilltops of energy above the upper energy bound A. Paths p and p' are homotopically equivalent, $p \sim p'$, while p and p'' are not homotopically equivalent. The product path pq exists, however products qp, pq' or qq'' do not exist. Path q'' is a circular path of extremity K_0.

Each path p on level set F is parametrized by the unit interval

$$I = [0,1] \tag{2}$$

and is regarded as a formal mapping from I to F:

$$p : I \rightarrow F \ . \tag{3}$$

Here $0 \in I$ corresponds to the origin (beginning) $L^*(p)$, and $1 \in I$ corresponds to the extremity (endpoint) $R^*(p)$ of the path p:

$$L^*(p) = p(0) \tag{4}$$

$$R^*(p) = p(1) \ . \tag{5}$$

The inverse path p^{-1} of p runs through the same points of F as does p, but in reversed order, for example,

$$p^{-1}(0) = p(1) \tag{6}$$

$$p^{-1}(1) = p(0) \ . \tag{7}$$

The inverse path p^{-1} represents a formally "reversed" chemical process within a classical model. The quantum chemical model is based on families of such paths, that is, collections of paths replace individual paths as fundamental entities.

Family [p] is the homotopy class of all paths in F which paths have common origin $L^*(p)$ and common extremity $R^*(p)$ with path p and which paths are homotopic to p, that is, which paths are continuously deformable into p within level set F:

$$[p] = \{p' : L^*(p')=L^*(p),\ R^*(p')=R^*(p),\ p' \sim p\} \tag{8}$$

Here

$$p' \sim p \tag{9}$$

denotes homotopic equivalence of p' and p. Each homotopy equivalence class [p] of (infinitely many) paths represents a formal reaction mechanism between the two nuclear configurations $L^*(p)$ and $R^*(p)$. Since $L^*(p)$ and $R^*(p)$ are points in the nuclear configuration space M, a formal reaction mechanism [p] still retains some features of a classical mechanical model. A family [p] is not yet fully compatible with the topological model of reactions, where chemical structures are represented by open sets of the nuclear configuration space M, rather than by individual points of M. However, the above "classical mechanical remnant" of reference to individual points of M can be removed easily when considering the fundamental group of reaction mechanisms (vide infra), since this group is invariant to the choice of the endpoints of the paths in any connected level set F.

2.3. The algebraic structure of reaction mechanisms

The algebraic structure leading to the fundamental group of reaction mechanisms is based on the definition of a product for reaction paths p, and on the subsequent definition of a product for "configuration-to-configuration" reaction mechanisms [p]. The product

$$r = pq \tag{10}$$

of two reaction paths p and q is the continuation of p by q, which product exists only for some ordered pairs of reaction paths, the condition for the existence of pq being

$$R^*(p) = L^*(q). \tag{11}$$

The product of two equivalence classes (reaction mechanisms) [p] and [q] is the reaction mechanism [pq] that is, the equivalence class containing the product path pq:

$$[p][q] = [pq]. \tag{12}$$

Homotopy equivalence class [pq] exists if and only if pq exists.

With this product the family

$$\Pi = \{[p]\} \tag{13}$$

of all equivalence classes on F is a groupoid: an associative algebraic structure with left units, right units and inverse defined as

$$L([p]) = [L^*(p)] \tag{14}$$

$$R([p]) = [R^*(p)] \tag{15}$$

and

$$[p]^{-1} = [p^{-1}] \tag{16}$$

respectively, but with no closure property in general. Hence Π is only a groupoid but not a group[5,6].

This groupoid Π, the fundamental groupoid of level set F of the potential energy hypersurface E(K), does, however, contain a group. For any given point $K_0 \in F$ the family Π_1 of homotopy classes of all closed reaction paths (loops) with endpoints

$$L^*(p) = R^*(p) = K_0 \tag{17}$$

is a stable subset and a subgroupoid of Π:

$$\Pi_1 = \{[p]: [p] \in \Pi,\ L^*(p) = R^*(p) = K_0 \in F\}. \tag{18}$$

The above endpoint condition ensures that within this subgroupoid Π_1

the product exists for any two homotopy classes. That is, the closure property is fulfilled for π_1, hence subgroupoid π_1 with the unique neutral element $[p_o]$, for which

$$p_o(I) = K_o \tag{19}$$

(that is, with the class of paths contractible to K_o), is also a group, the fundamental group of reaction mechanisms.

The elements of π_1 are the fundamental reaction mechanisms. Any reaction mechanisms within F can be represented as a segment of a fundamental reaction mechanism. In turn, each fundamental reaction mechanism is generated as a product of some generators of group π_1. These group generators are called the generator mechanisms, themselves being sufficient to generate any reaction mechanism on F. A typical group on a potential surface is a finitely generated free group[5]. π_1 as a group is invariant to the choice of K_o in any arcwise connected level set F, hence all reference to individual points of the nuclear configuration space can be removed. Group π_1 is a topological invariant of level set F, hence the fundamental group of reaction mechanisms, as an abstract group, is the same for all level sets F of the same energy hypersurface, or even of different energy hypersurfaces of different chemical systems, as long as these level sets have the same topology, that is, if they are related by some homeomorphism. Group π_1 is also invariant to changes in the energy bound A between critical levels, and it may change only if the potential surface has a critical point with energy A. Considering all possible energy changes in energy bound A, and regarding the subgroup relation as the ordering principle, the family of all groups for all bounds A has an algebraic structure of its own: these groups form a lower semilattice[5]. Various properties of this semilattice are discussed in reference [5].

2.4. The role of critical points and catchment regions in the algebraic structure of reaction mechanisms

Reaction mechanisms on a potential energy hypersurface can also be represented by the sequence of chemical species participating in the reaction. These chemical species, on the other hand, are represented by basins of various critical points, e.g. by basins of potential energy minima of the hypersurface. It has been shown[6] that such a representation of general reaction mechanisms has a rather direct connection with the homotopy equivalence class representation of reaction mechanisms. It is relatively easy to show that by restricting the analysis to paths interconnecting critical points of the potential energy hypersurface, an algebraic structure, analogous to groupoid π can be obtained. Hence, techniques applicable to the analysis of groupoid π are also applicable to this algebraic structure that is more directly related to the chemist's conventional concept of chemical reaction. Reaction paths leading from critical point to critical point reflect the (oversimplified, nevertheless useful) classical notion that chemical species can be represented by stationary points of potential surfaces.

As we have seen, homotopy equivalence of reaction paths, relative to fixed endpoints (denoted by $p \sim q$) introduces the equivalence classes $[p]$ and groupoid $\Pi = \{[p]\}$ into the set

$$\mathbf{p} = \{p\} \tag{20}$$

of all reaction paths in level set F. Consider an important subset of $\mathbf{p}$:

$$\mathbf{p}_{cc} \subset \mathbf{p} \tag{21}$$

where $\mathbf{p}_{cc}$ is the collection of all such paths p which have critical points as extremities:

$$\mathbf{p}_{cc} = \{p_{cc} : L^*(p_{cc}) = c(\lambda, i),\ R^*(p_{cc}) = c'(\lambda', i'), p_{cc} \in \mathbf{p}\}. \tag{22}$$

Here $c(\lambda,i)$ and $c'(\lambda',i')$ are critical points of index λ and λ', respectively; i and i' are indices of ordering. In the strict sense $L^*(p_{cc})$ and $R^*(p_{cc})$ themselves are formal "degenerate", "single point" paths, hence they are mappings. Nonetheless, we shall use the same notation for their actual point-set image, i.e., for the result of these mappings, as in eq. (22). If we denote the family of $\sim$ homotopy classes in $\mathbf{p}_{cc}$ by Π_{cc},

$$\Pi_{cc} = \{[p_{cc}]\}, \tag{23}$$

then evidently

$$\Pi_{cc} \subset \Pi\ . \tag{24}$$

Furthermore, Π_{cc} is a stable subset of Π, that is, the product of any two $[p_{cc}]$ classes, if it exists, is in Π_{cc}; the inverse $[p_{cc}^{-1}]$ and left and right units $L([p_{cc}])$, $R([p_{cc}])$ are in Π_{cc}. Consequently, Π_{cc} is a subgroupoid of fundamental groupoid Π.

Groupoid Π_{cc} describes the algebraic structure of the family of all those equivalence classes of reaction paths which lead from critical point to critical point. Since the topological concept of chemical structure is associated with basins of attraction, that is, with catchment regions of critical points, subgroupoid Π_{cc} is more directly relevant to reaction topology T_C than the fundamental groupoid Π [6]. The algebraic problem of reaction mechanisms can be further simplified by introducing further equivalence relations into set $\mathbf{p}_{cc}$. One such relation, denoted by **b**, serves to identify those paths (as <u>mappings</u>) which produce the same path-image in level set F. In the Appendix the definition and some properties of this relation **b** are given. Based on relation **b** it can be shown[6] that a common algebraic structure exists for reaction mechanisms, defined as homotopy classes of reaction paths, and for reaction mechanisms defined as open sets of reaction topology T_C (relative to the common level set F).

This result establishes the relation between the "homotopy class approach"[4,5] and the "open set approach"[6] to the problem of quantum chemical reaction mechanisms. This relation is not merely of theoretical interest, since the actual computational effort required for the determination of such open sets is usually much less than that needed for homotopy classes, especially if a basis for topology T_C is known[3].

APPENDIX

We shall prove that in any given level set $F \subset M$ the relation **b** among reaction paths[6] is an equivalence relation. We shall use the notations of ref [6] throughout. In particular, **a** denotes the homotopy of paths having critical points as endpoints and passing through the critical point $c(\lambda,i)$ of each catchment region $C(\lambda,i)$ entered, on each occasion when such penetration occurs. Relation **a** is a homotopy relative to these critical points. That is, two paths p_{ccc} and p'_{ccc} belong to the same $[p_{ccc}]_a$ relative homotopy class if and only if for some fixed elements

$$0 = \alpha_1, \alpha_2, .. \alpha_k = 1 \in I \tag{A1}$$

of the unit interval I and for some fixed sequence of critical points

$$c_1 = c(\lambda_1,i_1),\ c_2 = c(\lambda_2,\ i_2)\ ,..\ c_k = c(\lambda_k,i_k) \tag{A2}$$

the following k relations are satisfied:

$$p_{ccc}(\alpha_j) = p'_{ccc}(\alpha_j) = c_j \in F, \quad j=1,2,...k\,. \tag{A3}$$

The introduction of relation **b** is necessitated by the actual definition of relative homotopy as a continuous transformation between paths, as mappings, and not between the paths as the actual images (point sets in F) produced by these mappings. Our purpose is to interrelate reaction-path homotopy classes to point sets in topological space (M,T_C), and it is the point-set image of mappings p_{ccc} which is of primary interest to us. For this reason, all those $[p_{ccc}]_a$ equivalence classes within which common path images occur are collected into larger classes $[p_{ccc}]_b$, defined by a new equivalence relation **b**

$$p_{ccc}\mathbf{b}q_{ccc} \Leftrightarrow \text{exist } p'_{ccc},\ q'_{ccc}: p'_{ccc}\mathbf{a}q'_{ccc} \tag{A4}$$

where for the point-set images in F

$$p_{ccc} = p'_{ccc}, \quad p_{ccc}, p'_{ccc} \subset F, \tag{A5}$$

$$q_{ccc} = q'_{ccc}, \quad q_{ccc}, q'_{ccc} \subset F. \tag{A6}$$

The symmetry and reflexivity of relation **b** follow directly from this definition. In the following we shall give a direct proof for the transitivity of relation **b**, that implies that **b** is, indeed, an equivalence relation.

First we show that if

$$p_{ccc} \ \mathbf{b} \ q_{ccc} \tag{A7}$$

then there must exist some p''_{ccc} path with

$$p_{ccc} = p''_{ccc} \subset F \tag{A8}$$

for the point set images, such that

$$p''_{ccc} \ \mathbf{a} \ q_{ccc} \tag{A9}$$

Assume that p_{ccc}, as a mapping, assigns the

$$0 = \alpha_1^p, \alpha_2^p, \ldots \alpha_k^p = 1 \in I \tag{A10}$$

points, whereas mapping q_{ccc} assigns the

$$0 = \alpha_1^q, \alpha_2^q, \ldots \alpha_k^q = 1 \in I \tag{A11}$$

points of the unit interval I to the critical points $c_1, c_2, \ldots c_k$, respectively. There must exist some path p''_{ccc} with the same point set image as p_{ccc}, that assigns the (A11) sequence of points to the critical points. Such a path p''_{ccc} can be constructed as follows. Take $p_{i,i+1}$ (i=1,...k-1) to denote the ith segment of p_{ccc}, leading from c_i to c_{i+1}. Each path $p_{i,i+1}$ is a mapping

$$p_{i,i+1}: I \rightarrow F \tag{A12}$$

$$p_{i,i+1}(0) = c_i \tag{A13}$$

$$p_{i,i+1}(1) = c_{i+1} \tag{A14}$$

Then, the product path

$$p''_{ccc} = p_{1,2} \cdot \ldots \ p_{i,i+1} \cdot \ldots \ p_{k-1,k}$$

defined as the mapping

$$p''_{ccc}(\alpha) = \begin{cases} p_{1,2}\left((\alpha-\alpha_1^q)/(\alpha_2^q-\alpha_1^q)\right) & \text{if } \alpha_1^q \leq \alpha \leq \alpha_2^q \\ p_{i,i+1}\left((\alpha-\alpha_i^q)/(\alpha_{i+1}^q-\alpha_i^q)\right) & \text{if } \alpha_i^q \leq \alpha \leq \alpha_{i+1}^q \\ p_{k-1,k}\left((\alpha-\alpha_{k-1}^q)/(\alpha_k^q-\alpha_{k-1}^q)\right) & \text{if } \alpha_{k-1}^q \leq \alpha \leq \alpha_k^q \end{cases} \tag{A15}$$

has the required property, as it assigns the (A11) sequence of elements of I to the critical points, by definition (A15), and it also has the same point set image as p_{ccc}:

$$p''_{ccc} = p_{ccc} \subset F \qquad \text{(A16)}$$

From (A7) and (A16) the point set equivalences

$$p''_{ccc} = p_{ccc} = p'_{ccc} \subset F \qquad \text{(A17)}$$

and

$$q_{ccc} = q'_{ccc} \subset F \qquad \text{(A18)}$$

follow, and (A7) implies

$$p'_{ccc} \; \mathbf{a} \; q'_{ccc} \qquad \text{(A19)}$$

where p'_{ccc} and q'_{ccc} are paths fulfilling the conditions of definition (A4). Hence, point set equivalences (A17), (A18) and the fact that (A11) is the sequence common for p''_{ccc} and q_{ccc}, imply that

$$p''_{ccc} \; \mathbf{a} \; q_{ccc} \; . \qquad \text{(A20)}$$

In the possession of relation (A20), subject to (A7) and (A8), we prove now transitivity of relation **b**. That is, we show that from

$$p_{ccc} \; \mathbf{b} \; q_{ccc} \qquad \text{(A21)}$$

and

$$q_{ccc} \; \mathbf{b} \; r_{ccc} \qquad \text{(A22)}$$

the relation

$$p_{ccc} \; \mathbf{b} \; r_{ccc} \qquad \text{(A23)}$$

follows.

Evidently, the critical point sequence (A2) must be common for paths p_{ccc}, q_{ccc} and r_{ccc}, as it follows from (A21) and (A22). In the following a common letter symbol p, q or r implies common point set image. The definition of relation **b** implies that there must exist path pairs p'_{ccc}, q'_{ccc} and q''_{ccc}, r'_{ccc} such that relations

$$p'_{ccc} \; \mathbf{a} \; q'_{ccc} \qquad \text{(A24)}$$

and

$$q''_{ccc} \; \mathbf{a} \; r'_{ccc} \qquad \text{(A25)}$$

are fulfilled. In fact, property (A20), just proven, implies that both p'_{ccc} and r'_{ccc} can be chosen with the same (A11) point sequence of q_{ccc}, hence one may write

$$q'_{ccc} = q''_{ccc} = q_{ccc} \tag{A26}$$

for the mappings as well. Substitution into (A24) and (A25) gives

$$p'_{ccc} \ \mathbf{a} \ q_{ccc} \tag{A27}$$

and

$$q_{ccc} \ \mathbf{a} \ r'_{ccc} \tag{A28}$$

and the transitivity of equivalence relation **a** implies

$$p'_{ccc} \ \mathbf{a} \ r'_{ccc} \tag{A29}$$

Hence, (A23) is satisfied and **b** is indeed an equivalence relation.
Q.E.D.

ACKNOWLEDGEMENTS

This work was supported by a research grant from the Natural Sciences and Engineering Research Council of Canada.

REFERENCES

1. K. Fukui, 'A Formulation of the Reaction Coordinate', J. Phys. Chem., **74**, 4161-4163 (1970).
2. See e.g. K. Fukui, A. Tachibana and K. Yamashita, 'Toward chemodynamics', Int. J. Quant. Chem. Symp., **15**, 621-632 (1981), and references therein.
3. See e.g. P.G. Mezey, 'Reaction Topology: Manifold theory of potential surfaces and quantum chemical synthesis design', in R.B. King (Ed.), Chemical applications of topology and graph theory, Elsevier, Amsterdam, 1983, and references therein.
4. P.G. Mezey, 'The Topology of Energy hypersurfaces III. The fundamental group of reaction mechanisms on potential energy hypersurfaces', Theor. Chim. Acta, **00**, 000 (1984).
5. P.G. Mezey, 'The Topology of energy hypersurfaces IV. Generator sets for the fundamental group of reaction mechanisms and the complete set of reaction paths', Theor. Chim. Acta, **00**, 000 (1984).
6. P.G. Mezey, 'The algebraic structure of quantum chemical reaction mechanisms', Internat. J. Quantum Chem. Symp., **18**, 000 (1984).
7. M. Morse and S.S. Cairns, 'Critical Point theory in Global Analysis and Differential Topology', Academic Press, New York, 1969.
8. E.H. Spanier, Algebraic Topology, McGraw-Hill, New York, 1966.
9. M. Greenberg, Lectures on Algebraic Topology, Benjamin, New York, 1977.

10. J.N. Murrell and K.J. Laidler, 'Symmetries of Activated Complexes', Trans. Faraday Soc., **64**, 371-377 (1968).
11. J.N. Murrell and G.L. Pratt, 'Statistical Factors and the Symmetry of Transition States', Trans. Faraday Soc., **66**, 1680-1684 (1970).
12. L. Salem, 'Narcissistic Reactions: Synchronism vs. Nonsynchronism in Automerizations and Enantiomerizations', Acc. Chem. Res., **4**, 322-328 (1971).
13. R.E. Stanton and J.W. McIver, 'Group Theoretical Selection Rules for the Transition states of Chemical Reactions', J. Amer. Chem. Soc., **97**, 3632-3646 (1975).
14. T.D. Bouman, C.D. Duncan and C. Trindle, 'Group Theory and Reaction Mechanisms: Permutation Theoretic Prediction and Computational Support for Pseudorotation Modes in $C_2H_5^+$ and $C_5H_5^+$ Rearrangements', Int. J. Quantum. Chem., **11**, 399-413 (1977).
15. J. Maruani and J. Serre (Eds.), Symmetries and Properties of Non-rigid Molecules: A comprehensive survey, Elsevier, Amsterdam, 1983.
16. S.L. Altmann, Induced representations in crystals and molecules, Academic Press, New York, 1977.
17. Hs. H. Gunthard, A. Bauder and H. Frei, Topics in Current Chemistry, Springer Verlag, Berlin, 1979.
18. K. Balasubramanian, 'The symmetry groups of nonrigid molecules as generalized wreath products and their representations', J. Chem. Phys., **72**, 665-677 (1980).
19. L. Lathouwers and P. van Leuven, 'Generator coordinate theory of nuclear motion in molecules', Adv. Chem. Phys., **49**, 115-189 (1982).

REACTIONS IN DISORDERED MEDIA

A. Blumen
Lehrstuhl für Theoretische Chemie
Technische Universität München, D-8046 Garching, West Germany

G. Zumofen
Physical Chemistry Laboratory
ETH-Zentrum, CH-8092 Zurich, Switzerland

J. Klafter
Corporate Research Science Laboratories
Exxon Research and Engineering Company
Clinton Township, Annandale, N.J. 08801, USA

ABSTRACT. In this work we study A+A → P and A+B → P reactions. Our concern is the influence of the diffusive motion of the particles in establishing the classical kinetic behavior. We consider the time-dependence of the density of particles both in the presence and in the absence of diffusion and focus on pseudounimolecular A(0) << B(0), and on strictly bimolecular A(0) = B(0) initial conditions. In the absence of diffusion, the pseudounimolecular decays are highly nonexponential, and depend on the interaction range and on the dimensionality of the embedding reaction space. From a general approach we derive the decays for one-, two- and three-dimensional geometries and also the forms which obtain for fractals, structures which display non-integer dimensions. The introduction of diffusion wipes out the effects of different long-ranged interactions. However, diffusion is unable to completely mask the differences due to dimensionality, and thus the reaction kinetics still do not follow the simple "well-stirred-reactor" behavior. Here we show how these aspects may be accounted for through Smoluchowski-type forms. Moreover, fluctuations in the initial particle distribution get enhanced through the reaction, so that even the Smoluchowski-type extensions may fail at longer times: the strictly-bimolecular reaction A+B → P is such an example.

1. INTRODUCTION

The subject of reaction kinetics under diffusion limited conditions is an old topic [1,2], but only in recent years has one started to analyze in a systematic fashion the role of fluctuations [3-6] on the dynamics of the reactions and on the deviations from simple kinetic schemes.

R. Daudel et al. (eds.), Structure and Dynamics of Molecular Systems, 71–86.

When several molecular species are simultaneously present, site fluctuations arise naturally. Furthermore the reaction process enhances by itself the disorder (vide infra). In this work we focus on both pseudounimolecular and on bimolecular reactions and we establish for several models the different possible deviations from the classical, kinetic behavior. As it will become apparent in the following, a large role is played by the dimensionality of our systems. Hence, we will also consider reactions on fractal objects [7], which allow us to mimic disorder and to work in spaces of noninteger dimensions.

Here we start from a simple kinetic scheme in which A and B molecules react (for simplicity, irreversibly) to a product P. The usual kinetic reaction form is

$$A + B \xrightarrow{k} P \tag{1}$$

In the following we will disregard any influence (sterical or chemical) which the P-molecules may have on the further course of the reaction. Thus, we will view Eq. (1) as being:

$$A + B \xrightarrow{k} 0 \tag{2}$$

which corresponds to the annihilation of the A and B-particles.

Eqs. (1) and (2) obtain from the "well-stirred reactor" model, by which, due to an external ("stirrer") or internal (diffusion) agent, the concentration of A and B particles is kept homogeneous throughout the reaction volume. In this case the corresponding differential equation has the usual form:

$$\frac{dA}{dt} = -k\, A(t)\, B(t) \tag{3}$$

which expresses the decay in the number of particles by means of the reaction rate k. In the following we will also need the generalization of Eqs. (1) to (3) for time dependent rates, $k(t)$:

$$\frac{dA}{dt} = -k(t)\, A(t)\, B(t) \tag{4}$$

with a corresponding relation for $\frac{dB(t)}{dt}$. Since, obviously $\frac{dA(t)}{dt} = \frac{dB(t)}{dt}$, the expression $B(t)-A(t)$ is a constant of motion, $B(t) - A(t) = \delta$. Inserting this relation into Eq. (4) allows the separation of variables [8]:

$$\xi(t) \equiv \int_0^t k(t')dt' = - \int_{A_0}^{A} \frac{dA'}{A'(A'+\delta)} = \frac{1}{\delta} \left[\ln(1+\delta/A)-\ln(1+\delta/A_0)\right] \tag{5}$$

where A_0 is the original number of A particles. For the well-stirred reactor, which leads to time-independent rates, $\xi(t)$ is simply kt.

From Eq. (5) known special cases follow in a straightforward manner. If A_0 is much smaller than the initial number of B-particles, B_0, i.e. $A_0 \ll B_0$, then $\delta/A(t) \gg 1$ and $B(t) \simeq \delta$ during the whole course of the reaction. From Eq. (5) with $\delta/A(t) >> 1$ it follows:

$$A(t) \simeq A_0\ e^{-\delta\xi(t)} \simeq A_0 e^{-B_0\xi(t)} \tag{6}$$

Evidently, Eq. (6) turns into an equality for the reaction scheme $A + B \xrightarrow{k(t)} B$.

For a strictly bimolecular reaction $A_0 = B_0$ and $\delta = 0$. Taking the limit $\delta \to 0$ in Eq. (5) we obtain [8]:

$$A(t) = \frac{A_0}{1+A_0\xi(t)} \tag{7}$$

For very short times $A(t) \simeq A_0\ e^{-A_0\xi(t)} = A_0 e^{-B_0\xi(t)}$, i.e. the result follows the law Eq. (6). At longer times $A_0\xi(t) >> 1$, so that

$$A(t) \simeq 1/\xi(t) \tag{8}$$

The same results, Eqs. (7) and (8) hold for the reaction $A + A \to P$, when we replace the rate by $k(t)/2$, to take care of the fact that now two A particles disappear at each reaction step.

Focusing now on the well-stirred reactor case, $\xi(t) = kt$, we retrieve from Eqs. (6) and (7) well-known decay patterns: an exponential

$$A(t) \simeq A_0 e^{-B_0 kt} \tag{9}$$

for the pseudounimolecular (first-order) reaction, and

$$A(t) = \frac{A_0}{1+A_0 kt} \tag{10}$$

with $A(t) \simeq (kt)^{-1}$ at longer times for the $A + A \to P$ and the strictly bimolecular $A + B \to P$, $A_0 = B_0$ reactions.

Equations (9) and (10) are very general and do not depend on many aspects of interest, such as dimensionality or the form of the microscopic interactions. This is due to the fact that their use is based on a homogeneous spatial distribution of the particles during the whole course of the reaction, which precludes disorder effects. It is, however, evident that non-homogeneous conditions are widespread: First, the initial conditions may be already non-homogeneous (surface catalysis [9], pair-creation in irradiation tracks [10], energy transfer in doped systems [11,12]). Second, even under homogeneous initial conditions, the microscopic reactions create by themselves non-homogeneities and enhance already existing density fluctuations. Diffusion can only partly wipe out such disorder [4,6] but only when the diffusion length is large compared to the mean interparticle distance: At low particle densities diffusion (or stirring) cannot create a homogeneous background.

In this paper we will analyze in how far the deviations from the forms Eqs. (9) and (10) are amenable to a description in terms of a generalized rate k(t), Eqs. (6) and (7), and when such an approach may fail.

The work is structured as follows: In Sec. 2 we consider pseudounimolecular reactions A+B → P , $A_0 << B_0$ both in the absence (Sec. 2.1) and in the presence (Sec. 2.2) of diffusion, when the molecules react through distance-dependent interactions. As will become apparent in the following, in the static case the decay laws display both effects of interaction form and of dimensionality; it is precisely the latter aspect which allows to determine experimentally the non-integer dimension of fractals [13,14]. The diffusion blurrs in general, the form of the long-range-interactions; the decays still show, however, the dependence on dimensionality. The effects of disorder can be obtained by random-walk models, whose treatment is reserved to the following article in this Volume [15]. In Sec. 3 we continue with the study of the bimolecular reactions: we present the decays for the A+A → P and for the A+B → P strictly bimolecular ($A_0 = B_0$) reactions, for which we determine the role played by dimensionality. While the first reaction admits a very good description in terms of Eq. (7) and is thus of Smoluchowski-type, the strictly bimolecular reaction shows a complex form due to fluctuations in the density of the particles. By comparison with simulation calculations we present the decay patterns for longer times. The paper closes with a summary of results and future directions in Sec. 4.

2. PSEUDOUNIMOLECULAR REACTIONS

Pseudounimolecular reactions A+B → P , $A_0 << B_0$, can occur in the absence of diffusion when long-range interactions between the molecules are effective. As an example [16], trapped electrons in glasses react with electron acceptors (scavengers) through distance-dependent [17-19] interactions: the form of the interactions is usually taken to be:

$$w(\vec{r}) = \alpha e^{-\gamma r} \tag{11}$$

Here r is the interparticle distance and $w(\vec{r})$ may be envisaged to arise from the overlap of the wavefunctions of the electron and the scavenger, or from tunneling through a potential barrier. Evidently, Eq. (11) is only an approximation, and sometimes more complex forms for $w(\vec{r})$ are considered [19]. In previous works we have often modelled $w(\vec{r})$ through algebraic forms [20,21]:

$$w(\vec{r}) = \alpha r^{-s} \tag{12}$$

with s large. In this Section we will analyze in how far the effect of long-range interactions is reflected in the time-dependence of the density of A-particles. We begin with the static (diffusionless) case, and consider in the second part of this Section the influence of diffusion.

2.1. Reactions in the absence of diffusion

Since the number of A-particles is much smaller than that of B-particles, one can focus on a single A-particle, and then average its decay over all possible configurations of B-molecules surrounding it. Such a disordered situation may be modelled by assuming that all molecules occupy substitutionally with probability p positions on a given structure. For a fixed configuration $\underset{\sim}{K}$ of B-molecules the probability that the A-molecule did not react up to time t is exponential:

$$\Psi(\underset{\sim}{K};t) = \exp\left[-t \sum_{j\in \underset{\sim}{K}} w(\vec{r}_j)\right] = \prod_{j\in \underset{\sim}{K}} \exp\left[-t\, w(\vec{r}_j)\right] \tag{13}$$

Here $w(\vec{r})$ denotes the long-range interactions, Eqs. (11) or (12).

The quantity of experimental interest is not $\Psi(\underset{\sim}{K};t)$ but its ensemble average over all possible configurations $\underset{\sim}{K}$ of molecules distributed on the lattice, $\Psi(t)$, where $\Psi(t) = \langle\Psi(\underset{\sim}{K};t)\rangle_{\underset{\sim}{K}}$. We now introduce the stochastic variables ζ_j, so that $\zeta_j=1$ if site j is occupied and $\zeta_j=0$ otherwise. As stated the sites are randomly occupied with probability p; in a finite volume V containing B_0 B-molecules and N_T sites $p \simeq B_0/N_T$. From Eq. (13) one has:

$$\begin{aligned}\Psi(t) &= \langle \prod_i{}' \exp\left[-t\zeta_i w(r_i)\right]\rangle_{\{\zeta_i\}} \\ &= \prod_i{}' \langle \exp\left[-t\zeta_i w(\vec{r}_i)\right]\rangle_{\zeta_i} \\ &= \prod_i{}' \{1-p+p\exp\left[-tw(\vec{r}_i)\right]\}\end{aligned} \tag{14}$$

Here the product extends over all sites of the structure with the exception of the origin. This means that the ensemble average reproduces the details of the structure (it may also depend on the origin, if the structure is not translationally invariant). In the process of the reaction the whole lattice is being sampled. Eq. (14) is exact and is valid for all values of p, p<1, and for all times. It is general for any geometry (finite or infinite, with or without translational symmetry).

For a low density of B molecules, p << 1, an approximate form to $\Psi(t)$ can be derived from Eq. (14) [22]. In the continuum description one introduces the density $\rho(\vec{r})$ of sites, and has to first order in the concentration of B particles:

$$\Psi(t) \simeq \exp\left(-p \int d\vec{r}\, \rho(\vec{r})\{ 1-\exp\left[-tw(\vec{r})\right]\}\right) \tag{15}$$

Let us first exemplify the procedure for regular lattices in d-dimensions (d=1,2, or 3), where $\rho(\vec{r}) = \rho = N_T/V$. We start from Eq. (11) and insert it into Eq. (15), obtaining:

$$\Psi(t) \simeq \exp\left[-c\, p\rho\, g_d(\alpha t)\right] \tag{16}$$

Here $p\rho$ is the density of B-molecules, $c = \pi^{d/2}\gamma^{-d}/\Gamma(1+d/2)$ and $g_d(z)$ is an analytic function

$$g_d(z) = d\int_0^\infty dx\, x^{d-1}\left[1-\exp(-ze^{-x})\right] \qquad (d\neq 0) \tag{17}$$

For large z $g_d(z) \sim \ell n^d z$. Hence, at longer times, in 3 dimensions one obtains for the decay law under the interaction $w(\vec{r}) = \alpha \exp(-\gamma r)$

$$A(t) \simeq A_0 \exp\left[-c(B_0/V)\, \ell n^3 (\alpha t)\right] \quad , \tag{18}$$

where we used $p=B_0/N_T$ and $\rho=N_T/V$. The decay is clearly nonexponential and mirrors in the logarithmic time dependence the interaction law, Eq. (11). If instead of using this interaction we start from Eq. (12) we obtain:

$$\Psi(t) \simeq \exp\left[-\tilde{c}\, p\rho\, (\alpha t)^{d/s}\right] \tag{19}$$

with $\tilde{c} = \pi^{d/2}\,\Gamma(1-d/s)/\Gamma(1+d/2)$. In three dimensions the form corresponding to Eq. (18) is

$$A(t) \simeq A_0 \exp\left[-\tilde{c}(B_0/V)(\alpha t)^{3/s}\right] \tag{20}$$

where $w(r) = \alpha r^{-s}$. Again Eq. (20) is nonexponential and reflects the form of the interaction, Eq. (12). To obtain the decay laws in the case in which the underlying lattice is fractal one has to insert the relation $\rho(r) \sim r^{(\bar{d}-d)}$ into Eq. (15) where now $\bar{d}$ is the fractal (Hausdorff) dimension [7]. This amounts to replacing d in Eqs. (18) and (20) by $\bar{d}$ [13]. Although Eqs. (18) and (20) are very different from the exponential decay of Eq. (9), they belong to the class of Eq. (6) when we identify $\xi(t)$ with $c\,\ell n^3(\alpha t)/V$ or with $\tilde{c}\,(\alpha t)^{3/s}/V$, respectively. From these forms we also obtain effective time-dependent rates:

$$k(t) = 3c\, \ell n^2(\alpha t)/(Vt) \tag{21}$$

or

$$k(t) = 3\tilde{c}\alpha(\alpha t)^{-1+3/s}/(Vs) \tag{22}$$

Eqs. (21) and (22) can also be directly determined from Eq. (4), rewritten as:

$$k(t) = B_0^{-1}\, \dot{\Psi}(t)/\Psi(t) = -B_0^{-1} \frac{d}{dt}\ell n[\Psi(t)] \tag{23}$$

Applications of this analysis to experimental situations are provided in Refs. 8, 16 and 23.

2.2. Reactions in the presence of diffusion

As already mentioned, diffusion acts as an internal stirrer. Under diffusion the number of A-particles decays quicker than Eqs. (18) and (20) and the dependence on the microscopic interaction forms becomes weaker. Nevertheless, since d=2 is a marginal dimension, deviations from the exponential decay of Eq. (9) occur in low-dimensions.

In order to exemplify these statements, we follow first the analysis of Ref. 24. Let us denote by D_A and D_B the diffusion coefficients of the A- and B-particles, and set $D = D_A + D_B$. Then for integer-dimensional spaces, Eq. (15) gets replaced by:

$$\Psi(t) \simeq \exp\{-p\rho \int d\vec{r}\, [1-E(t,\vec{r})]\} \tag{24}$$

where $E(t,\vec{r})$ is the solution of the reaction-diffusion equation:

$$\frac{\partial E(t,\vec{r})}{\partial t} = [D\nabla^2_{\vec{r}} - w(\vec{r})]\, E(t,\vec{r}) \tag{25}$$

with initial condition $E(0,\vec{r}) = 1$. Using Eq. (23) the rate coefficient for isotopic interactions w(r) is:

$$k(t) = \frac{\tilde{\tilde{c}}}{V} \int_0^\infty \frac{\partial E(t,r)}{\partial t}\, r^{d-1}\, dr \tag{26}$$

Here $\tilde{\tilde{c}} = 2\pi^{d/2}/\Gamma(d/2)$ and we neglect finite volume effects by integrating from zero. The rate k(t) approaches a constant at longer times, if the time-independent equation

$$[D\nabla^2_r - w(r)]\, E_s(r) = 0 \tag{27}$$

admits a bounded, non-trivial solution $E_s(r)$. The constant is:

$$k = (\tilde{\tilde{c}}D/V) \lim_{r\to\infty} r^{d-1} \frac{\partial E_s(r)}{\partial r} \tag{28}$$

Consider the algebraic interaction form, $w(r) = \alpha r^{-s}$, Eq. (12), for which Eq. (27) can be solved in terms of modified Bessel-functions $I_\lambda(z)$ and $K_\lambda(z)$. One obtains [24]

$$E_s(r) = [c_1 K_{|\lambda|}(z) + c_2 I_{|\lambda|}(z)] z^\lambda \tag{29}$$

Here $\lambda = (d-2)/(s-2)$, $z = [2/(s-2)] (\alpha/d)^{1/2} r^{1-s/2}$, and the constants c_1 and c_2 are to be determined from appropriate boundary conditions. We focus on $r\to\infty$, i.e. on $z\to 0$. Since $z^\lambda I_{|\lambda|}(z)$ is bounded in the neighborhood of $z=0$, only the c_1-term matters. Now, $z^\lambda K_{|\lambda|}(z)$ is bounded for $\lambda > 0$, i.e. for $d > 2$ and diverges for $\lambda \leq 0$, i.e. for $d \leqslant 2$. Hence, a stationary, bounded, nontrivial $E_s(r)$ exists only for $d > 2$, and thus only above two dimensions does one obtain for $\Psi(t)$ an exponential regime at longer times.

The numerical solutions of Eq. (25) are rather cumbersome, and rely sometimes on numerically unstable Laplace-transforms or on quite arbitrary Padé approximations, as discussed in Ref. 25. In our experience we found that random-walk simulations of the motion allow to determine the pseudounimolecular decays $\Psi(t)$ accurately over many orders of magnitude [26,27].From random-walk models approximating analytical forms are also readily obtainable. Furthermore, diffusion models sometimes mask important features: thus $E(t,\vec{r})$ depends only on D but not on D_A and D_B, separately, whereas at very long times $\Psi(t)$ depends on whether the B-particles move or are static [3,4]. Random walk models are the topic of the following article in this Volume [15].

To exemplify the previous statements we present here, following Ref. 15, some random-walk results. Denoting by S(t) the mean number of distinct sites visited by a walker in time t one has as an approximate decay law [1,28]

$$A(t) \simeq A_0 \exp[-B_0 S(t)/N_T] \tag{30}$$

Expressing S(t) as a function of the number of steps taken in time t, $n \simeq t/\tau$, where τ^{-1} is the stepping frequency, one obtains for not too small n:

$$S_n \sim a_1\sqrt{n} + a_2/\sqrt{n} + \ldots \qquad (d=1) \qquad (31a)$$

$$S_n \sim a_1 n/\ell n(a_2 n) + \ldots \qquad (d=2) \qquad (31b)$$

$$S_n = a_1 n + a_2\sqrt{n} + \ldots \qquad (d=3) \qquad (31c)$$

and, in general, for fractals with spectral dimension $\tilde{d}$:

$$S_n \sim a_1 n^{\tilde{d}/2} + \ldots \qquad (\tilde{d}<2) \qquad (31d)$$

The forms of Eqs. (31) are strongly dimension-dependent, and the values of the parameters a_i depend on the lattice structure. Note that only in the case of (31c) (d=3) does Eq. (30) lead to the exponential behavior predicted by Eq. (9) for a fixed reaction rate k. In this case the correction terms from (31c) lead to Smoluchowski-expressions for k(t) [1,29]

$$k(t) = -B_0^{-1}\dot{A}(t)/A(t) = N_T^{-1}\frac{dS(t)}{dt} = (a_1 + a_2/2\sqrt{t\tau})/N_T \qquad (32)$$

Equation (30) is thus of Smoluchowski-type [1]. We now turn our attention to the bimolecular reactions.

3. BIMOLECULAR REACTIONS

In this Section we consider both the A + A → P and the A + B → P reactions: as stressed in the Introduction, pseudounimolecular reactions are the special case of A + B → P when $A_0 \ll B_0$. Here we consider the case of strict equality of the initial number of particles $A_0 = B_0$, and we compare the results to Smoluchowski-type approximations.

We start from the decay of the pseudounimolecular form (30) and follow our approach, Ref. 8, Eqs. (25)ff. Using now Eq. (32) as an approximation to the true rates one has from Eq. (7)

$$\Phi_{Sm}^{AB}(t) = \frac{A(t)}{A_0} = \frac{1}{1 + (A_0/N_T)S(t)} = \frac{1}{1 + pS(t)} \qquad (33)$$

for the strictly bimolecular (A + B → P, $A_0 = B_0$) reaction. For the reaction A + A → P two particles get annihilated at each reaction step and microscopically the annihilation proceeds by a factor of two quicker, but the particle density is twice as large and enters the rate quadratically. Hence, the decay should obey approximately the same form:

$$\Phi_{Sm}^{AA}(t) = \frac{1}{1 + pS(t)} \tag{34}$$

It now remains to establish in how far Eqs. (33) and (34) are valid.

We begin with the A + A → P reaction, since it will turn out to be <u>less</u> influenced by fluctuations than the strictly bimolecular A + B → P, $A_0 = B_0$, reaction. This reaction is of importance in energy transfer problems, since it describes exciton up-conversion and annihilation processes [12, 30-32]. We model the reaction in a random-walk picture: We consider both fractals (Sierpinski-gaskets) [7] and also regular lattices, and we start from configurations containing some $2\cdot10^4$ molecules. For $p = A_0/N_T$ we take $p = 0.1$, so that $N_T = 2\cdot10^5$. Periodic boundary conditions are used, so that the systems appear infinite. The A molecules are placed randomly on the structures, following a binomial (yes-no) distribution. At each reaction step the molecules are moved one by one to neighboring sites. Any two molecules that during this process happen to occupy the same site are immediately removed.

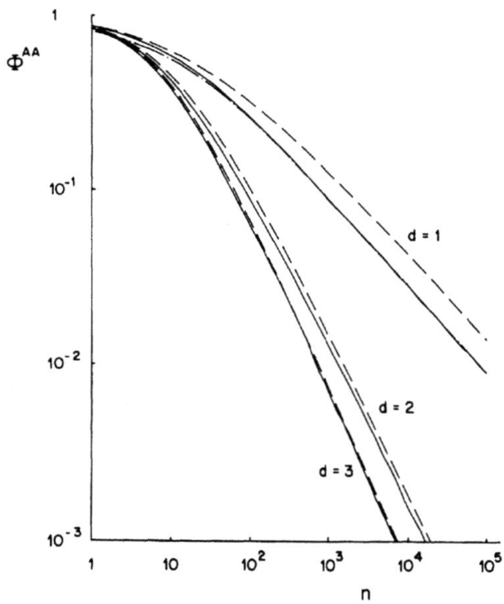

Figure 1. Decays due to the bimolecular reaction A + A → P. The full lines are the simulation results for the linear chain (d=1), for the square (d=2), and for the simple cubic lattice (d=3). The dashed lines are the approximate forms, Eq. (34) and the dash-dotted line is the continuum-model result of Ref. 33.

In Figure 1 we present the results for the decay following the A + A → P reaction on the linear chain, the square lattice and the simple cubic lattice. The full lines give the simulation results obtained by averaging over 100 distinct initial conditions and walks each. We choose to plot $\ell n\ (A(t)/A_0)$ vs. $\ell n\ t$. In these scales, according to Eqs. (34) and (31), at long times the decays should follow straight lines. From the figure, this behavior is clearly apparent. Moreover, it turns out that the Smoluchowski form is a good approximation when we identify $S(t)$ with $S_{2n}-1$ in Eq. (34). These forms are given in Fig. 3 as dashed-lines. The difference between the results is rather small.

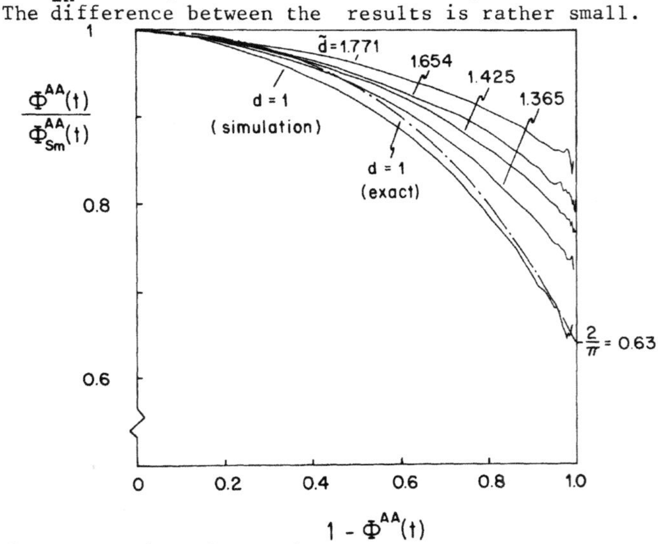

Figure 2. The ratios of the simulated over the Smoluchowski-form results for the bimolecular reaction A + A → P. For the linear chain (d=1) and for fractals (Sierpinski gaskets) of different spectral dimension $\tilde{d}$ the ratios (full lines) are plotted as functions of the progress of the reaction $1 - \Phi$. The dash-dotted line and the analytical result $2/\pi$ are taken from the continuum model in Ref. 33.

As shown by Torney and McConnell, [33] the one-dimensional A + A → P problem admits in the continuum, diffusion-equation limit, an exact solution. They find that the relative difference between the two forms $\Phi^{AA}_{exact}/\Phi^{AA}_{Sm}$ goes monotonously from 1 to $2/\pi$ in the 1-d case. We find that with increasing dimension, the differences between the exact decay and the Smoluchowski form become less and less significant. In Fig. 2 we display these differences for the range of dimensions between 1 and 2, by considering annihilation processes on Sierpinski-gaskets with spectral dimensions $\tilde{d}$ = 1.365, 1.425, 1.654 and 1.771. We plot Φ_{Sm}/Φ_{exact} as a function of the reaction completion coordinate $1-\Phi$, so as to faciliate the comparison to Ref. 33.

The main conclusion of our study of the A + A → P reaction is that the Smoluchowski-approach provides a good approximation over the range of the decay studied. We view Eq. (34) as a very good tool for comparison to experimental findings.

For the A + B → P, $A_0=B_0$ reaction the situation is different, due to spatial fluctuations, which get enhanced by the chemical reaction [4,6,34,35]. This finding is especially interesting, since some recent works have argued otherwise [36,37]. The reason for this effect is that at longer times, due to the progress of the reaction, large regions containing only A or only B molecules appear. Then the diffusion does not provide an efficient stirring anymore, and the reactions proceed more slowly, since mostly only molecules at the boundaries of these regions can react. The expected decay at longer times follows then [4,6,34,35] $\Phi^{AB}_{exact} \sim t^{-d/4}$ for regular lattices and $\Phi^{AB}_{exact} \sim t^{-\tilde{d}/4}$ for fractals ($\tilde{d} < 4$).

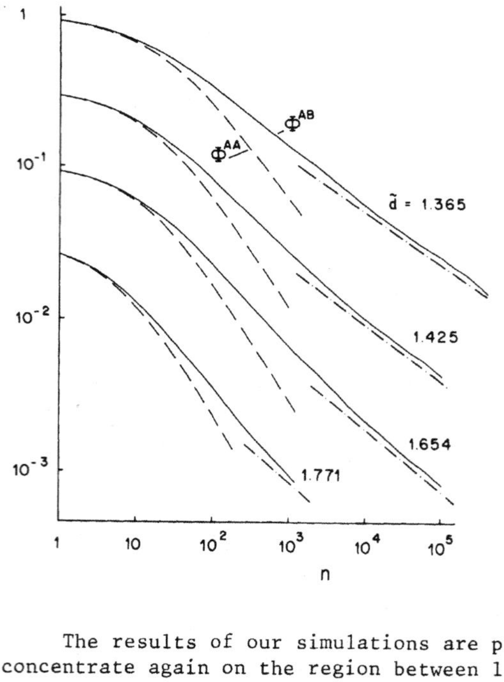

Figure 3. Simulated decays due to bimolecular reactions on fractals (Sierpinski gaskets) of different spectral dimensions $\tilde{d}$. Full lines for the reactions A + B → P, $A_0=B_0$ with expected slopes at longer times denoted by dash-dotted lines. Dashed lines for the reactions A + A → P. For both types of reactions the initial concentrations were chosen to be p=0.1.

The results of our simulations are presented in Fig. 3. Here we concentrate again on the region between 1 and 2 dimensions, for which the crossover from the Smoluchowski-form (which goes as $t^{-\tilde{d}/2}$ at longer times) to the $t^{-\tilde{d}/4}$ decay becomes clearly evident. The full lines give the decay due to the A + B → P reaction, and are the result of some 10 realizations of initial conditions and walks involving some 10^5 particles placed on fractals, such that the initial concentrations are p=0.1 in every case. The spectral dimensions are $\tilde{d}$ = 1.365, 1.425, 1.654 and 1.771. The dashed lines are the results for the A + A → P reaction. Fig. 3 shows clearly that the two reactions (A + A and A + B) obey different laws at longer times. Again, we chose to plot $\ln \Phi_{exact}$ vs. $\ln t$ so that in these scales the expected long time form $t^{-\tilde{d}/4}$ is a straight line of slope $-\tilde{d}/4$. The dash-dotted lines in the Figure indicate this slope.

From a theoretical point of view it is interesting to show how the change in the time dependence of $\Phi^{AB}(t)$ enters. We let ourselves be led by the form of Eq. (33) and plot in Fig. 4 the function $[\Phi^{AB}_{exact}(t)]^{-1}-1$ vs. $\ln t$. As long as an approximation such as Eq. (33) is correct, this function should parallel $S_{2n}-1$. In the Figure the two behaviors are indicated, the simulation results being given as full lines and the $S_{2n}-1$ forms as dashed lines. It is again apparent that after a short time period the description in terms of sites visited becomes unsatisfactory and that the full lines cross over to power-law dependencies of the form $t^{\tilde{d}/4}$: These slopes are indicated in Fig. 4 as dash-dotted lines. We remark that the higher the dimension, the larger the region in which the Smoluchowski form is adequate, and the later the crossover effect.

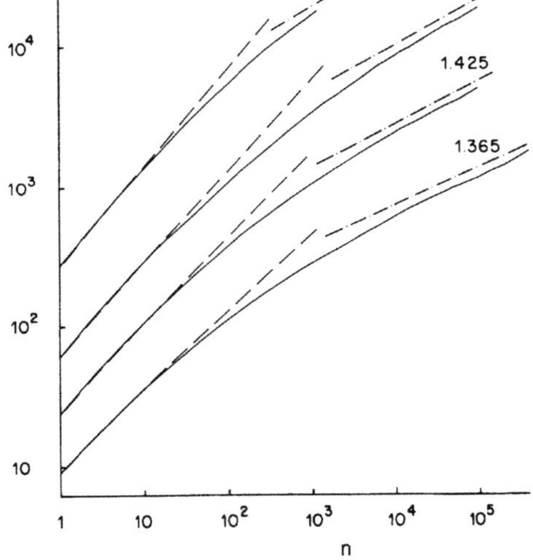

Figure 4. Decays due to bimolecular reactions A + B →P on fractals (Sierpinski gaskets) of different spectral dimensions $\tilde{d}$. The simulation results, given as $(\Phi^{AB})^{-1}-1$, are denoted by full lines, the dashed lines are the $(A_0/N_T)(S_{2n}-1)$ forms, and the dash-dotted lines indicate the expected slopes at longer times.

4. SUMMARY AND CONCLUSIONS

In this paper we have analyzed the $A + A \rightarrow P$ and the $A + B \rightarrow P$ reactions and have modelled them through random walks on regular and on fractal lattices. One goal of our study was to find out in how far classical reaction kinetics is a good description of the decay in the presence of disorder and when, because of density fluctuations, the standard kinetical schemes cease to hold.

In order to be able to distinguish disorder effects from simple dimensional aspects, we have analyzed these reactions in the kinetic scheme, incorporating time-dependent rates. We have shown now the limiting cases of pseudounimolecular ($A_0 << B_0$) and of strictly bimolecular ($A_0=B_0$) decays follow from exact solutions for arbitrary initial concentrations A_0 and B_0. Dimensionality aspects are readily accounted for by relating the time-dependent rates to the number of distinct sites visited by random-walkers. Such relations give then Smoluchowski-type forms, both for regular lattices and also for fractal structures.

For the pseudounimolecular reactions the decay rates show in the absence of diffusion both the influence of the microscopic interactions and also that of the spatial dimension. The inclusion of the diffusion renders the decays more regular, without, however, leading to exponential forms for dimensions equal or less than two. We have shown, using an exactly solvable equation that d=2 is indeed the marginal dimensionality of the problem. Pseudounimolecular decay forms may be accurately obtained using random-walk simulations.

We have simulated both on regular lattices and on fractals the $A + A \rightarrow P$ and the $A + B \rightarrow P$, $A_0 = B_0$ strictly bimolecular reactions. For the first we find agreement with the Smoluchowski form for the whole decay range studied, whereas the second obeys such a form only in the initial time domain and crosses over to a slower decay $t^{-\tilde{d}/4}$ ($\tilde{d} < 4$) at longer times. Thus, reactions display a richer behavior than what standard kinetic approaches predict, a behavior which may be experimentally accessible. Here a time-dependent monitoring of the reaction process is needed, perhaps along lines similar to the measurements of energy transfer in condensed media.

ACKNOWLEDGEMENTS

Our thanks are due to Professor K. Dressler for helpful discussions. We acknowledge a grant of computer time from the computer center of the ETH-Zurich and the support of the Deutsche Forschungsgemeinschaft and of the Fonds der Chemischen Industrie. The hospitality of Exxon Research and Engineering Co. is gratefully acknowledged by A. B.

REFERENCES

1. S. Chandrasekhar, Rev. Modern Physics 15, 1 (1943) and ref. mentioned, such as M. von Smoluchowski, Phys. Z. 17, 557; 585 (1916); Z. Phys. Chem. (Leipzig) 92, 129 (1917).
2. D.F. Calef and J.M. Deutch, Ann. Rev. Phys. Chem. 34, 493 (1983).
3. B. Ya Balagurov and V.G. Vaks, Zh. Exp. Theor. Fiz. 65, 1939 (1973) [English translation Sov. Phys. JETP. 38, 968 (1974)]
4. A.A. Ovchinnikov and Ya.B. Zeldovich, Chem. Phys. 28, 215 (1978).
5. M.D. Donsker and S.R.S. Varadhan, Commun. Pure Appl. Math. 28, 525 (1975); 32, 721 (1979).
6. D. Toussaint and F. Wilczek, J. Chem. Phys. 78, 2642 (1983).
7. B.B. Mandelbrot, 'The Fractal Geometry in Nature' (Freeman, San Francisco, 1982).
8. A. Blumen, J. Klafter and G. Zumofen, Phys. Rev. B27, 3429 (1983).
9. J.M. Smith, 'Chemical Engineering Kinetics' (McGraw-Hill, Kogakusha, Tokyo, 1981).
10. T.R. Waite, Phys. Rev. 107, 463 (1957); 107, 471 (1957).
11. J. Klafter and A. Blumen in: 'Random Walks and their Applications in the Physical and Biological Sciences', ed. M.F. Shlesinger and B.J. West (American Institute of Physics, N.Y., 1984) p. 173.
12. P.W. Klymko and R. Kopelman, J. Phys. Chem. 87, 4565 (1983).
13. J. Klafter and A. Blumen, J. Chem. Phys. 80, 875 (1984).
14. U. Even, K. Rademann, J. Jortner, N. Manor and R. Reisfeld, Phys. Rev. Lett. 52, 2164 (1984).
15. G. Zumofen, A. Blumen and J. Klafter, this Volume.
16. J. Klafter, A. Blumen and G. Zumofen, J. Phys. Chem. 87, 191 (1983).
17. J.R. Miller, J. Chem. Phys. 56, 5173 (1972); J. Chem. Phys. 79, 1070 (1975).
18. R.K. Huddleston and J.R. Miller, J. Phys. Chem. 85, 2292 (1981); 86, 200, 1347 (1982).
19. A.B. Doktorov, R.F. Khairutdinov and K.I. Zamaraev, Chem. Phys. 61, 351 (1981).
20. A. Blumen and R. Silbey, J. Chem. Phys. 70, 3707 (1979).
21. A. Blumen, J. Klafter and R. Silbey, J. Chem. Phys. 72, 5320 (1980).
22. A. Blumen, Nuovo Cimento 63B, 50 (1981).
23. F. Willig, A. Blumen and G. Zumofen, Chem. Phys. Lett. 108, 222 (1984).
24. K. Allinger and A. Blumen, Chem. Phys. Lett. 82, 76 (1981).
25. K. Allinger and A. Blumen, Chem. Phys. Lett. 79, 494 (1981).
26. G. Zumofen and A. Blumen, Chem. Phys. Lett. 88, 63 (1982).
27. J. Klafter, G. Zumofen and A. Blumen, J. Physique Lett. 45, L49 (1984).
28. H.B. Rosenstock, Phys. Rev. 187, 1166 (1969); SIAM J. Appl. Math. 27, 457 (1974).
29. R.C. Powell and Z.G. Soos, J. Lumin. 11, 1 (1975).
30. P. Evesque, J. Physique (Paris) 44, 1217 (1983).

31. R. Kopelman, P.W. Klymko, J.S. Newhouse and L.W. Anacker, Phys. Rev. B29, 3747 (1984).
32. D.C. Torney and H.M. McConnell, Proc. R. Soc. London A387, 147 (1983)
33. D.C. Torney and H.M. McConnell, J. Phys. Chem. 87, 1941 (1983).
34. P. Meakin and H.E. Stanley, J. Phys. A17, L173 (1984).
35. K. Kang and S. Redner, Phys. Rev. Lett. 52, 955 (1984).
36. R.M. Noyes and W.C. Gardiner, J. Phys. Chem. 85, 594 (1981).
37. V.V. Antonov-Romanovskii, Fiz. Tverd. Tela 25, 3291 (1983). [English transl. Sov. Phys. Solid State 25, 1896 (1983)].

RANDOM WALKS ON FRACTALS

G. Zumofen
Physical Chemistry Laboratory
ETH-Zentrum, CH-8092 Zurich, Switzerland

A. Blumen* and J. Klafter
Corporate Research Science Laboratories
Exxon Research and Engineering Company
Clinton Township, Annandale, N.J. 08801, USA

ABSTRACT. We study the pseudounimolecular reaction $A+B \rightarrow P$, with $A(0) \ll B(0)$ and we model it through a random-walk picture. Two distinct cases are considered: the trapping problem, in which the minority, A-particles, move while the B-molecules are stationary, and the target problem, in which the A molecules act as fixed targets and only the B-particles move. As underlying structures we consider both translationally-symmetric lattices and also fractals (Sierpinski gaskets). Fractals are a large class of geometrical objects; apart from being interesting by themselves, fractals also provide a way to mimic disorder. Although both the target and the trapping problem can be described by the same reaction-diffusion equation, their decays are different: It is the trapping which shows at longer times the slower decay and the larger deviations from standard kinetic decay laws. Both, for the trapping and target problems, the deviations get more pronounced if the dimensionality of the underlying structure decreases; the decay laws found for Sierpinski gaskets are intermediate between the decays for one- and two-dimensional regular lattices.

1. INTRODUCTION

Random walks have a long history and, as standard stochastic processes, have been successfully used to model chemical reactions [1-3]. The difference between the reaction-diffusion description of chemical kinetics and the random-walk picture consists in the discretization of the space over which the walk takes place. As is well-known by now,

---

* On leave from Lehrstuhl für Theoretische Chemie, Technische Universität München, D-8046 Garching, West Germany

R. Daudel et al. (eds.), Structure and Dynamics of Molecular Systems, 87–100.

many results do not depend on the continuous or discrete picture of the reaction space [1], so that for a wide range of problems both methods can be used, especially when performing numerical calculations. In addition, a large body of exact analytical solutions are known for random walks [3], fact which enhances their usefulness.

In this paper we study in the framework of random-walks the $A+B \rightarrow P$ reaction for the case that the concentrations of A- and B-molecules are very different, say $A_0 \ll B_0$, where A_0 and B_0 are the initial numbers of particles. As discussed in the previous paper [4] this situation corresponds to a pseudounimolecular reaction. We will consider here two different random-walk models for this reaction: In Sec. 2 we start with the trapping problem, in which the A-molecules move, whereas the B particles are stationary and act as static traps. Then, in Sec. 3 we focus on the target problem, in which the B-molecules move, whereas the A-molecules are fixed in space and play the role of targets. Both models have a wide range of applications since random walks on lattices with randomly distributed reactive centers have been extensively studied [3,5]. Of main concern here are mobile defects in crystals [6], electronic and vibrational energy transfer [7-10] and nuclear magnetic resonance [11]. Recently, models of self-attracting polymers [12] and of the Williams-Watts form of dielectric relaxation [13] have also been mapped onto a random walk picture.

In the past most studies were limited to regular lattices in one-, two- and three-dimensions. In nature, however, many structures obey a weaker symmetry than translational invariance: they are dilatationally invariant (self-similar), and belong to the class of geometrical objects called fractals [14] . Many stochastically disordered structures, like percolation clusters at criticality [15,16] linear and branched polymers [17], aggregates constructed by diffusion-limited growth [21] may be viewed as fractals. It is thus of interest to extend the random-walk picture to fractal structures.

In this work we consider trapping and target problems on fractal structures. Comparing the results to those which obtain for regular lattices we find that the properties of the gaskets mirror their intermediate (non-integer) state. As fractals we use the Sierpinski gaskets [14], which are deterministic: they can be built according to an iterative prescription, which determines unambiguously which points belong to the lattice. Deterministic fractals can be constructed in a wide variety [22]: See Sec. 2 for an example of a Sierpinski-gasket.

2. THE TRAPPING PROBLEM

We start our study of the pseudounimolecular reaction $A+B \rightarrow P$, $A_0 \ll B_0$, with the trapping problem, in which the B-molecules are static. It is now sufficient to consider the fate of a single A-molecule, which performs a random walk over a given structure. Here the B-molecules are distributed randomly, occupying the sites of the structure with probability p. For a finite, relatively large system of N_T sites, $p \simeq B_0/N_T$. At each step the walker moves, with equal probability, to one of its neighboring sites. The reaction is assumed to occur instantaneously, when the A-molecule lands on a site occupied by a B-molecule. Thus the B-molecules act as traps.

In order to obtain the survival law of the A-molecules, an average over all random-walk realizations and all distributions of B-molecules is needed. We proceed as in our previous works [10,23]. For a particular realization of the random walk on the structure we let R_n denote the number of distinct sites visited in n steps, where we set $R_0=1$. Since we also consider fractals, the stochastic variable R_n may depend both on the starting point on the structure and also on the sequence of directions of the steps. We note that for a regular lattice the starting point is irrelevant. For the same realization of the walk we let F_n denote the probability that trapping has not occured up to the nth step: [10,23,24]

$$F_n = (1 - p)^{R_n - 1} \tag{1}$$

assuming the origin of the walk not to be a trap. The measurable survival probability after n steps is $\tilde{\Phi}_n$, the average of F_n over all realizations of the random walk [7,10,25,26] and over all starting points

$$\tilde{\Phi}_n = \langle F_n \rangle = \langle (1-p)^{R_n - 1} \rangle \tag{2}$$

Introducing $\lambda = - \ln (1-p)$, Eq. (2) turns into:

$$\tilde{\Phi}_n = e^{\lambda} \langle e^{-\lambda R_n} \rangle \tag{3}$$

which allows the following expansion in the cumulants $K_{j,n}$ of the distribution of R_n:

$$\tilde{\Phi}_n = e^{\lambda} \exp \left[\sum_{j=1}^{\infty} K_{j,n} (-\lambda)^j / j! \right] \tag{4}$$

The first two cumulants are for instance:

$$K_{1,n} = \langle R_n \rangle \equiv S_n \tag{5}$$

and

$$K_{2,n} = \langle R_n^2 \rangle - \langle R_n \rangle^2 \equiv \sigma_n^2 \tag{6}$$

where S_n and σ_n^2 denote the mean and the variance of R_n.

From the knowledge of all cumulants $\tilde{\Phi}_n$ can be determined exactly using Eqs. (3) and (4). In general the distribution of R_n is not known in great detail, and one has from the first N (known) cumulants the approximate form:

$$\tilde{\Phi}_{N,n} \simeq e^{\lambda} \exp\left[\sum_{j=1}^{N} K_{j,n} (-\lambda)^j/j!\right] \tag{7}$$

The expression for N = 1

$$\tilde{\Phi}_{1,n} = \exp\left[-\lambda (S_n - 1)\right] \tag{8}$$

has been advanced in many areas [7,9,27]: In the random-walk field it corresponds to the first-passage-time approximation [9]. In the fractal field it was recently used by de Gennes [27]. For N = 2 one obtains from Eq. (7) the form [10,28]

$$\tilde{\Phi}_{2,n} = \exp\left[-\lambda(S_n - 1) + \lambda^2\sigma_n^2/2\right] \tag{9}$$

Now, for regular lattices one has for not too small n [2,3,5]:

$$(d=1) \quad S_n = a_1\sqrt{n} + a_2/\sqrt{n} + \ldots \; ; \; \sigma_n^2 \sim 4(\ell n2 - 2/\pi)n \tag{10a}$$

$$(d=2) \quad S_n = a_1 n/\ell n(a_2 n) + \ldots \; ; \; \sigma_n^2 \sim n^2/\ell n^4 n \tag{10b}$$

$$(d=3) \quad S_n = a_1 n + a_2\sqrt{n} + \ldots \; ; \; \sigma_n^2 \sim n \; \ell n \; n \tag{10c}$$

where the a_i are constants which depend on the lattice structure.

With these expressions for S_n and σ_n^2 we found for three dimensional regular systems Eq. (8) to be a good approximation over the main portion of the decay [28]. Furthermore, Eq. (9) which includes the variance, approximates the survival probability of the A-molecule very well over several orders of magnitude in its decay, both for three- and for two-dimensional systems [10]. For the linear chain, Eq. (7) converges very slowly, so that in order to describe two orders of magnitude in the decay, at least four terms (N = 4) in Eq. (7) are needed [10].

The expressions for $\tilde{\Phi}_n$ lead now to the decay rates $k_n = B_o^{-1}(\tilde{\Phi}_{n-1} - \tilde{\Phi}_n)/\tilde{\Phi}_n$, the discrete analog of $k(t) = -B_o^{-1}\dot{\tilde{\Phi}}(t)/\tilde{\Phi}(t)$. From Eqs. (3) to (9) it is obvious that k_n is not exactly proportional to $p \simeq B_o/N_T$. However the proportionality gets better for lower p and higher dimensionalities. In the region where the proportionality holds one obtains as in the previous work [4], the rates:

$$k_n = N_T^{-1} \frac{dS_n}{dn} \tag{11a}$$

or, for continuous times, with $t \simeq n\tau$, where τ is the stepping time:

$$k(t) = N_T^{-1} \frac{dS(t)}{dt} \tag{11b}$$

We now look at fractals, and choose as prototypes the Sierpinski gaskets. A member of this class of geometrical objects which can be embedded in the plane is shown in Fig. 1.

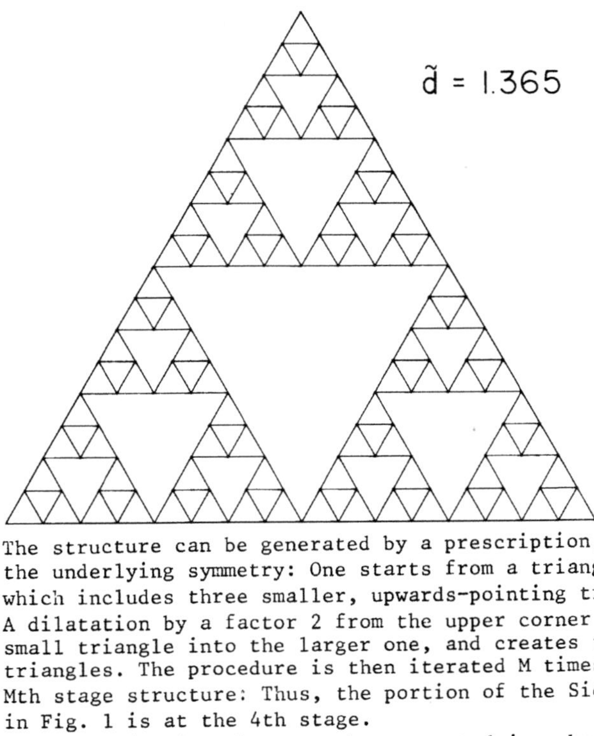

Figure 1. The Sierpinski gasket at the 4th stage. The spectral dimension is $\tilde{d}$ = 1.365.

The structure can be generated by a prescription, which renders clear the underlying symmetry: One starts from a triangle of sidelength 2, which includes three smaller, upwards-pointing triangles of sidelength 1. A dilatation by a factor 2 from the upper corner transforms the upper small triangle into the larger one, and creates two additional, larger triangles. The procedure is then iterated M times, and leads to the Mth stage structure: Thus, the portion of the Sierpinski gasket depicted in Fig. 1 is at the 4th stage.

Sierpinski gaskets can be generated in embedding spaces of arbitrary dimension d, by starting with the corresponding hypertetrahedrons [14,22]. Also, related fractals of more general shape may be constructed along similar lines of thought [22].

Distinct from regular lattices, whose random-walk properties are mainly determined by their embedding spaces, see Eqs. (10) as a typical example, properties of fractal lattices are describable through several, mostly non-integer parameters, which play roles similar to the dimension. For us, here two major parameters are of concern: The fractal

(Hausdorff)-dimension $\bar{d}$, which is related to the density of sites, and the spectral (fracton) dimension $\tilde{d}$, which appears in connection with dynamical properties (such as heat conduction, wave propagation and also reaction and diffusion) on the fractals. Thus, for Sierpinski-gaskets embedded in d-dimensional spaces one has [14]

$$\bar{d} = \ell n\ (d+1)/\ell n 2 \tag{12a}$$

and

$$\tilde{d} = 2\ \ell n\ (d+1)/\ell n\ (d+3) \tag{12b}$$

Therefore, for the Sierpinski gasket of Fig. 1 $\bar{d} = 1.584$ and $\tilde{d} = 1.365$. In general, the relation $\tilde{d} \leqslant \bar{d} \leqslant d$ holds; furthermore for Sierpinski gaskets $1 < \tilde{d} < 2$, as may be seen from Eq. (12b).

In Ref. 29 we have determined S_n and σ_n^2 through random-walk simulations on Sierpinski gaskets. The changes to Eqs. (10) are as follows:

$$(\tilde{d} < 2) \qquad S_n \simeq a_1 n^{\tilde{d}/2} \quad ; \qquad \sigma_n^2 \sim n^{\tilde{d}} \tag{13}$$

The results both for S_n and σ_n^2 could be very well fitted by Eq. (13) for several gaskets considered [29]. Eq. (13) interpolates between the one- and the two-dimensional forms, Eqs. (10a) and (10b). One should note that Eqs. (13) and also the survival probability $\tilde{\Phi}_n$, to be discussed below, depend only on $\tilde{d}$, but not on $\bar{d}$. This distinguishes random-walk decays from reactions mediated through long-ranged interactions in the absence of diffusion (see the previous article, Ref. 4, and also Ref. 30), which depend on $\bar{d}$.

In Fig. 2 we present the survival probabilities $\tilde{\Phi}_n$ for several values of the concentrations of B-molecules, which are randomly distributed on the Sierpinski-gasket of Fig. 1 ($\tilde{d} = 1.365$, p = 1%, 3%, 10% and 30%). The $\tilde{\Phi}_n$ are plotted logarithmically versus n; the full curves correspond to the exact decay, whereas the broken lines are the approximating forms $\tilde{\Phi}_{1,n}$ and $\tilde{\Phi}_{2,n}$, Eqs. (8) and (9). For the plot Eq. (13) for S_n and σ_n^2 was used, in order to have approximate expressions with a minimal number of parameters. For all B-concentrations the decays are clearly <u>nonexponential</u>. We find a behavior much reminiscent of survival probabilities for walkers on the square lattice: the mean number of sites visited, S_n, does not describe the decay well, whereas inclusion of the variance σ_n^2 considerably improves the agreement [10]. In Ref. 29 we found by analysing several Sierpinski-gaskets that the agreement is better for the higher $\tilde{d}$ values. This corroborates nicely with the parallel trend for regular lattices.

Dynamical quantities for Sierpinski-gaskets such as the moments of the distribution of distinct sites visited and also the survival probability show a behavior which is situated between that which obtains for one- and for two-dimensional regular lattices. The unifying aspect is the <u>compact</u> exploration of the geometrical structures by the walker, a concept stressed by de Gennes [27]. As in the one-dimensional case, this leads to relations between the moments of the distributions, re-

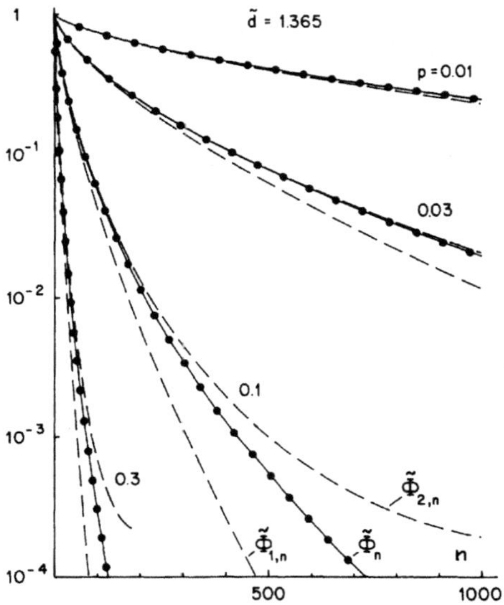

Figure 2. The survival probability $\tilde{\Phi}_n$ for various trap concentrations p on a Sierpinski gasket with $\tilde{d} = 1.365$ where n is the number of steps. The dots and full lines denote the simulation data whereas the dashed lines are the $\tilde{\Phi}_{1,n}$ and $\tilde{\Phi}_{2,n}$ approximations Eqs. (8) and (9), respectively.

lations which appear from the numerical simulations [29].

Concluding this section on the trapping problem, we emphasize that the survival probability is not necessarily an exponential function of time. This is exemplified by Eqs. (7) to (9), together with the appropriate forms for the cumulants, Eq. (10) and (13), and also by Fig. 2. These results hold for short and moderately long times. As discussed by us elsewhere [31] at long times the asymptotic decay laws on fractals are

$$\tilde{\Phi}_n \sim \exp\left[-c\,\lambda^{2/(\tilde{d}+2)} n^{\tilde{d}/(\tilde{d}+2)}\right] \tag{14a}$$

which are an extension of the forms which hold for regular lattices [32,33]

$$\tilde{\Phi}_n \sim \exp\left[-c\,\lambda^{2/(d+2)} n^{d/(d+2)}\right] \tag{14b}$$

Equations (14) are again, nonexponential. Furthermore they are nonanalytical with respect both to λ and to n.

3. THE TARGET PROBLEM

In this section we consider the target problem, $A+B \rightarrow P, A_0 << B_0$, in which the B-molecules move, whereas the A molecules are static and act as targets. We focus on a single A-particle, assumed to be at the origin of the coordinate system. The B-molecules occupy the remaining sites of the structure. We find it convenient to allow for a multiple occupancy of sites by B molecules. As initial conditions we analyze three kinds of distributions [34]: the Poisson distribution, by which the number of B-particles at each site follows a Poisson-law, the binomial distribution, under which each site may be occupied with probability p by exactly one B-particle, and the multinomial distribution, by which M B-molecules are distributed over N structure sites [34].

Let us now set the origin of the coordinate system at the location of a target and denote by $F_m(\vec{r})$ the probability that a random walker starting from $\vec{r}$ reaches the origin $\vec{0}$ for the first time in the mth step. For regular lattices, because of the symmetry of the walk $F_m(\vec{r})$ is also the first-passage time from $\vec{0}$ to $\vec{r}$, as defined by Montroll and Weiss [35]. In general, we denote by $H_n(\vec{r})$ the probability that a first passage from $\vec{r}$ to $\vec{0}$ occured in the first n steps and have:

$$H_n(\vec{r}) \equiv \sum_{m=1}^{n} F_m(\vec{r}) \tag{15}$$

The probability therefore that a walker from $\vec{r}$ did not reach $\vec{0}$ in the first n steps is thus

$$\Phi_n(\vec{r}) = 1 - H_n(\vec{r}) \tag{16}$$

Using Eq. (16) we obtain the survival probability of the A-molecule by appropriately weighing products of the $\Phi_n(\vec{r})$-functions. We take the Poisson-distribution, given by:

$$g(j) = e^{-p}\, p^j/j! \tag{17}$$

where g(j) is the normalized probability of having j B-particles at one site, and $p \simeq B_0/N_T$. The decay law of the target is then:

$$\Phi_n^P = \prod_{\vec{r}}{}' \{\sum_j g(j)[\Phi_n(\vec{r})]^j\} \tag{18}$$

where the product extends over all structure sites, with the exception of the origin, and the index P in Φ_n^P denotes the Poisson-distribution. Inserting Eqs. (16) and (17) into Eq. (18) leads to:

$$\Phi_n^P = \prod_{\vec{r}}{}' \{\sum_j e^{-p}[p\Phi_n(\vec{r})]^j/j!\}$$
$$= \prod_{\vec{r}}{}' \exp[-p+p\,\Phi_n(\vec{r})] = \exp[-p \sum_{\vec{r}}{}' H_n(\vec{r})] \tag{19}$$

For regular lattices Eq. (19) may be further simplified since, according to Eqs. (III.2) and (III.3) of Ref. 35 one has

$$\sum_{\vec{r}}{}' F_m(\vec{r}) = S_m - S_{m-1} \qquad (m \geq 1) \tag{20}$$

where again S_m is the mean number of distinct sites visited by a random walker in m steps, with $S_0=1$. Introducing now Eqs. (15) and (20) into (19) one has exactly:

$$\Phi_n^P = \exp\left[-p(S_n-1)\right] \tag{21}$$

For fractals the connection between Eqs. (19) and (21) is only approximate, since $H_n(\vec{r})$ depends then explicitly on the location of the trap (origin), and one has to first average over all possible trap locations. Eq. (21) corresponds then to using the first term of the cumulant expansion of Eq. (19). We expect it to hold well for quasihomogeneous fractals, such as the Sierpinski gaskets.

We note that Eq. (21) corresponds to the first-passage-time approximation of the trapping problem, Eq. (8), with λ replaced by p. For S_n, Eqs. (10) for regular lattices and (13) for fractals can be used. As before, only in three-dimensions does one obtain a quasiexponential decay. Different from the trapping problem, in which at extremely long times a crossover to the forms (14) takes place, here the form (21) persists for regular lattices even at very long times.

To display deviations from Eq. (21) we now turn to the target problem when the initial distribution of the B-molecules is binomial. Then Eq. (17) gets replaced by the distribution $(1-p)\,\delta_{j,0}+p\delta_{j,1}$, where $\delta_{j,i}$ is the Kronecker-delta symbol, and from Eq. (18) one has

$$\Phi_n^B = \prod_{\vec{r}}{}' \left[1-p + p\,\phi_n(\vec{r})\right] = \prod_{\vec{r}}{}' \left[1-pH_n(\vec{r})\right] \tag{22}$$

where B denotes the binomial distribution and where we made use of Eq. (16). We notice that Eq. (22) is amenable to an exact treatment in the framework of the generating-functions formalism [35,36], and Eq. (21) is its leading term in the low-density expansion, $p \ll 1$. To demonstrate the last statement it suffices to take logarithms of both sides of Eq. (22) and to use the Taylor-series expansion:

$$\ell n\Phi_n^B = \sum_{\vec{r}}{}' \,\ell n\left[1-pH_n(\vec{r})\right] = -\sum_{j=1}^{\infty} (p^j/j) \sum_{\vec{r}}{}' \left[H_n(\vec{r})\right]^j \tag{23}$$

To first order in p one recovers Eq. (19). Noting that all the $H_n(\vec{r})$ terms are non-negative, one also finds $\Phi_n^B \le \Phi_n^P$ for all n. Since the $H_n(\vec{r})$ are bounded from above by unity, a lower bound for Φ_n^B is easily obtained by replacing in Eq. (23) $[H_n(\vec{r})]^j$ by $H_n(\vec{r})$. Thus setting

$$\lambda \equiv \sum_{j=1}^{\infty} p^j/j = -\ell n\,(1-p) \tag{24}$$

one has

$$\Phi_n^P(\lambda) \le \Phi_n^B(p) \le \Phi_n^P(p) \tag{25}$$

which allows to bound Φ_n^B through Φ_n^P (or equivalently, Φ_n^P through Φ_n^B).

It follows that the qualitative behavior of Φ_n^B and of Φ_n^P is the same: The correction terms do not change drastically the decay behavior.

We now consider the multinomial distribution, starting directly from B_o-molecules distributed on N_T sites. The decay is then:

$$\Phi_n^M = N_T^{-B_o} \sum \frac{(B_o)!}{\ell_1!\dots\ell_{N_T}!} [\Phi_n(\vec{r}_1)]^{\ell_1} \dots [\Phi_n(\vec{r}_{N_T})]^{\ell_{N_T}} \tag{26}$$

where the sum runs over all $\{\ell_i\}$ sets such that $\ell_1 + \dots + \ell_{N_T} = B_o$. Eq. (26) is the obvious restriction of Eq. (18) to fixed numbers of B-molecules and sites. Using the multinomial formula the summation is immediate:

$$\Phi_n^M = N_T^{-B_o} \left[\sum_{i=1}^{N_T} \Phi(\vec{r}_i)\right]^{B_o} \tag{27}$$

and, as in the equations following Eq. (19)

$$\Phi_n^M = (1 - S_n/N_T)^{B_o} \tag{28}$$

The last expression has been advanced by Hamill and Funabashi [37] and by Shlesinger and Montroll [13]. In the thermodynamic limit, $N_T \to \infty$, $B_o \to \infty$, such that $p = B_o/N_T$, one has $\Phi_n^M \to e^{-pS_n}$. Here again Eq. (21) provides the leading behavior. Note the order in which the limiting process is to be performed: one first chooses the value of the maximal number n of steps and then takes N_T to be much larger than this value.

To summarize the results of this and of the previous section, we compare in Figs. 3 and 4 the decay laws of the trapping and of the target problems. To exemplify, we start from the simple cubic lattice (Fig. 3) and extend then our considerations to the Sierpinski gasket with $\tilde{d} = 1.365$ (Fig. 4). In both cases we plot the results for two concentrations (p = 10% and 50%). In Fig. 3 the forms (21) are given as broken lines: They provide the exact solution of the target problem for a Poisson distribution of walkers, Φ_n^P, and for these forms only S_n is needed.

For the cubic lattice we calculated S_n exactly through the very accurate (non-stochastic) series inversion technique of Ref. 25. In Fig. 3 we plotted both $\Phi_n^P(p)$ and $\Phi_n^P(\lambda)$, with $\lambda = -\ln(1-p)$, since these give upper and lower bounds for $\Phi_n^B(p)$, Eq. (25). The target decay for a binomial distribution Φ_n^B is plotted as a full line. This form is also exact, Eq. (23), but it involves additional information, since not only the S_n, but also the $H_n(\vec{r})$ are needed. We obtained these through an iterative (non-stochastic) procedure. We find that the relations which bound Φ_n^B through Φ_n^P are correct, so that these forms are very similar.

The solution of the trapping problem, Eq. (3) involves an average over all random-walk realizations, which is not in general known in closed form. Thus, in order to obtain the decay we simulated some 10^4 walks, following the procedure of Ref. 10. The simulation results are indicated as dots, and the dash-dotted line is an analysis of the decay in terms of the first two cumulants, Eq. (9). It is evident that the simulation works very well in this range, and that the $\tilde{\Phi}_n$ is bounded from below by the form $\Phi_n^P(\lambda)$, Eq. (21).

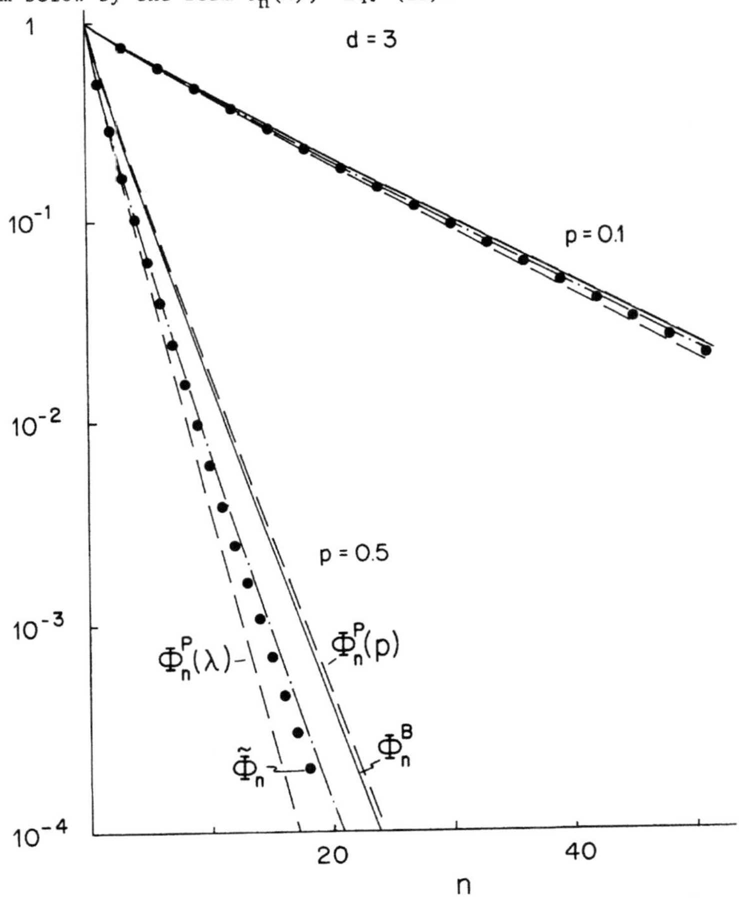

Figure 3. Pseudounimolecular decays on a simple cubic lattice. The exact decays of the target problem are given as full lines, Φ_n^B, for an initial binomial distribution, and as dashed lines, Φ_n^P, for an initial Poisson distribution of walkers. The dots and the dash-dotted lines denote the simulation calculations of the decay $\tilde{\Phi}_n$ due to trapping. The concentrations of the majority species are p=0.1 and p=0.5.

In Fig. 4 we present the corresponding results for random walks on the Sierpinski gasket. Here we have determined $H_n(\vec{r})$ iteratively for a given origin on the fractal lattice and we have used mean values of S_n

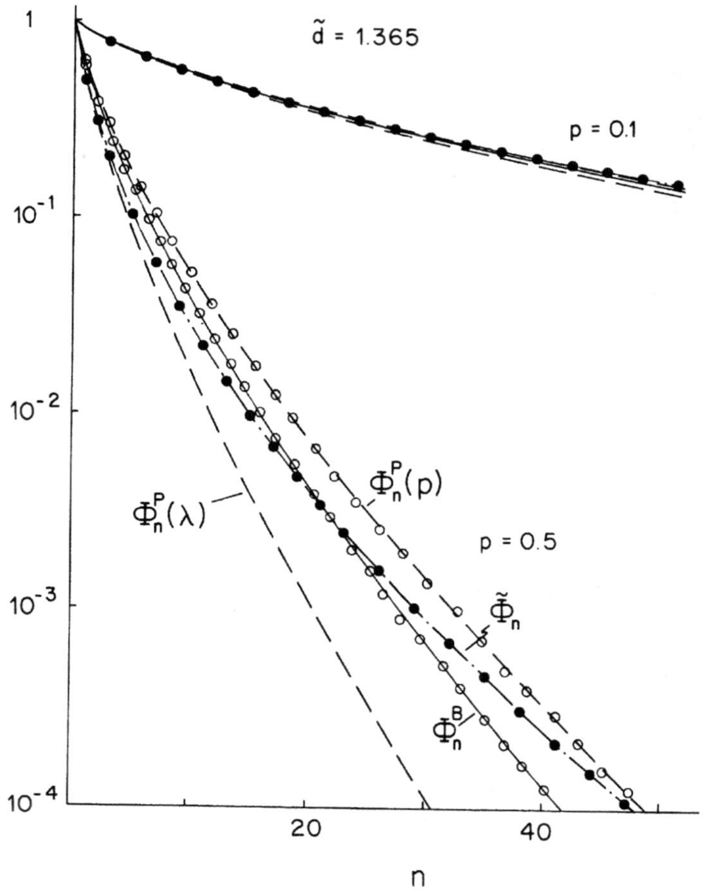

Figure 4. Pseudounimolecular decays on a d=2 Sierpinski gasket. The notations are as in Fig. 3. The dashed and full lines are the approximate decays corresponding to Eqs. (21) and (23), and the open circles are simulation calculations for the target problem.

in Eq. (21) and of $\sum_{\vec{r}}[H_n(\vec{r})]^j$ in Eq. (23) by averaging appropriately over different origins. The solution of the trapping problem is taken from Fig. 2.

Comparing Figs. 3 and 4 we find that in three dimensions Eq. (21) is a much better approximation both for the target $\Phi_n^B(p) \simeq \exp[-p(S_n-1)]$ and also for the trapping $\tilde{\Phi}_n \simeq \exp[-\lambda(S_n-1)]$ decay. In all cases, this approximation works better for smaller concentrations ($p \ll 1$) and at shorter times. As a rule [see also Eqs. (14)], at sufficiently long times the target decay is more rapid than the trapping.

4. SUMMARY AND CONCLUSIONS

In this article we have studied the pseudounimolecular $A+B \rightarrow P$, $A_o \ll B_o$ reaction in the framework of random walks, and we have considered both regular and fractal lattices. We have focussed on the trapping problem, in which the A-molecules but not the B-molecules move and on the target problem, in which the A-molecules are static and the B-particles move.

In the trapping problem approximations based on the cumulants are qualitative in the short-time and low concentration limits, but at longer times deviations from such forms become increasingly important. The role of the dimensionality is very important: The survival probability for A-molecules moving on Sierpinski gaskets shows a behavior intermediate between that found for one- and two-dimensional walks.

In Sec. 3 we have presented exact decay laws for the target problem on regular lattices. These forms are also good approximations for the target-decay on quasihomogeneous fractals. We are able to incorporate in the exact forms three distinct initial conditions, by letting the B-molecules follow Poisson, binomial and multinomial distributions. These differences show up in the decays which, however, follow quite closely Eq. (21). We infer that the target problem is well represented by kinetic forms, when the dimensional effects are included. Thus the motions of the majority and of the minority species are not interchangeable; the trapping problem leads to slower decays in the long-time regime than the target problem.

ACKNOWLEDGEMENTS

Our thanks are due to Professor K. Dressler for helpful discussions. The hospitality of Exxon Research and Engineering Co. is gratefully acknowledged by A. B. where part of this work was carried out. We acknowledge a grant of computer time from the computer center of the ETH-Zurich and the support of the Deutsche Forschungsgemeinschaft and of the Fonds der Chemischen Industrie.

REFERENCES

1. S. Chandrasekhar, Rev. Modern Physics **15**, 1 (1943).

2. M.N. Barber and B.W. Ninham, 'Random and Restricted Walks; Theory and Applications'(Gordon and Breach,New York, 1970).
3. G.H. Weiss and R.J. Rubin, Adv. Chem. Phys. 52, 363 (1983).
4. A. Blumen, G. Zumofen and J. Klafter, this Volume.
5. E.W. Montroll and B.J. West, in: 'Fluctuation Phenomena', E.W. Montroll and J.L. Lebowitz eds. (North-Holland, Amsterdam, 1979).
6. M. Koiwa ans S. Ishioka,J. Stat. Phys. 30, 477 (1983).
7. H.B. Rosenstock, Phys. Rev. 187, 1166 (1969); SIAM J. Appl. Math. 27, 457 (1974).
8. D.L. Huber in: 'Laser Spectroscopy of Solids', W.M. Yen and P.M. Selzer eds. (Springer-Verlag, Berlin, 1981).
9. J. Klafter and R. Silbey, J. Chem. Phys. 74, 3510 (1981).
10. G. Zumofen and A. Blumen, Chem. Phys. Lett. 88, 63 (1982).
11. P.M. Richards, J. Stat. Phys. 30, 497 (1983).
12. S. Redner and K. Kang, Phys. Rev. Lett. 51, 1729 (1983).
13. M.F. Shlesinger and E.W. Montroll, Proc. Natl. Acad. Sci. U.S.A. 81, 1280 (1984).
14. B. Mandelbrot, 'The Fractal Geometry in Nature' (W.H. Freeman, San Francisco, 1982).
15. S. Alexander and R. Orbach, J. Phys. Lett. 43,L625 (1982).
16. R. Rammal and G. Toulouse, J. Phys. Lett. 44,L13 (1983).
17. S. Havlin and D. Ben-Avraham, J. Phys. A: Math Gen. 15, L311 (1982).
18. T. Witten and L.M. Sander, Phys. Rev. Lett. 47, 1400 (1981).
19. S. Alexander, C. Laermans, R. Orbach and H.M. Rosenberg, Phys. Rev. B28, 4615 (1983).
20. D. Avnir, D. Farin and P. Pfeifer, J. Chem. Phys. 79, 3566 (1983).
21. U. Even, K. Rademann, J. Jortner, N. Manor and R. Reisfeld, Phys. Rev. Lett. 52, 2164 (1984).
22. R. Hilfer and A. Blumen, J. Phys. A17, L537 (1984).
23. G. Zumofen and A. Blumen, J. Chem. Phys. 76, 3713 (1982).
24. J. Klafter, A. Blumen and G. Zumofen, J. Stat. Phys. (Sept. 1984).
25. A. Blumen and G. Zumofen, J. Chem. Phys. 77, 5127 (1982).
26. G.H. Weiss, Proc. Nat. Acad. Sci. U.S.A. 77, 4391 (1980).
27. P.G. de Gennes, C. R. Acad. Ser. II 296, 881 (1983).
28. G. Zumofen and A. Blumen, Chem. Phys. Lett. 83, 372 (1981).
29. A. Blumen, J. Klafer and G. Zumofen, Phys. Rev. B28,6112 (1983).
30. J. Klafter and A. Blumen, J. Chem. Phys. 80, 875 (1984).
31. J. Klafter, A. Blumen and G. Zumofen, J. Phys. Lett. 45, L49 (1984).
32. B. Ya. Balagurov and V.G. Vaks, Zh. Exp. Teor. Fiz. 65, 1939 (1973) [English translation; Sov. Phys. Jetp 38, 968 (1974)].
33. M.D. Donsker and S.R.S. Varadhan, Comm. on Pure and Appl. Math. 28, 525 (1975); 32, 721 (1979).
34. W. Feller, 'An Introduction to Probability Theory and Its Applications' (Wiley, New York 1971) Vol. I. Chap. 6.
35. E.W. Montroll and G.H. Weiss, J. Math. Phys. 6, 167 (1965).
36. E.W. Montroll, J. Math. Phys. 10, 753 (1969).
37. W.H. Hamill and K. Funabashi, Phys. Rev. B16, 5523 (1977).

LOCALIZATION OF EXCITONIC EIGENSTATES IN MOLECULAR CRYSTALS

J.-P. Lemaistre and A. Blumen *
Centre de Mécanique Ondulatoire Appliquée, 23, rue du Maroc
F-75019 Paris, France

ABSTRACT. A study of the influence of the structural disorder on the localization properties of excitonic eigenstates is presented. The effects of the diagonal and off-diagonal disorders on the exciton wave-functions are analyzed through the calculations of inverse participation ratios (IPR) for short-ranged (exchange,octupolar) and long-ranged (dipolar) types of interactions. Configurationally averaged calculations are performed and provide the density and the energy distribution of the IPR at various trap concentrations. The method is then used to analyze the localization behavior of both the guest and host eigenstates of isotopically mixed crystals over the full range of trap concentrations. The computed values of the IPR for guest and host exciton states provide a quantum picture of the mixed crystal and allow to follow the gradual transition from small aggregates to the crystal when the concentration increases from low to high values. Applications are made to naphthalene and anthracene triplet states.

1. INTRODUCTION

The problem of a transition between localized and delocalized quantum states in disordered systems has been the center of considerable attention [1-3]. The question is of importance in many areas, such as electronic processes in non-crystalline materials [4,5] and the energy transfer in matrix-isolated species and in mixed molecular crystals[6-8]. In the classical Anderson-transition approach one focuses on the diagonal disorder which obtains when, on a periodic lattice, the site-energies are statistically distributed. Recently the important role of the structural inhomogeneities and of their influence on the localization properties has been investigated [8-13]. Stuctural disorder which must be distinguished from the substitutional disorder [9] arises when active particles are randomly placed on an underlying lattice, thus forming random patterns. In this case the off-diagonal terms are highly correlated, since the absence of an active particle from a site disrupts completely all interactions with that site. Furthermore, for a low occupancy of sites by particles (high disorder), restricting the interactions to

R. Daudel et al. (eds.), Structure and Dynamics of Molecular Systems, 101–117.

nearest neighbors is not justified in general, since then longer-ranged distance-dependent interactions become effective and have to be taken into account. Thus, the parameters which determine the structural disorder are the range of the interparticle potential $V(\vec{r})$ and the concentration of particles. The role of the potential $V(\vec{r})$ is analyzed by considering short-range (exchange and octupolar) interactions and long-range (dipolar) interactions. We use as a localization criterium for the excitonic eigenstates the inverse participation ratios [8-10,14]. The density of excitonic eigenstates and the distribution in energy of their IPR are numerically evaluated. In our numerical calculations we keep the number of particles constant and, in order to overcome the statistical fluctuations, we perform configurational averagings. Contrarily to the pure substitutionally off-diagonal disorder which leads to delocalization [3,9,15] the calculations show that the structural off-diagonal disorder localizes strongly [8]. In our study we also include the diagonal disorder. Thus both effects of diagonal and off-diagonal structural disorders enhance the localization of the excitonic eigenstates.

The calculation method is then applied to study the localization properties of the triplet excitonic eigenstates of isotopically mixed crystals. For these systems the sites are occupied by either protonated or deuterated molecules. We first start by considering a deuterated crystal (host) in which a fraction of sites is randomly substituted by protonated molecules (guests) acting as traps. All the trap sites are coupled either directly if they are in nearest-neighbor positions, or indirectly by superexchange through the virtual states of the host crystal. Since the triplet interactions are mainly two-dimensional [16] we use as a model a square lattice. The edge effects due to the finite size of the lattice are removed by closing the plane edge to edge [17]. Thus the diagonalization of the full Hamiltonian of the guest-host system gives the eigenstates in the energy domains of both the guest and host states and allows to obtain the corresponding localization indices and their distribution. Using the symmetry properties of the crystal and those of the triplet states [18-20] the phosphorescence intensities can be easily calculated and the lineshapes compared to experimental data [21-23]. Such calculations are out of the scope of the present study.

In section 2 we outline the theoretical approach of the localization problem. The numerical calculations of the localization indices of the excitonic eigenstates are provided in section 3. The effects of the structural off-diagonal and diagonal disorders are illustrated over the full range of concentrations for short-ranged and long-ranged types of interactions. In section 4 we use our numerical method to describe the guest and host triplet eigenstates of isotopically mixed crystals with application to naphthalene and anthracene. Finally, in section 5 we summarize the main results of our study.

2. THEORETICAL APPROACH

2.1. Hamiltonian and eigenstates

We consider a regular cubic array of N sites, which features a three dimensional crystal. In a particular configuration of randomly distributed particles some M sites, $M = cN$, where c denotes the concentration, are occupied. The total Hamiltonian associated to this configuration of M interacting particles is

$$H = \sum_i E_i |i\rangle\langle i| + \sum_{i,j}' V(\vec{r}_{ij}) |i\rangle\langle j| \quad (1)$$

In Eq. (1) the summation runs over all the occupied lattice sites and $\vec{r}_{ij}$ are their mutual distances. The interactions $V(\vec{r}_{ij})$ between the randomly distributed particles define the off-diagonal disorder. As typical $V(\vec{r}_{ij})$ dependencies (which occur often in molecular crystals) we have chosen short-ranged (exchange and octupolar) as well as long-ranged (dipolar) interactions, namely :

$$V(r) = V_0 \exp(-\alpha(r-1)) \quad (2a)$$

and

$$V(r) = V_0 r^{-s} \qquad (s = 3 \text{ or } 7) \quad (2b)$$

Here $r \equiv |\vec{r}|$ is expressed in lattice units and V_0 is the interaction strength between nearest-neighbor sites. In the following we choose V_0 as energy unit and we assume that the interactions are isotropic $V(r) \equiv V(\vec{r})$.

For the diagonal disorder which is due to the fluctuations in the values E_i of the site energies we follow the prescription of the Anderson-model : we take the E_i to be randomly and uniformly distributed in the interval $[-W/2, +W/2]$ around the mean value E_0. For simplicity we set E_0 at the origin of the energy scale, $E_0 = 0$.

For a given configuration of interacting particles the total Hamiltonian of Eq. (1) is diagonalized. One obtains thereby all eigenvalues and eigenvectors $H|\psi_k\rangle = E_k|\psi_k\rangle$. The index k indicates the different eigenstates, which in the local basis $|i\rangle$ are

$$|\psi_k\rangle = \sum_i C_i^k |i\rangle \quad (3a)$$

where the coefficients C_i^k fulfill the normalization condition

$$\sum_i |C_i^k|^2 = 1 \quad (3b)$$

The computed E_k values give the density of states $\rho(E)$.

2.2. Localization behavior of eigenstates

From the coefficients C_i^k we obtain readily the quantity

$$L_k \equiv \sum_i |C_i^k|^4 \quad (4)$$

which defines the inverse participation ratio commonly used as localization criterium [8-10] . Obviously L_k is close to unity for a state localized on a single site only and drops to zero as $1/n$ for a state evenly delocalized over n sites. The numerical procedure used to calculate the eigenstates of the Hamiltonian, Eq. (1) allows to obtain, in addition to the density of states, the distribution of the L_k as a function of energy. Furthermore we can also study the distribution of the localization indices by deriving the number of states $n(L)\,dL$ whose L_k fulfill the condition $L \leq L_k \leq L + dL$. An easy-to-view description for the distribution of the L_k and useful in the study of its concentration dependence is provided by $\lambda(L)$, the cumulative density of eigenstates whose localization indices lie below L. The function $\lambda(L)$ is thus

$$\lambda(L) = \int_0^L n(L')\, dL' /M \qquad (5)$$

3. NUMERICAL CALCULATIONS

In the numerical calculations the localization indices of the eigenstates are evaluated for several three-dimensional systems of randomly distributed particles. Both short-ranged (exponential and octupolar) and long-ranged (dipolar) types of couplings are used. The diagonalization of the Hamiltonian, Eq. (1), is performed by taking the number of particles M to be around 100. Thus the finite lattice size depends on the concentration and is fixed so that $cN \simeq 100$. In order to account well for the statistical fluctuations we average over a fair number of distinct configurations. All the results presented here have been performed for 30 configurations. The interactions $V(r)$ are typical for triplet and singlet states of aromatic crystals such as naphthalene and anthracene. The diagonal disorder, whenever introduced, corresponds to the inhomogeneous linewidths observed in the phosphorescence and fluorescence spectra of these systems.

Let us start by considering the off-diagonal disorder and then compare the results to those obtained by adding the diagonal disorder. Thus in Fig. 1 we display the plots of $\lambda(L)$, the relative number of eigenstates with a localization index less than L, for exponential, octupolar and dipolar interactions. The drawings correspond to calculations performed in the absence ($W/V_0 = 0$) and in the presence ($W/V_0 \neq 0$) of diagonal disorder and were performed for various concentrations of particles ranging from $c = 0.01$ to $c = 0.9$. The calculations show that with increasing concentration the states become more and more delocalized. Indeed an increase in c corresponds to spatial configurations which show a higher degree of order. We also notice that the delocalization of the eigenstates occurs gradually when the concentration grows. From Fig. 1 we infer that the inclusion of the diagonal disorder acts localizing and leads to a drastic effect for the short-range interactions. We can also notice that this effect is more pronounced for not very delocalized states, whereas delocalized states are less influenced. Comparing the three series of curves allows to see that the exchange and octupolar interactions, both being short-ranged, are similar. In general the octupolar interaction is intermediate between the exchange and the long-ranged dipolar interaction.

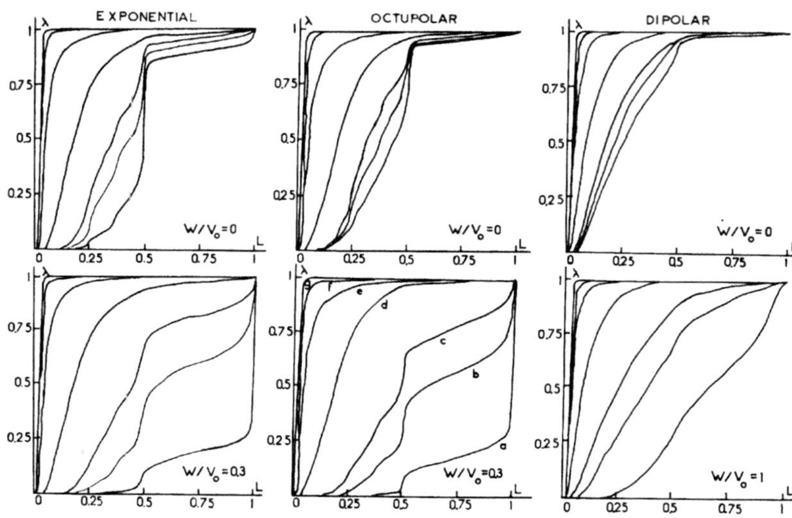

Figure 1. Plots of $\lambda(L)$ for exponential, octupolar and dipolar interactions. The concentrations are c = 0.01(a); 0.05(b) ; 0.1(c) ; 0.25(d); 0.5(e) ; 0.75(f) ; 0.9(g).

3.1. Short-ranged interactions

In Figs.2 and 3 we display calculations for exponential interactions which were performed in the absence (a) and in the presence (b) of diagonal disorder ($W/V_0 = 0.3$). The number density n (L) of eigenstates (bottom) and their energy distribution (top) are given in Fig. 2 for c = 0.05 and in Fig. 3 for c = 0.1. From these figures we observe that there are no states present below $L_k \leq 0.1$ and that the number of states with $0.1 < L_k \leq 0.2$ is very low. If we choose the value 0.2 as a boundary between the localized and delocalized behavior this implies for our system that most states are localized. In the absence of the diagonal disorder Figs 2 and 3 show a prominent peak at L = 0.5 meaning that in many cases two particles couple strongly with each other. Now if the perfect energetic resonance between all particles is disrupted by the inclusion of the diagonal disorder one observes a broadening in the density ρ (E) of eigenstates. Furthermore the peak at L = 1 in the n (L) histograms gains in intensity at the detriment of the peak at L = 0.5. We associate the remaining states at L = 0.5 with dimer states, i.e. pairs in nearest neighbor position. We note that both the diagonal and the off-diagonal structural disorder favor the localization and their combined effect is cooperative. For short-ranged interactions (exchange and octupolar) the effects of the diagonal disorder are well observable for concentrations up to 0.25. Beyond this concentration the states become delocalized and the influence of the diagonal disorder is less pronounced. The delocalization behavior in the range above 0.25 is illustrated in Figs. 4 and 5. In fig. 4 we show the energy distribution ρ (E) of eigenstates at high concentrations. We consider here the off-diagonal disorder only. The histograms of the number density of eigenstates n (L) are depicted in Fig. 5.

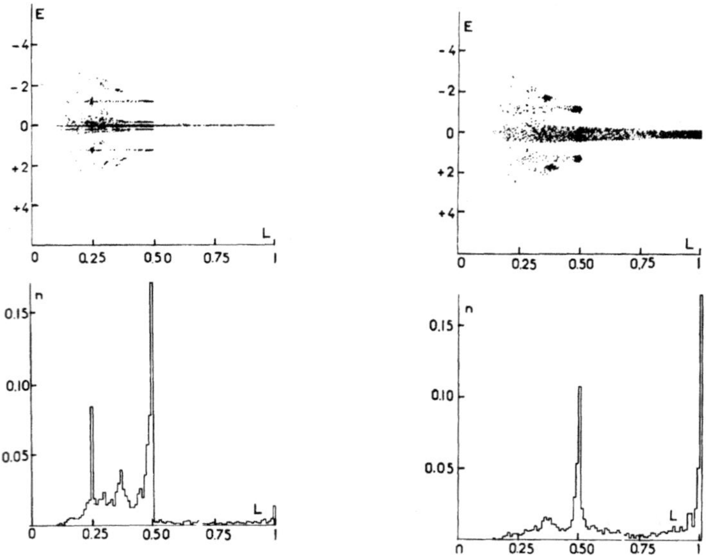

Figure 2. The number density n (L) of the localization indices and the energy distribution ρ (E) for exponential interactions in the absence (left) and in the presence (right) of diagonal disorder ($W/V_0 = 0.3$) at a concentration of c = 0.05.

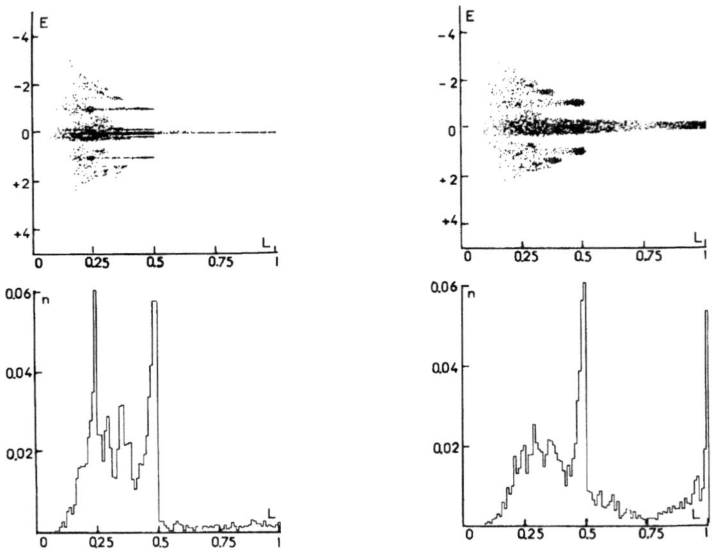

Figure 3. Same as in Fig. 2 at a concentration of c = 0.1.

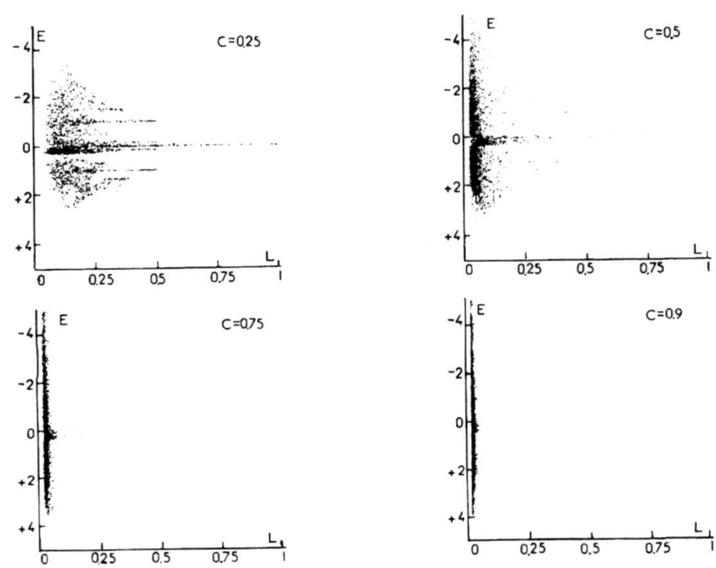

Figure 4. The energy distributions ρ(E) for exponential interactions at concentrations c = 0.25, 0.5, 0.75 and 0.9.

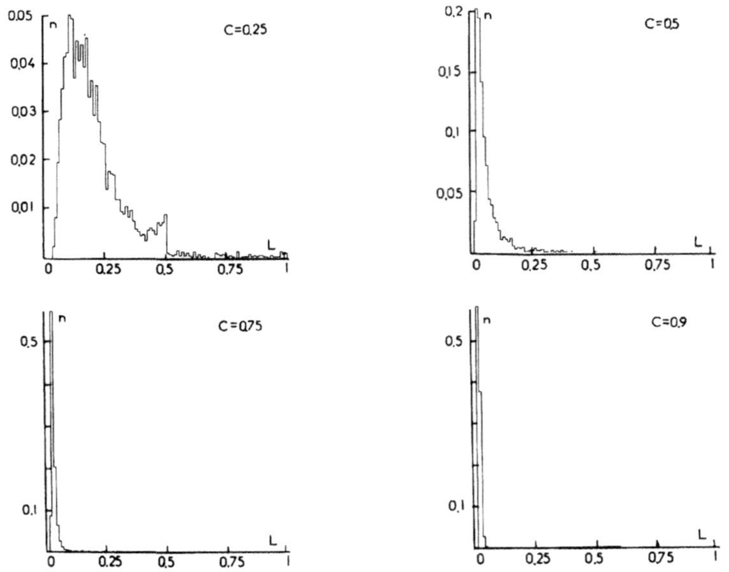

Fig. 5. The number density n (L) for exponential interactions at concentrations c = 0.25, 0.5, 0.75 and 0.9.

3.2. Long-ranged interactions

Let us now consider the dipolar case, in which the interactions V (r) are long-ranged. In the absence of the diagonal disorder most of the states are delocalized, even at low concentration (c = 0.01) ; this is illustrated in Fig. 1c. As in the previous discussion we display the energy distribution ρ (E) of the eigenstates together with the number density n (L). The results presented in the figures 6, 7 and 8 at concentrations c = 0.01, 0.05 and 0.1 respectively have been performed with and without the diagonal disorder, taking in the former case $W/V_0 = 1$. For long-ranged dipolar interactions the inclusion of the diagonal disorder leads, in addition to the broadening of the distributions ρ (E) and n (L), to delocalization over small size cluster states with $0.5 < L_k < 1$. For the value of the parameter W used in our calculations the effect of diagonal disorder is clearly observable up to 0.1. At higher concentrations the off-diagonal disorder is dominant. The concentration dependence of the delocalization behavior of the eigenstates is illustrated in Fig. 9 for ρ (E) and in Fig. 10 for n (L).

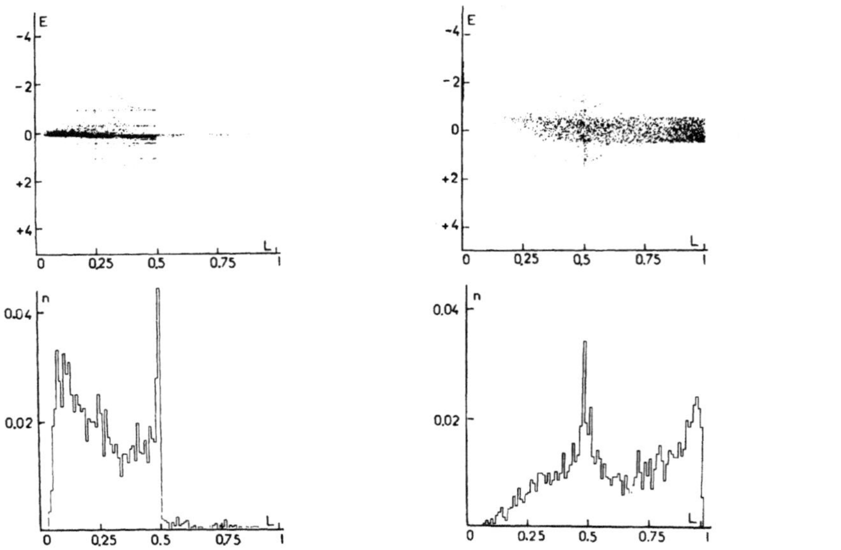

Fig. 6. The number density n(L) of the localization indices and the energy distribution ρ(E) for dipolar interactions in the absence (left) and in the presence (right) of diagonal disorder ($W/V_0 = 1$) at a concentration of c = 0.01.

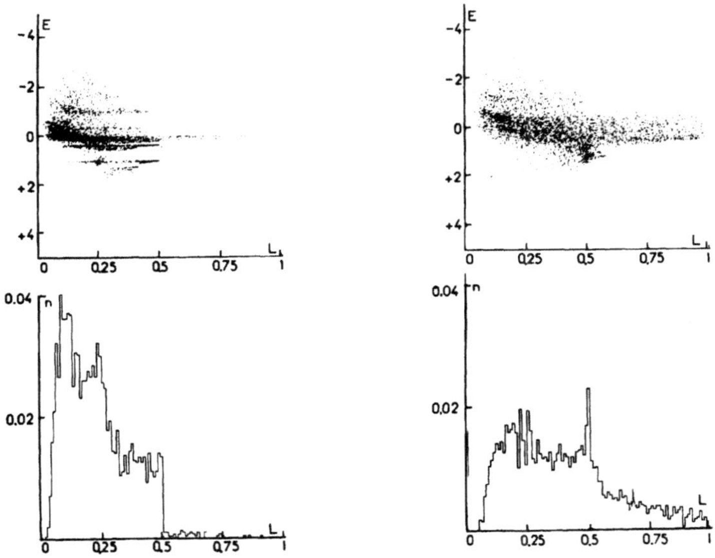

Fig. 7. Same as in Figure 6 at concentration c = 0.05.

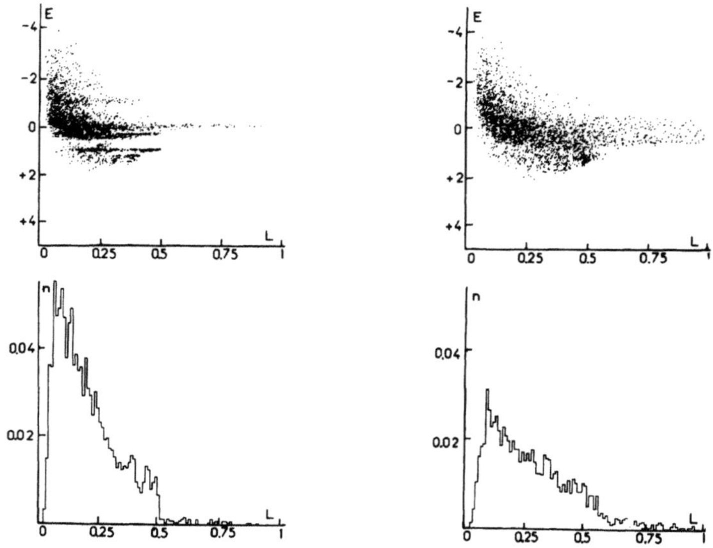

Figure 8. Same as in Fig. 6 at concentration c = 0.1.

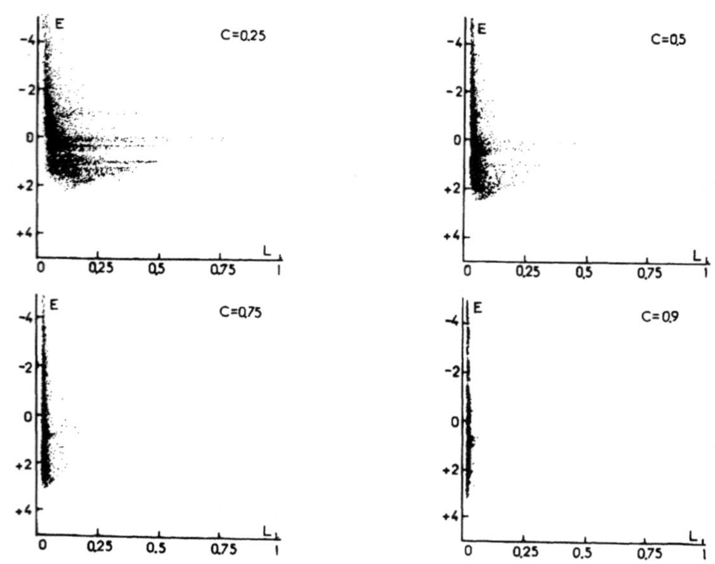

Figure 9. The energy distributions $\rho(E)$ for dipolar interactions at concentrations c = 0.25, 0.5, 0.75 and 0.9.

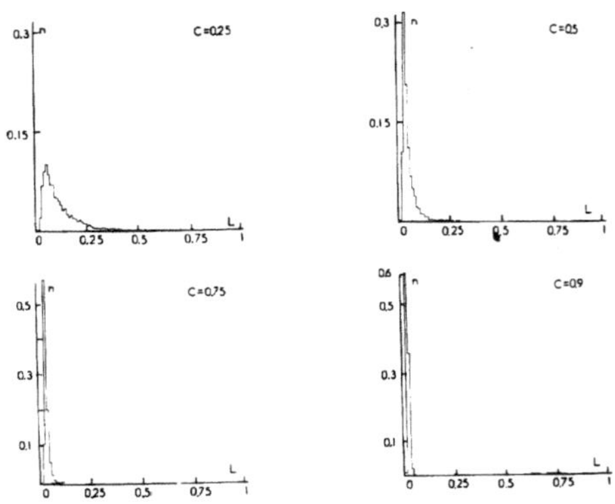

Figure 10. The number density n (L) for dipolar interactions at concentrations c = 0.25, 0.5, 0.75 and 0.9.

4. APPLICATION TO MOLECULAR CRYSTALS

Molecular crystals and more particularly isotopically mixed molecular crystals provide interesting systems for the study of the localization of excitonic eigenstates. In general the crystal may be viewed as representing a perfect structure in which disorder is introduced substitutionally by doping the material with molecules which differ in their isotopic constitution (traps). As it is well known the positions and the orientations of the isotopic traps are often those of the substituted molecules of the crystal. Furthermore, as for the triplet states of naphthalene and anthracene, the states of the host crystal are energetically well separated from the trap states. As may be evident the localized or delocalized behavior of the trap and host eigenstates strongly depends on the nature and the dimensionality of the interactions, on the trap concentrations and also on the energy splittings between the trap and host states. At low concentrations (say less than 0.1) the trap states are those of small molecular clusters, a situation which corresponds to localized wavefunctions. On the other hand at high concentrations of traps, these form clusters which evolve into extended domains, associated with delocalized bands. In the intermediate concentration range, where neither the site representation nor the band-states description is adequate, the identification of the numerous geometrical configurations of clusters becomes an inextricable problem and the nature of the corresponding eigenstates is not easy to assign. In order to overcome this difficulty we calculated numerically the localization indices associated with the eigenstates of both trap and host crystal molecules according to the method presented in section 2.

4.1. Hamiltonian of the mixed crystal

We consider an isotopically mixed crystal in which the sites are occupied by either protonated or deuterated molecules. We start from a deuterated crystal (host) in which a fraction, c, of protonated molecules (guests), of lower energy, is randomly substituted. The total Hamiltonian is

$$H = \sum_i E_i |i\rangle\langle i| + \sum_{j>i} V_{ij} |i\rangle\langle j| \qquad (6)$$

In Eq. (6) the sums run over all sites. We assume that the guest (E_g) and the host (E_h) energies are not affected by the disorder so that $E_i = E_g$ or $E_i = E_h$ depending on the occupation of each site. We also assume that the guest-guest, host-host and guest-host interactions are identical since the triplet bandwidths of pure protonated and pure deuterated crystals are similar. The numerical diagonalization of the full Hamiltonian of Eq. (6) gives the eigenenergies E_k and the eigenvectors $|\psi_k\rangle$ of both the guest and host states. The index k is used here to distinguish the different electronic eigenstates. Obviously, in the case of a perfect crystal, the index k is a vector of the reciprocal lattice. The direct diagonalization procedure for coupled guest and host eigenstates overcomes the difficulty of identifying and counting the numerous configurations of clusters. From the knowledge of the eigenvectors we evaluate the localization indices L_k for each eigenstate and their distribution between L = 0 (delocalized states) and L = 1 (localized states). We thus obtain the normalized density of localization indices n (L) and the cumulative number of states λ (L) (see section 2.2) over the energy range of the guest and the host excitons.

4.2. Naphthalene and anthracene triplet states

As a model for naphthalene or anthracene crystal, we use a square arrangement of N = 81 molecules since, for both materials, the triplet interactions are mainly two-dimensional [16]. In order to partially overcome the edge effects due to the finite size of the model the square is closed edge to edge. The two systems are quite similar. Compared to naphthalene, in anthracene the interactions are stronger and the guest-host energy separation is smaller. Thus, the energy separation between guest and host is E_h-E_g = 95 cm^{-1} for naphthalene (Nh_8) and 60 cm^{-1} for anthracene (Ah_{10}). In our calculations we take as nearest neighbor triplet interactions the values V_{ij} = 1.2 cm^{-1} for Nh_8 [20] and V_{ij} = −2.55 cm^{-1} [22,23] for Ah_{10} when i and j are translationally inequivalent; V_{ij} = +0.7 cm^{-1} for Nh_8 and V_{ij} = −2 cm^{-1} for Ah_{10} when i and j are translationally equivalent along the b crystal axis.

The numerical diagonalization of the Hamiltonian of Eq. (6) provides the eigenenergies and the eigenvectors of both guest and host states. For each eigenstate we calculate the localization indices L_k according to Eq. (4). In Figs 11 and 12 we display the histograms of the eigenstates populations n (L) at various guest concentrations c relevant for comparison to experimental data available on Nh_8 [21] and Ah_{10} [23] respectively. At c = 0.15 the host eigenstates are delocalized states while the guest eigenstates are localized over the various

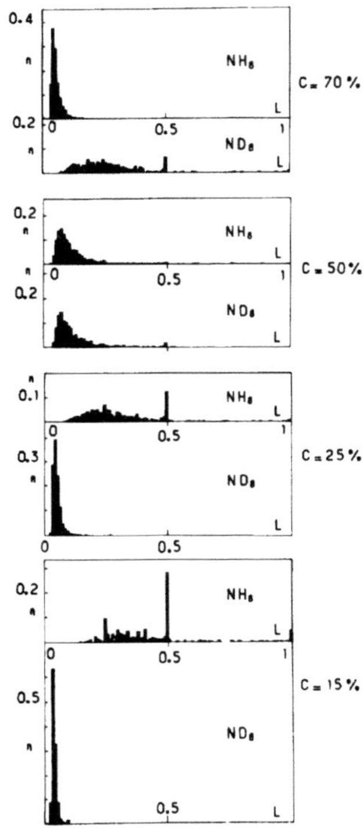

Figure 11. The number density n (L) for naphthalene(Nh_8 and Nd_8) triplet eigenstates at the Nh8 concentrations c = 0.15, 0.25 0.5 and 0.7.

configurations of clusters. At higher concentrations(c = 0.25, 0.30), we observe an increase of the populations of such clusters. At c = 0.5 the histograms associated to guest and host eigenstates are quite similar. At high concentrations (c = 0.7, 0.85) we note the symmetric roles played by the guest and host eigenstates. Furthermore these calculations emphasize the important role of pair eigenstates having a localization index L_k equal to 0.5. These pairs may be assigned to dimers (i.e. two guest molecules coupled through nearest neighbor interactions) or pairs (i.e. two guest molecules separated by several host molecules and coupled through superexchange interactions). As mentioned in the previous section this class of molecular pairs which disappear by adding some diagonal disorder may be seen as possible centers for the triplet-triplet annihilation process between traps at the trap energy (i.e. within the inhomogeneous linewidth of the monomer line). For this process, it turns out that the energies of the clusters are out of resonance with those of the monomers. In Figs 13 and 14 we display the cumulative number of states $\lambda(L)$ with localization index L_ν below a certain value Lin the energy domains of guest and host excitons for Nh_8 and Ah_{10} respectively. The concentration dependence of λ shows that the transition between the localized and the delocalized behavior occurs gradually when the concentration increases.

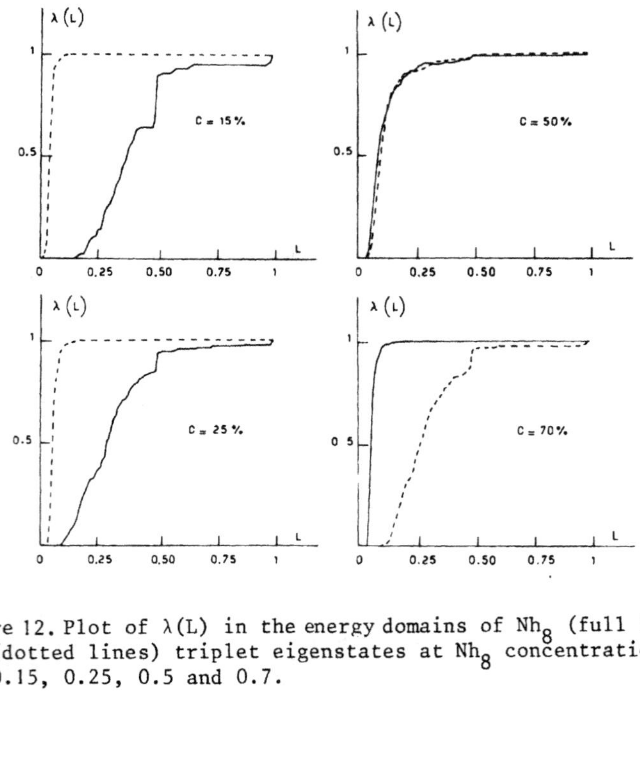

Figure 12. Plot of λ(L) in the energy domains of Nh_8 (full lines) and Nd_8 (dotted lines) triplet eigenstates at Nh_8 concentrations of $c \cong 0.15$, 0.25, 0.5 and 0.7.

5. CONCLUSION

We have presented a complete study of the influence of the off-diagonal and diagonal structural disorders on the localization properties of excitonic eigenstates in structurally disordered media. The IPR method has been used to define the localization behavior of the exciton wavefunctions associated to randomly distributed particles coupled by short-ranged and long-ranged types of interactions. We have performed numerical calculations, configurationally averaged, of the localization indices over the full range of concentrations. From the calculations of λ(L) we have shown that the delocalization of the eigenstates occurs gradually when the particle concentration increases. The energy distribution ρ (E) together with the density n (L) of localization indices

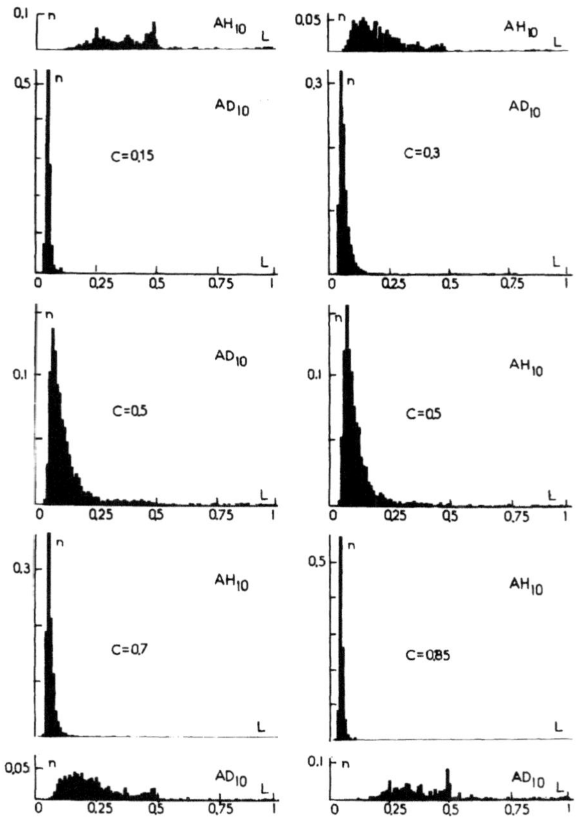

Figure 13. The number density n (L) for anthracene (Ah_{10} and Ad_{10}) triplet eigenstates at Ah_{10} concentrations c = 0.15, 0.3, 0.5 and 0.85.

allow us to illustrate the drastic localization effect of the diagonal disorder for short-ranged interactions at low concentrations. Furthermore, and from the n (L) histograms, we observe at low concentrations a large population of pair eigenstates delocalized over two sites. The resonance interaction between these pairs can easily be destroyed by

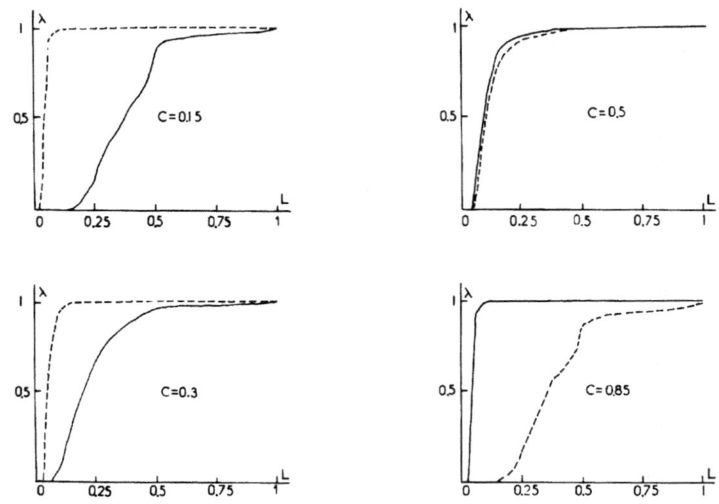

Figure 14. Plot of λ(L) in the energy domains of Ah_{10} (full lines) and Ad_{10} (dotted lines) triplet eigenstates at Ah_{10} concentrations c = 0.15, 0.3, 0.5 and 0.85.

adding a small diagonal disorder (inhomogeneous broadening) thus leading to eigenstates localized over one site (L = 1).

We have then applied the IPR method to provide a quantum picture of isotopically mixed crystals at trap concentrations ranging from large dilution to highly concentrated systems. The localization indices of both the guest and the host system have been performed for two-dimensional short-ranged triplet interactions. From the n(L) histograms and the λ(L) function we have shown the gradual delocalization (localization) of the eigenstates associated to guest (host) molecules when the guest concentration increases.

Finally, from our work, we found the direct numerical procedure to be a very useful method to analyze the quantum properties of structurally disordered systems over a large domain of concentrations and for various types of interactions. The calculation of these basic quantities is a fundamental step before describing dynamical processes in condensed media.

ACKNOWLEDGEMENTS

The supports of the Deutsche Forschungsgemeinschaft and of the Fonds der Chemischen Industrie are gratefully acknowledged by A.B.

REFERENCES

⋆ On leave from Lehrstuhl für Theoretische Chemie, Technische Universität München, Lichtenbergstrasse 4, D-8046 Garching, West Germany.

1- P.W. Anderson, Phys. Rev. 109 (1958) 1492.
2- D.J. Thouless, in *Ill-Condensed Matter*, R. Balian, R. Maynard and G. Toulouse, eds. (North-Holland, Amsterdam, 1979), p.1.
3- Y. Nagaoka and H. Fukuyama, eds, *Anderson Localization* (Springer Verlag, Berlin, 1982).
4- N.F. Mott and E.A. Davis, *Electronic Processes in Non-Crystalline Materials*, Second ed. (Clarendon, Oxford, 1979).
5- E.N. Economou, *Green's Functions in Quantum Physics*, Second ed. (Springer Verlag, Berlin, 1983).
6- A.H. Francis and R. Kopelman, *Laser Spectroscopy of Solids*, W.M. Yen and P.M. Selzer eds. (Springer Verlag, Berlin, 1981), p 241.
7- J. Klafter and J. Jortner, J. Chem. Phys. 71 , 2210 (1979).
8- J.P. Lemaistre and A. Blumen, Chem. Phys. Lett. 99. 291 (1983).
9- F. Yonezawa, J. Non-Cryst. Solids 35/36 , 29 (1980).
10- W.Y. Ching and D.L. Huber, Phys. Rev. B25 , 1096 (1982) ; B26 , 5596 (1982).
11- S. Stafström, R. Ricklund and K.A. Chao, Phys. Rev. B27, 6158 (1983).
12- S.N. Evangelou, Phys. Rev. B27 , 1397 (1983).
13- R. Raghavan, Phys. Rev. B29 , 748 (1984).
14- P. Dean, Proc. Phys. Soc. (London) 73, 413 (1959) ; Proc. Phys. Soc. (London) A254, 507 (1960) ; Rev. Mod. Phys. 44, 127 (1972).
15- P.D. Antoniou and E.N. Economou, Phys. Rev. B16, 3768 (1977).
16- R. Brown, J.-P. Lemaistre, J. Megel, Ph. Pee, F. Dupuy and Ph. Kottis, J. Chem. Phys. 76 , 5719 (1982).
17- Ph. Pee, J.-P. Lemaistre, F. Dupuy, R. Brown, J. Megel and Ph. Kottis, Chem. Phys. 64 , 389 (1982).
18- H. Sternlicht and H.M. McConnell, J. Chem. Phys. 35 , 1793 (1961) ; R.M. Hochstrasser, J. Chem. Phys. 47 , 1015 (1967).
19- J.-P. Lemaistre and Ph. Kottis, J. Chem. Phys. 68 , 2730 (1978) ; J.-P. Lemaistre and A.H. Zewail, J. Chem. Phys. 72 , 1055 (1980).
20- J.-P. Lemaistre, Ph. Pee, M. Beguery, F. Dupuy, J. Megel, R. Brown and Ph. Kottis, Chem. Phys. Letters 89 , 207 (1982).
21- J.-P. Lemaistre, A. Blumen, F. Dupuy, Ph. Pee, R. Brown and Ph. Kottis, J. Phys. Chem. 88 , 4655 (1984).
22- H. Port, D. Rund and H.C. Wolf, Chem. Phys. 60 , 81 (1981).
23- D. Rund and H. Port, Chem. Phys. 78 , 357 (1983).

FUSION AND TRANSFER OF TRIPLET EXCITONS IN ISOTOPICALLY MIXED NAPHTHALENE CRYSTALS

Ph. Pée, R. Brown, F. Dupuy, Y. Rebière, Ph. Kottis
Centre de Physique Moléculaire Optique et Hertzienne
Université de Bordeaux I and CNRS - F 33405 TALENCE

and J.-P. Lemaistre
Centre de Mécanique Ondulatoire Appliquée - F 75019 PARIS

ABSTRACT. Results on temperature and concentration dependences of triplet exciton fusion and triplet energy transfer in isotopically mixed naphthalene crystals are presented. Exciton fusion in the trap states is investigated by delayed fluorescence emission. Energy transfer efficiency is analyzed by the relative transfer yield to a supertrap acting as a sensor. The temperature dependences observed on delayed fluorescence and on energy transfer are similar, with a threshold at T $\sim$ 5K, suggesting a common origin for both processes. Numerical solutions of a master equation in a two-dimensional crystal are provided. The results may be understood by considering a two optical phonon mechanism (with a phonon energy equal to 40 cm^{-1}) and by taking into account the long-range dipolar interaction between traps due to the spin-orbit coupling of triplet states to upper singlet states.

1. INTRODUCTION

The dynamics of excitons in molecular crystals, both their transfer and their fusion, has been the subject of numerous experimental and theoretical studies {1-12}. But why molecular crystals in particular ? In these solids the intermolecular couplings (J,K $\sim$ 10 cm^{-1}) are very small compared to the site excitation energies (E $\sim$30000 cm^{-1}). As a first approximation the crystal may be described by the "oriented gas" model of independent molecules. Nevertheless the weak intermolecular couplings are sufficient enough at low temperature to create bands of collective excited states. In the case of pure crystals these collective states are described by means of Bloch waves (linear combinations of N resonant site localized excitations).

$$|E_k,\vec{k}> = (1/\sqrt{N}) \{ \sum_{m=1}^{N} \exp(i\, \vec{k}.\vec{r}_m)\, |m>\} \exp(-iE_k t/\hbar) \qquad (1a)$$

$$\text{where} \quad |m> = \phi^{*}_m \prod_{n=1}^{N}{}' \; \phi^{o}_n \qquad (1b)$$

R. Daudel et al. (eds.), Structure and Dynamics of Molecular Systems, 119–136.

But in real crystals the translational symmetry required breaks down because of the presence of inevitable impurities or crystal defects. In this case it may be difficult to draw an experimental distinction between exciton dynamics and processes such as scattering or trapping. The problem may be simplified by creating well defined and efficient localized states with

i) finite spatial extent

ii) readily measured energy levels and polarisations.

The most commonly applied method is isotopic substitution of some crystal sites. Guest molecules have the same size as host molecules, thus limiting the distorsion of the matrix. Furthermore the guest-guest, guest-host and host-host interactions are slightly similar and the substituted molecules have the same orientation. Other simplifications also depend on the system chosen.

This work concerns the first triplet states of host deuterated naphthalene crystals ($C_{10}D_8$ or ND_8) containing protonated naphthalene substitutional traps ($C_{10}H_8$ or NH_8). This system has the following advantages.

i) A long triplet state lifetime ($\sim$2s for the guest system).

ii) Triplet levels in the visible spectrum ($\sim$4700 Å or 21200 cm^{-1}) are easily populated with a tunable dye laser.

iii) The NH_8 triplet level is $\Delta \simeq 100$ cm^{-1} below the ND_8 exciton band. This is a typical example of the "deep trap limit" since $|V/\Delta| << 1$ where V ($\sim$1.2 cm^{-1}) denotes the interaction between nearest neighbor sites. The crystal may thus be considered from an energy point of view as composed of two independent quantum systems. Moreover at T=1.6K, there is no thermal activation from the NH_8 band to the ND_8 band. It is possible therefore to speak of a truly selective excitation of the NH_8 traps.

iv) Triplet state interactions in naphthalene are to first approximation of the short-ranged exchange type thus limited to nearest neighbor sites.
Furthermore the short-ranged nature of exchange interaction allows to ignore the interactions along the $\vec{c}$ crystal axis. A naphthalene crystal excited to its first triplet state may be considered as an ensemble of independent $(\vec{a},\vec{b})$ planes {13,14}.

The conditions required for a proper understanding of the propagation and the interaction of triplet excitons are :

i) A proper spectroscopic study of the crystal triplet states energy levels and polarisations.

ii) A well defined topology : Spatial distribution of the NH_8 traps and their couplings.

iii) Rigourously defined initial conditions : Selective laser excitation of the trap triplet states at very low temperature (T=1.6K).

In this paper, we give results concerning two indicators of exciton propagation : Delayed fluorescence due to triplet state fusion and transfer to non-resonant acceptor traps present at trace concentrations. We shall point out that similarities between these indicators (constant values at T < 4K and strongly increased yields at T $\sim$5K) may be of common origin. The sharp thresholds at T $\sim$5K are explained by reference to exciton-phonon coupling. The high yields at low trap concentrations are evidence that we *must* take into account the weak but long-range dipolar coupling in the triplet state, which is due to spin orbit coupling to higher singlets.

In what follows, part 2 describes the experimental system, by a spectroscopic and topological study of the mixed crystal NH_8/ND_8. Part 3 reviews our results on the dynamics of triplet excitons, both fusion resulting in delayed fluorescence and donor-acceptor transfer. These results are interpreted in the light of exciton-phonon coupling. The conclusion is presented in part 4.

2. DESCRIPTION OF THE EXPERIMENTAL SYSTEM

Experiments were done on host crystals of deuterated naphthalene (ND_8) containing protonated naphthalene (NH_8) at concentrations between 0.1 % and 15 % molar. The experimental conditions have been described else where {15}.

As pointed out in the introduction, the definition of the system means precise determination of the energy levels and of the topology (spatial distribution, couplings) of the mixed crystal NH_8/ND_8. This was achieved by recording phosphorescence excitation spectra at T = 1.6K in the spectral region of the traps {16}. Figure 1a shows a recording obtained when the laser has a linewidth of 0.8 cm^{-1} : We observe a strong deformed line with a weaker satellite at +1.7 cm^{-1}.

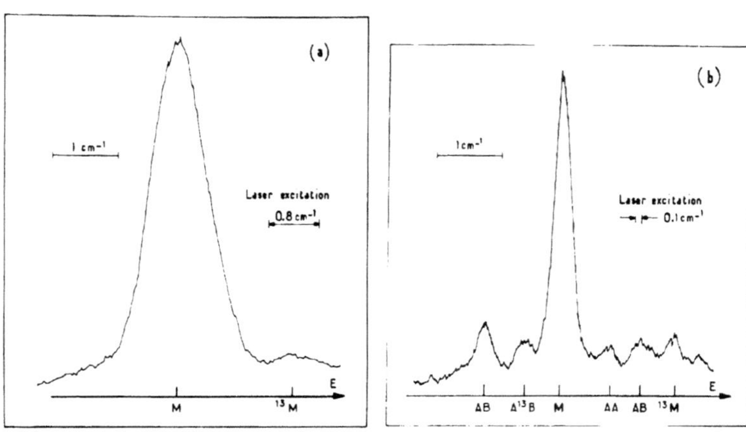

Figure 1. Phosphorescence excitation spectra of the trap states in an ND_8 crystal containing 2 % NH_8 recorded at T = 1.6K with a laser resolution of 0.8 cm^{-1} (a) and 0.1 cm^{-1} (b).

In Figure 1b, the spectrum with a laser resolution of 0.1 cm^{-1} reveals a strong central line, attributed to isolated NH_8 molecules or monomers and other weaker lines, assigned to associations of two or more NH_8 molecules (dimers, trimers,...). These structures, quite different from what one would expect on the assumption of a uniform distribution (no satellite lines) of the NH_8 in the ND_8 matrix, show the microscopic inhomogeneous disordered nature of the mixed crystal.

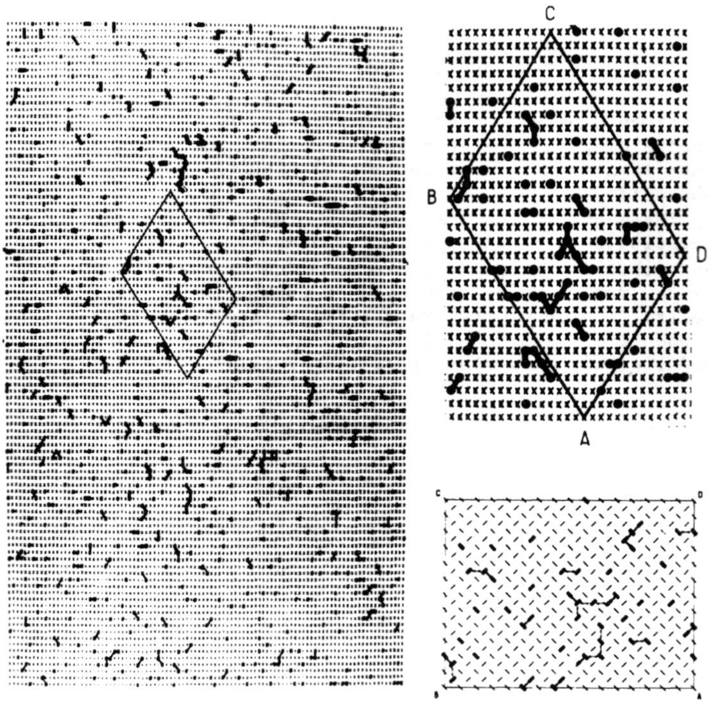

Figure 2. Computer generation of a trap distribution at c = 10 % in a two dimensional square lattice (left). A part of this grid (upper right) and the same transposed into the crystal lattice (bottom right). Note the presence of random clusters.

Figure 2 shows a simulation of the traps distribution in a two dimensional square lattice (which is isomorphous with the $(\vec{a},\vec{b})$ crystal plane). This simulation obtained with a pseudo-random number generator emphasizes the microscopic inhomogeneous nature of the traps distribution in the mixed crystal. Averaging over a large number of similar grids the concentrations and the separations of the random aggregates may be evaluated {18}.

Diagonalization of the interaction matrix written on the site basis yields the energy levels and polarisations of the aggregates. The edge effects were removed by joining together opposite sides {17}. Once the concentrations, the energy levels and the transition moments are known, a phosphorescence spectrum may be calculated (Fig. 3). For the calculation of the transition moments we used the spin-orbit coupling mechanism to the upper singlet states {19}.

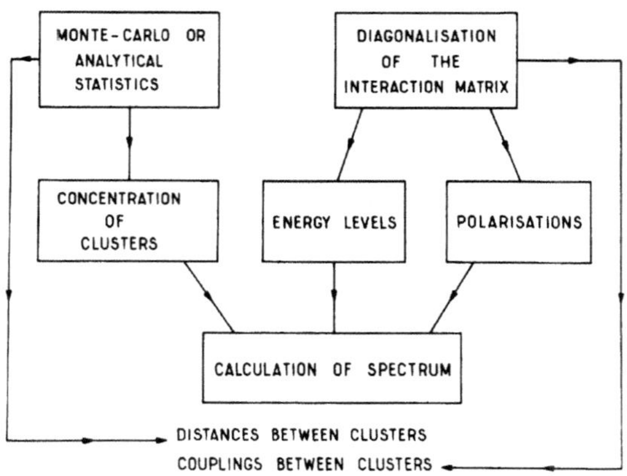

Figure 3. A scheme illustrating the different steps involved in the calculation of the excitation spectra.

The satisfactory agreement between experimental and theoretical spectra (Fig. 4) justifies the assumption of random inhomogeneities. Detailed results on the topology of the mixed crystal have been given elsewhere {18}. In particular there is a wide spread of distances between traps leading to an even wider spread of couplings. These fluctuations in the traps distribution are important when one considers the dynamics of triplet excitons.

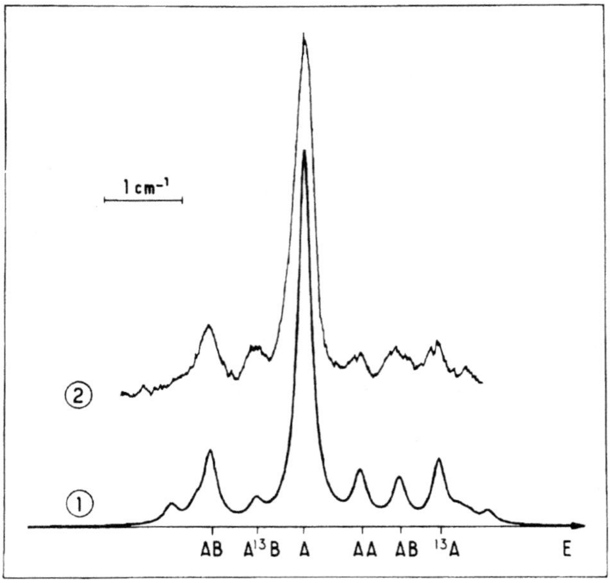

Figure 4. Theoretical (1) and experimental (2) phosphorescence excitation spectrum of the naphthalene mixed crystal at a NH_8 concentration c = 2 %. The calculation has been performed by taking into account the spin-orbit coupling mechanism. The side peaks are due to various cluster states arising from strongly coupled resonant sites and from aggregates containing a non-resonant site (here ^{13}C at +1.7 cm^{-1}).

3. EXCITON DYNAMICS : FUSION AND TRANSFER OF TRIPLET EXCITONS

Migration of triplet excitons may be analyzed by

i) Delayed fluorescence, following the fusion of two excitons.

ii) Sensitized phosphorescence due to quasi-resonant transfer between donor states (traps) followed by non-resonant transfer to an acceptor (super-traps).

3.1. Experimental results

In this section we shall present our experimental results on exciton fusion and exciton transfer in isotopically mixed naphthalene crystals. A detailed study of the temperature dependences of both mechanisms is provided for various trap concentrations.

3.1.1. Fusion of triplet excitons. Delayed fluorescence, following the interaction of two triplet excitons (figure 5), has the same spectral composition as prompt fluorescence in the same conditions (temperature, concentration,...) and has a long lifetime, like that of the triplet state which gives rise to it. The commonly supposed mechanism {5} is akin to the Auger effect (spontaneous ionisation due to the interaction of excited orbitals).

$$\begin{array}{lll} T^{*} + T^{*} \rightarrow & S^{**} & + S_0 \\ & \downarrow & \\ & S^{*} & + \text{phonons} \\ & \downarrow & \\ & S_0 & + h\nu \end{array}$$

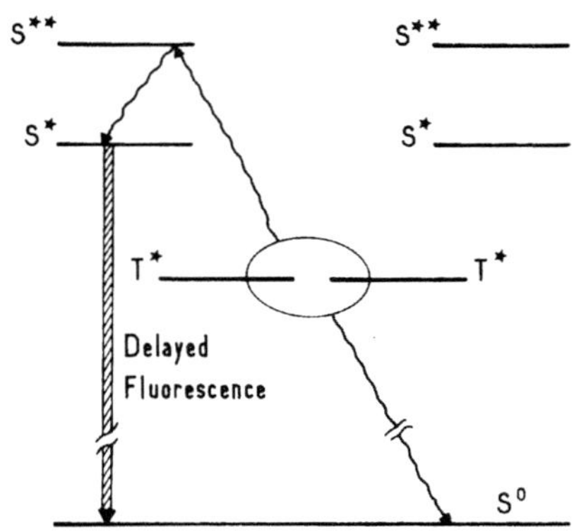

Figure 5. A commonly used scheme showing the triplet fusion mechanism leading to delayed fluorescence.

We measured the total intensity of delayed fluorescence as a function of temperature for crystals with different trap concentrations : 0.6 %, 3 %, 9 % (figure 6). Four distinct temperature domains may be found, I to IV.

i) In the low temperature region, I, T < 4K, delayed fluorescence is independent of the temperature and increases with the trap concentration. We will write $I_{DF}(c,T) = I_o(c)$.

ii) At around T = 5K there is a thermal activation, where $I_{DF}(c,T)$ passes from the value $I_o(c)$ to a maximum $I_{Max}(c)$ at $T \sim 6K$. The activation is relatively more important at lower trap concentrations. Indeed, if c > 7 %, there is no activation. This temperature effect is described in two ways in figure 7. a) On plotting $\ell n\ (I_{DF}\ (c,T))$ vs. 1/T, we obtain straight lines with apparent activation energy vàrying between 28.4 cm^{-1} and 0 cm^{-1} as c increases from 0.6 % to 9 % ; ; b) on plotting $\ell n\ (I_{DF}(c,T) - I_o(c))$ vs 1/T i.e. by substracting the temperature independent contribution, we found an activation energy $\Delta E \sim 35\ cm^{-1}$ at any concentration.

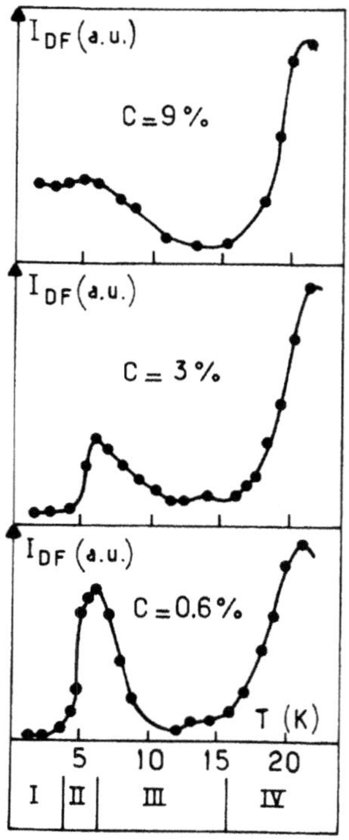

Figure 6. The temperature dependence of the delayed fluorescence I_{DF} after excitation of the ND_8 triplet band and observed at long time (see Ref. {12}) for three values of the trap concentration. The figure displays the four domains of interest (see text).

iii) In region III, $I_{DF}(c,T)$ decreases with increasing temperature. This may be seen as an efficient transfer to trace impurities, since as we shall see below the yield of transfer from NH_8 to βMN (2-methylnaphthalene a natural impurity in naphthalene) is close to 1.

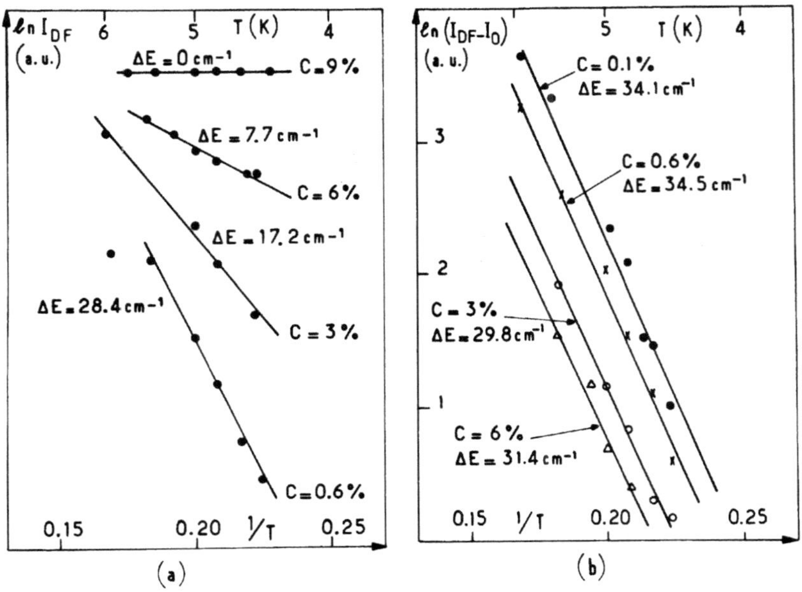

Figure 7. Thermal activation plots of the delayed fluorescence intensity observed in region II ; $\ell n\ (I_{DF})$ vs $1/T$ (a) and $\ell n\ (I_{DF}-I_o(c))$ vs $1/T$ (b). $I_o(c)$ is the constant (at concentration c) low temperature contribution.

iv) Above T = 18K (region IV) $I_{DF}(c,T)$ increases strongly with temperature. The observed activation energy of 90 cm^{-1}, equal to the trap depth shows the dominant role played by the host exciton band of the crystal in this region.

3.1.2. Transfer of triplet excitons. Direct, spatial, measurement of energy transfer between donors is difficult to carry out {20,21}. An indirect method is used here. A low, known concentration of an acceptor impurity is included in the mixed crystal. It is chosen to have an energy level lower than the donor and to easily substitute a host or a donor site. At low temperatures the emission of the acceptor (super-traps) may be used as an indicator of donor-donor transport (figure 8). In the isotopically mixed naphthalene crystal a suitable choice for the sensor is given by βMN (2-methylnaphthalene) at concentrations $C_A < 0.1$ %. At a given acceptor concentration (in fact $C_A \sim 0.03$ %) the distribution of donor acceptor distances may be calculated on the assumption of a randomly inhomogeneous distribution. At this concentration ($C_A = 0.03$ %) the average donor-acceptor distance is 20 unit cells, so that acceptor emission is a good indicator of donor-donor transfer. At low temperature, back transfer from acceptor to donor may be neglected. The yield of donor-acceptor transfer is expressed as a relative phosphorescence intensity.

$$r = I_A / (I_A + I_D)$$

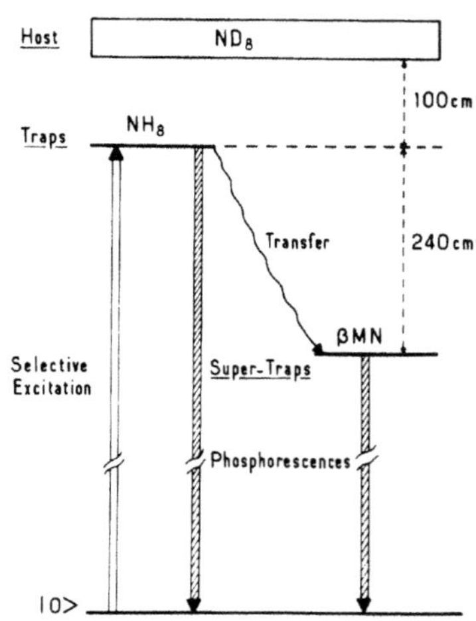

Figure 8. Energy levels of the ND_8/NH_8 mixed crystal involved in the transfer mechanisms. The NH_8 trap states are selectively excited and the phosphorescence of both the traps and the supertraps (βMN) are recorded.

Figure 9 shows the variation of r as a function of temperature for various donor concentrations, c = 2.5 %, 3.75 %, 5 % and 10 %. Three domains may be distinguished.

i) At low temperature the yield r is temperature independent and increases with concentration. Note that in this temperature region the phosphorescence intensity of the acceptor is approximately constant ($I_o(c)$).

ii) At $T \sim 5K$ we observe a thermal activation. We have plotted in Fig. 10 the acceptor phosphorescence intensity corrected for the constant low temperature value $I_o(c)$, $\ell n\ (I_{\beta MN}(c,T)-I_o(c))$ as a function of $1/T$. The activation energy obtained at any concentration is of the order of $\Delta E = 35\ cm^{-1}$.

iii) Above T = 10K, the yield r becomes constant.

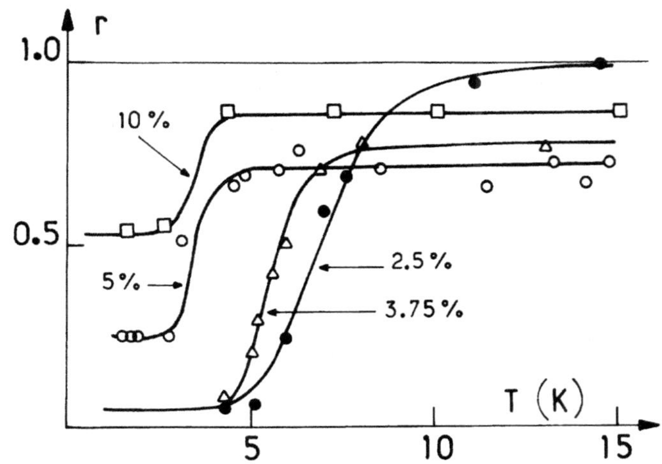

Figure 9. Temperature dependence of the transfer yield to acceptors $r = I_A/(I_A+I_D)$ at various donor concentrations. The solid lines are only visual guides.

3.2. Qualitative interpretation of experimental results

The temperature dependences and concentration effects observed on delayed fluorescence and on energy transfer are similar :

i) At low temperature, there is essentially no temperature dependence and increasing emissions with donor concentration.

ii) At T ∿ 5K delayed fluorescence and acceptor phosphorescence emission exhibit a thermal activation of $\Delta E = 35\ cm^{-1}$.

These similar dependences suggest a common origin for both processes. Due to the strong inhomogeneities in the trap distribution leading to the existence of aggregates the dynamical properties of the mixed crystal must be analyzed in terms of "dynamical aggregates" of relatively strongly coupled sites. These couplings, while being smaller than the spectral resolution are important in view of the long natural lifetime of triplet donors. The temperature independent emissions, delayed fluorescence or phosphorescence, observed at low temperatures may be due to fusion or trapping from eigenstates belonging to the same "dynamical aggregate". Then at T ∿ 5K the thermal activation (of energy $\Delta E \simeq 35\ cm^{-1}$) may be assigned to hopping between such aggregates. This assumption explains the variations of the yields of transfer and of delayed fluorescence as functions of temperature and concentration :

i) At very low temperature there is a quantum delocalisation over the aggregates, to an extent determined by the concentration. Transfer to acceptors occurs from any donor aggregate close to an acceptor. As the concentration rises the number and the extent of the aggregates increases, with corresponding improvement of the yields {9}. According to this model fusion will occur between excitons sitting on the same aggregate.

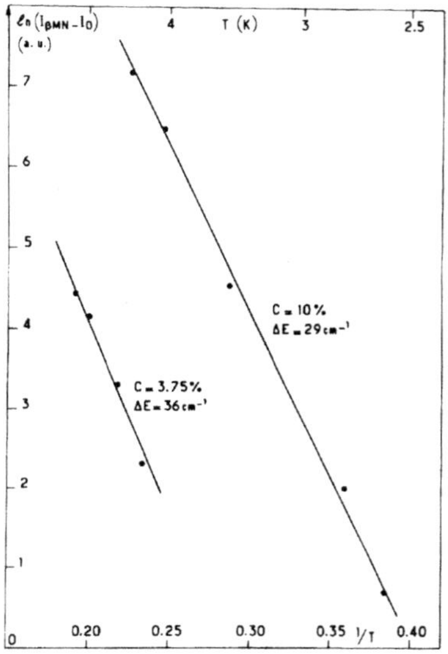

Figure 10. Thermal activation plot of the acceptor emission at donor concentration c = 3.75 % and c = 10 % corrected for the constant low temperature phosphorescence I_o : ℓn $(I_{\beta MN} - I_o)$ vs 1/T.

ii) When the temperature increases, there is a strong threshold due to incoherent motion between the "dynamical aggregates". The activation energy of $\Delta E \simeq 35$ cm^{-1} found for both transfer and fusion shows that it is the activation energy of a donor-donor transfer event. A model of exciton-phonon coupling, incorporating the random distribution of traps is presented below. It justifies the surprisingly large activation energy of the resonant donor-donor transfer.

iii) As the range of hopping increases, an increasing number of excitons becomes trapped. The acceptor emission rises while the delayed fluorescence drops off.

3.3. Model of transfer in the random system NH_8/ND_8

Consider a mixed crystal represented by the random grid (figure 2). At low concentrations the donor eigenstates are localized site states.

Hopping between them may be described by a master equation for the site occupations, p_i (t) :

$$\frac{dp_i}{dt} = -p_i/\tau_i - (\sum_{j\neq i} k_{ji})\, p_i + \sum_{j\neq i} k_{ij}\, p_j \quad , \; i=1,\ldots,M \qquad (2)$$

where τ_i = intrinsic lifetime of the molecules i = 1,...,M

k_{ij} = transfer rate from site j to site i.

k_{ij} may be written as a product of a distance dependent term $f(R_{ij})$ and a temperature dependent part ξ (T).

$$k_{ij} = f(R_{ij})\;\xi\,(T)$$

In hopping models, $f(R_{ij}) \propto |v(R_{ij})|^2$, where $v(R_{ij})$ is the electronic matrix element between localised states $|j>$ and $|i>$, e.g. superexchange $v(R_{ij}) \propto \exp(-\alpha R_{ij})$ or multipolar coupling $v(R_{ij}) \propto 1/R_{ij}^m$. Two typesof temperature dependence ξ(t) are exhibited :

i) Trapping, the final step of donor to acceptor transfer, involves the emission of a cascade of large phonons to conserve energy. In a system such as $\beta MN/NH_8/ND_8$, where $E_D - E_A \sim 240\ cm^{-1}$ and at low temperatures ($k_B T << h\omega$) the emission of these phonons is independent of temperature, ξ(T) is a constant, and is equal to zero if j is an acceptor (no back transfer).

ii) Several phonon assisted transfer mechanisms were reviewed to account for the activation of transfer between donors.

The rates between sites 1 and 2, coupled by a matrix element v and separated by an energy gap δ(< inhomogeneous linewidth) are {23}

i) One phonon process : This must be a long wavelength acoustic phonon since $\delta << E_{Debye}$

$$\Gamma_{21}^{(1)ac} \propto v^2\,\delta^2\,T \qquad (3a)$$

ii) two phonon processes : These are the absorption and the emission of two phonons such as $h\omega - h\omega' = \delta$

$$\Gamma_{21}^{(2)ac} \propto v^2\,T^7 \quad \text{for two acoustic phonons and} \qquad (3b)$$

$$\Gamma_{21}^{(2)opt} \propto v^2 \exp(-E_m/k_B T) \quad \text{for two optical phonons} \qquad (3c)$$

where E_m is the energy of the bottom of the phonon optical branches. These laws may be used to scale the numerical solutions of the master equation (Eq.2). As an illustration we show in Fig. 11 the site occupation times calculated from this model. This is in fact an illustration of a continuous time random walk on a random medium. In this calculation superexchange couplings were assumed and the temperature effect was scaled by the two optical phonon law.

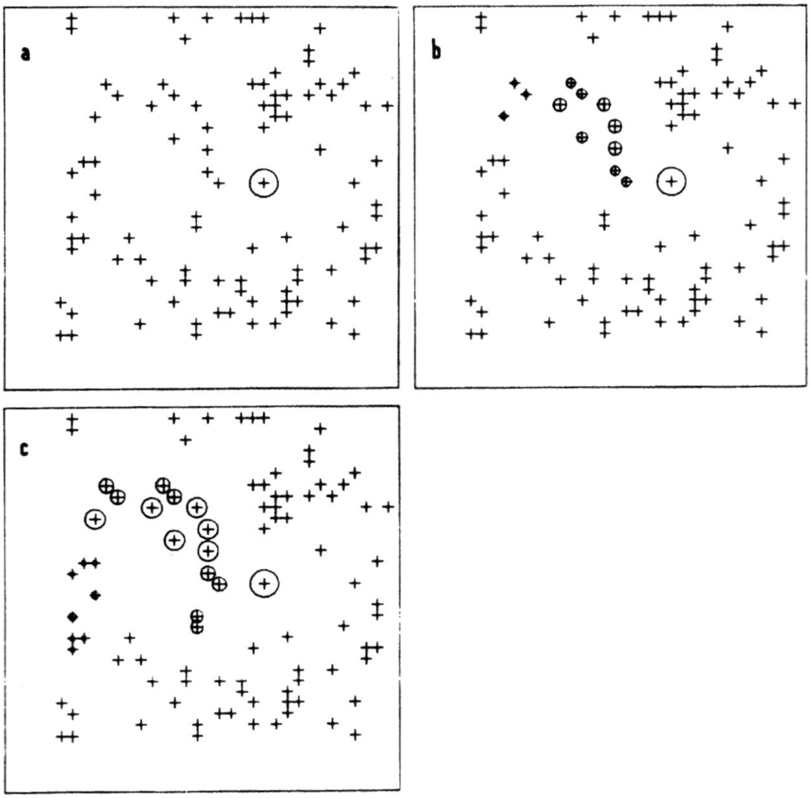

Figure 11. An illustration of the total time spent by excitons on different sites in a random walk terminated by their emission (on an acceptor not shown here). The areas of the circles are proportional to the logarithms of the site occupation times. Three increasing values of the temperature are shown (a, b and c respectively). Note the very uneven occupations of the sites.

Increasing values of the temperature T(a) < T(b) < T(c) lead to an increase of delocalization. Note also that the occupations are so inhomogeneous that for the purpose of illustration the areas of the discs are proportional to the logarithms of the occupations. The random grids ranged in size from 50x50 sites to 140x140 sites depending on donor concentration. Statistically significant results were obtained on averaging over 50 grids. As shown in Fig. 12 only the two optical phonon process accounts for the threshold at T $\sim$ 5K using a phonon energy of E_m=40 cm^{-1} in fair agreement with the experimental activation energy ($\Delta E \sim 35$ cm^{-1}).

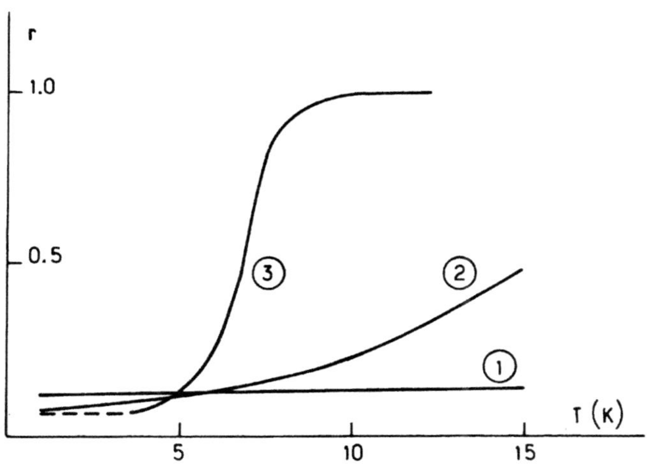

Figure 12. Typical activation curves of donor-acceptor transfer calculated for the three activation processes discussed in the text ; one acoustic phonon (1), two acoustic phonons (2), two optical phonons (3).

Two types of electronic interactions were tested.

i) Superexchange interactions calculated for the mixed naphthalene system {18}. In this case the yield, r, of energy transfer is slowly varying function of temperature at low donor concentrations, in contradiction with experiments (see Fig. 13a and also Fig. 9).

ii) Because of the spin-orbit coupling of the triplet state to upper singlet states which gives phosphorescence emission, the donor-donor interaction v includes a dipolar term {19}. Although this contribution is negligible at short distances (or at high concentration), it cannot be ignored at long distances (or at low concentrations). In the latter case the dipolar term is predominant over superexchange (for distances larger than 5 unit cells). On taking this term into account a better fit is obtained as shown in Fig. 13b.

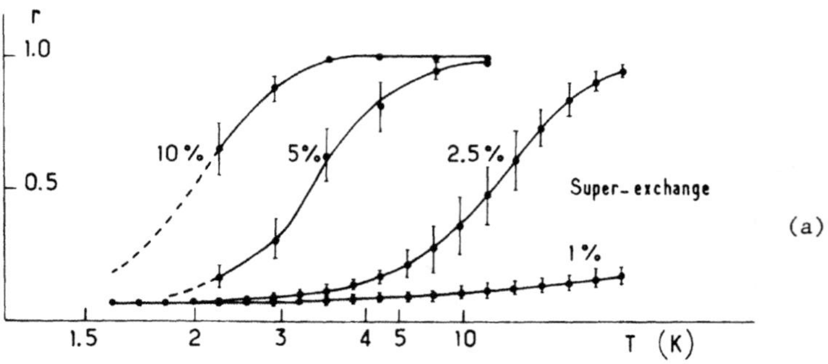

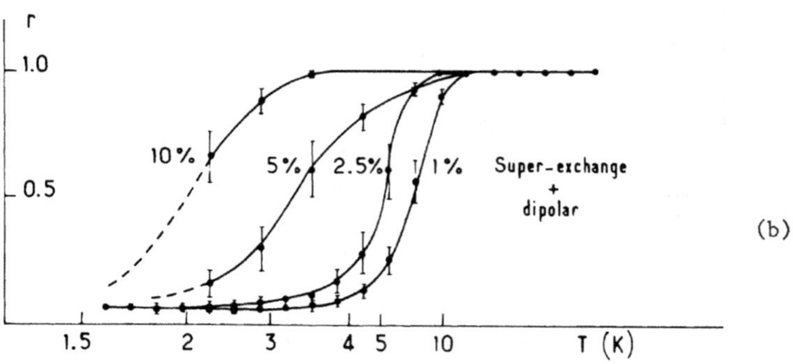

Figure 13. Calculated yield r of donor-acceptor transfer before (a) and after (b) the inclusion of dipolar coupling at long distances. Error bars are 95 % confidence limits.

Finally in Fig. 14 we show the final set of calculated curves taking into account the additional (to the superexchange) dipolar coupling and using the two optical phonon process.

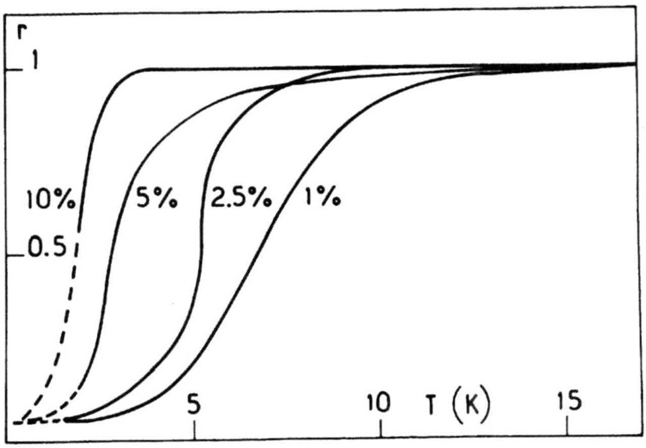

Figure 14. Calculated yield r of donor-acceptor transfer as a function of the temperature for various donor concentrations. The calculation has been performed by considering the superexchange + dipolar coupling and using the two optical phonons process ($E = 40\ cm^{-1}$). These curves have to be compared to those given in Fig. 9.

4. CONCLUSION

Comparison of results on delayed fluorescence and on energy transfer yields in isotopically mixed naphthalene crystals reveals identical mechanisms. At low temperature ($T < 4K$) the emissions are temperature independent so that a different approach based on quantum delocalization has to be developed. Between 4K and 20K incoherent motion prevails. All the observed and concentration dependences are due to donor-donor processes only. Above T = 20K the thermal activation to the host triplet band complicates the dynamics. We have shown that a simple two dimensional model of the triplet mixed crystal can describe both the static and the dynamics properties in the triplet state of a random mixed crystal of naphthalene. The use of special experimental methods, such as very low temperatures and selective excitation, yields results sufficiently "pure" to be rapidly related to the theory. Results on triplet energy transfer show the importance of a process of absorption and emission of two optical phonons. They also show that the singlet character of the triplet state, due to spin-orbit coupling, cannot be neglected since it introduces a long ranged dipolar coupling, effective at low concentrations. Finally, the present study may be extended to other systems, with different couplings and dimensions.

ACKNOWLEDGEMENT We would like to thank Dr. J. Mégel for kindly synthesizing and providing us with naphthalene crystals.

REFERENCES

{1} A. Suna, Phys. Rev. B : Solid State 1 (1970) 1716.
{2} S.E. Webber, C.E. Swenberg, Chem. Phys. 49 (1980) 231.
{3} V.M. Kenkre, Y.M. Wang, Chem. Phys. Letters 87 (1982) 263.
{4} V.M. Kenkre, Phys. Rev. B : Condensed Matter 22 (1982) 2089.
{5} H. Sternlicht, G.C. Nieman, G.W. Robinson, J. Chem. Phys. 38 (1963) 1326.
{6} P.W. Klymko, R. Kopelman, J. Chem. Phys. 86 (1982) 3686.
{7} R. Kopelman, E.M. Monberg, F.W. Ochs, Chem. Phys. 19 (1977) 413.
{8} F.B. Tudron, S.D. Colson, J. Chem. Phys. 65 (1976) 4184.
{9} R. Brown, J.P. Lemaistre, J. Mégel, Ph. Pée, F. Dupuy, Ph. Kottis, J. Chem. Phys. 76 (1982) 5719.
{10} A.H. Francis, R. Kopelman, Topics in applied physics 49 (1981) 241 and references therein.
{11} R. Silbey, in : *Modern problems in condensed matter sciences* vol.3 ed. V.M. Agranovich , R.M. Hochstrasser ; North-Holland, Amsterdam (1983).
{12} Ph. Pée, Y. Rebière, F. Dupuy, R. Brown, Ph. Kottis, J.P. Lemaistre, J. Phys. Chem. 88 (1984) 959.
{13} B.S. Sommer, J. Jortner, J. Chem. Phys. 50 (1969) 839.
{14} V.L. Broude, A.V. Leiderman, T.G. Tratas, Soviet Phys. Solid State 13 (1972) 3058.
{15} F. Dupuy, Ph. Pée, R. Lalanne, J.P. Lemaistre, C. Vaucamps, H. Port, Ph. Kottis, Mol. Phys. 35 (1978) 595.
{16} Complete definition of the system requires also knowlegde of the host band spectra {17}. However, in this low temperature work at T = 1.6K, the traps are the only dynamically active sites.
{17} Ph. Pée, J.P. Lemaistre, F. Dupuy, R. Brown, J. Mégel, Ph. Kottis, Chem. Phys. 64 (1982) 389.
{18} Ph. Pée, R. Brown, F. Dupuy, Ph. Kottis, J.P. Lemaistre, Chem. Phys. 35 (1978) 429.
{19} J.P. Lemaistre, Ph. Pée, M. Béguery, F. Dupuy, J. Mégel, R. Brown, Ph. Kottis, Chem. Phys. Letters 89 (1982) 207.
{20} P. Avakian, R.E. Merrifield, Phys. Rev. Letters 13 (1964) 541.
{21} V. Ern, P. Avakian, R.E. Merrifield, Phys. Rev. 148 (1966) 862.
{22} R. Kopelman, E.M. Monberg, F.W. Ochs, P.N. Prasad, J. Chem. Phys. 62 (1975) 292.
{23} R. Brown, Ph. Pée, F. Dupuy, Ph. Kottis, J. Physics C : Solid State Phys. to be published.

CONCENTRATION DEPENDENCES OF HOMOGENEOUS BROADENINGS AND CROSS-RELAXATION RATES IN CONDENSED MEDIA : THE ROLE OF PAIR DISTRIBUTIONS

J.-P. Lemaistre and J.-P. Korb
Centre de Mécanique Ondulatoire Appliquée, 23, rue du Maroc
75019, Paris, France

ABSTRACT. The concentration dependences of dipolar homogeneous broadenigs and cross-relaxation rates in condensed media are analyzed by means of effective reduced distributions. Elimination of interacting species due to either clustering or recombination processes, which do not participate to the processes under study, leads to define effective concentrations instead of sample concentrations. Numerical calculations of these effective concentrations are performed in 2D and 3D lattices. They are used to predict the concentration behavior of homogeneous broadenings and cross-relaxation rates over a large domain of concentrations.

1. INTRODUCTION

The study of dynamical processes in condensed media such as energy transfer (1), exciton fusion (2), cross relaxation(3-6)and electron-radical recombination (7) is a subject of current interest. The understanding of such mechanisms, especially when the experiments are performed at low temperatures, requires the knowledge of the topological distributions of the interacting species. Usually the species are assumed to be randomly distributed over the possible sites of a two-dimensional (2D) or three-dimensional (3D) lattice. In the limit of the large dilution of the species the radial distribution functions are well-known (8) and the evaluation of the related averaged quantities is easy to perform. However, the experiments are often performed at concentrations for which the analytical expressions are not applicable. Thus, until what concentrations are these theoretical models available and their description of experiments relevant ?

In the calculation of physical observables associated to a particular mechanism the number of interacting species involved is not simply related to the concentration of the sample. We shall illustrate this statement by considering the following typical cases. The first example is that of a molecular crystal (host) in which some sites (guests) are randomly distributed over the whole crystal and coupled by dipolar interactions. In the limit of large dilution the guest molecules are monomers and the homogeneous broadening resulting from their couplings

R. Daudel et al. (eds.), Structure and Dynamics of Molecular Systems, 137–147.

shows a linear dependence with concentration. Increasing the concentration leads to the formation of clusters (molecules in nearest-neighbor positions), and deviations to linear concentration behavior are then expected to occur (9). Thus the low-concentration approximation is no longer valid and the calculation of homogeneous linewidths which describes the guest sites as an ensemble of independent pairs is not justified. Nevertheless when the nearest-neighbor interactions are larger than the homogeneous linewidth involved the cluster eigenstates are out of resonance with those of the isolated guest sites. As a consequence the concentration-dependent broadening has to be calculated over the distribution reduced to isolated guest sites and whose effective concentration, denoted hereafter C_{eff}, is different from the guest concentration c of the sample (with $C_{eff} < c$). The idea of effective concentration (10), for a given process, may be useful in the interpretation of delayed fluorescence experiments at low temperature for the understanding of exciton fusion in molecular crystals (2) . The clusters with different configurations have non-degenerate eigenstates and their coresponding wavefunctions have different symmetries so that the interclusters fusion process is expected to be unlikely. Thus exciton fusion between excited isolated states (or energetically equivalent excited cluster states) may become the predominant mechanism. According to the previous assumption, the evaluation of the effective concentration of pairs, which are possible centers for the fusion process, is of primary interest. The last example, mentioned here, where the idea of effective concentration can be used, is provided by calculations of cross-relaxation rates between overlapping magnetic resonance lines of electrons and radicals when electron-radical recombination takes place (3-6). Theoretical expressions for these rates have been derived for two randomly distributed dilute spin species featuring trapped electrons and free radicals produced by γ-irradiation of organic glasses (4). Comparison with ELDOR experiments allows to make a distinction between the distance of closest approach and the average distance of dipolar interaction of the two species at high dilution. The fact that the closest approach distance increases with concentration has been qualitatively understood as a distance under which the highly paramagnetic species combine rather than a distance related to the steric hindrance of the species (4) . After recombination the paramagnetic species produced by irradiation become unobservable by magnetic resonance. Above a certain concentration range the efficiency of the recombination process can be very important thus leading to a deviation between the effective concentration C_{eff} of the remaining spin species and their concentration c before recombination.

The purpose of this study is to provide numerical calculations of the distributions of interacting species for two and three-dimensional lattices. Section 2 is devoted to the theoretical pair distributions at low concentration and the calculations of related averaged quantities. We outline in section 3 our numerical calculation method : we estimate the effective concentration, C_{eff}, of isolated species (i.e. the species with no nearest neighbors) as a function of the concentration c of the sample over the full range of concentration. The above method is then applied in section 4 to calculate the concentration dependence of homogeneous broadenings over the reduced distributions (i.e. distributions obtained after elimination of clusters). The numerical results giving the radial distribution of closest species pairs are compared to analytical expressions calculated for the effective concentrations derived above and not for the concentration of the sample. By considering the distributions of two different kinds of species featuring electrons and radicals our method allows us to obtain the number of recombination events, inside a circle or a sphere of radius Ro, and the distribution of the remaining species. The concentration dependence of the dipolar transfer time between the remaining species is then evaluated. Finally in section 5 we summarize the main conclusions of our study.

2. THEORETICAL PAIR DISTRIBUTIONS AT LOW CONCENTRATION

Let us consider a random distribution of species in a d-dimensional (d = 2 or 3) regular lattice and let us denote by c the concentration of occupied sites. The radial density of probability of finding a pair of species separated by a distance R is given at low concentration ($c<<1$) by

$$\Phi_d(R) = d\,\alpha_d\, R^d \exp(-\alpha_d R^d) \quad (1)$$

For two and three dimensional lattices $\alpha_2 = \pi c/s_o$ and $\alpha_3 = 4\pi c/v_o$ where the site area s_o or the site volume v_o equal to unity if R is expressed in latttice units. The density functions in Eq. (1) take maximum values for the distances

$$R_{Max,d} = (1/2\pi c)^{1/d} \quad (2)$$

For a given dimensionality d the $R_{Max,d}$ values only depend on the concentration c.

In what follows, we shall consider that the species are coupled by dipolar interactions. Thus, calculations of dipolar homogeneous broadenings and dipolar energy transfer rates involve averages of $1/R^s$, with s = 3 and 6 respectively, over the radial distribution $\Phi_d(R)$

$$<1/R^s>_d = \int_{R_o}^{+\infty} (1/R^s)\, \Phi_d(R)\, dR \quad (3)$$

In Eq. (3) R_o is the distance of minimal approach between two species. At high dilution ($c<<1$) a straightforward calculation gives for a 2D lattice

$$<1/R^3>_2 = 2c/R_o^3 - \sigma(c^2) \quad (4a)$$

$$<1/R^6>_2 = c/2R_o^6 - \sigma(c^2) \quad (4b)$$

and for a 3D lattice

$$<1/R^3>_3 = (c/R_o^3)\,(-\gamma_E - \ell n\ c) + \sigma(c^2) \quad (5a)$$

$$<1/R^6>_3 = (c/R_o^6) + \sigma(c^2) \quad (5b)$$

where $\gamma_E = 0.577..$ is Euler's constant.
At low concentration, $c \ll 1$, the homogeneous linewidths (4a,5a) and dipolar energy-transfer rates (4b,5b) are linearly concentration-dependent.

3. NUMERICAL EVALUATION OF EFFECTIVE CONCENTRATIONS

We first consider a regular (square or cubic) lattice made up of N sites. A fraction cN of these sites, where c denotes the concentration, is randomly occupied thus featuring a random distribution of coupled (here through dipolar interactions) species. At very low concentration the species form essentially monomers. Increasing the concentration leads to the appearance of clusters whose eigenstates have different energies from those of the isolated species. Note that these out of resonance states will not contribute to the homogeneous broadening of monomer lines. In our numerical method we eliminate all the species whose distance is less than a given reference distance Ro. We thus obtain the effective concentration C_{eff} of such isolated species and the concentration ratio, C_{eff}/c, of the number of monomers to that of the initially occupied sites for each concentration c and for various values of Ro. As we shall see later, and due to the sigmoïdal shape of the calculated curves, the ratio C_{eff}/c may be described by the following relationship

$$C_{eff}/c = (1 - c)^{Z(c)} \tag{6}$$

In Eq. (6) Z(c) can be seen as an "effective" coordination number which obviously depends on the dimensionality and also on the radius Ro. In the limit of large dilution, Z equals the coordination number of the lattice for the corresponding values of Ro. In order to calculate the average dipolar energy, $\Lambda\alpha\langle 1/R^3\rangle$ over the reduced distribution, and to obtain its concentration dependence, we perform the calculation of $1/R^3_{ij}$ over all pairs (i,j) whose distances are greater than Ro.

$$\Lambda(\alpha) = \sum_{j>i} (1/R^3_{ij})_{(\alpha)} \quad ; \; R_{ij} > R_o \tag{7}$$

Here, α stands for a given configuration of $C_{eff}N$ sites. The distance Ro is a parameter which can be chosen so that the energy splitting between the dimer states of the two species distant Ro exceeds the homogeneous linewidth of isolated species. Thus Λ is obtained by a cumulative averaging over a large number n of configurations α

$$\Lambda = \sum_{\alpha=1}^{n} \Lambda(\alpha) \tag{8}$$

Secondly and in order to feature the distributions of trapped electrons and radicals produced by γ-irradiation in organic glasses we randomly generate another set of occupied sites at the same concentration. We shall denote (o) for radicals and (★) for electrons. Our simulation model is built up so that a radical (o) and an electron (★) cannot be set on the same site. We assume that the recombination process between radicals and electrons occurs if their relative distance is less than Ro that we call the recombination radius. In our procedure the two corresponding sites are eliminated since the molecular species created by recombination cannot be "seen" by electron spin resonance spectroscopy. The electron-radical pair distributions and the effective concentration (C^{o}_{eff} and $C^{\star}_{eff}$) of the species remaining after recombination are then evaluated. It is precisely these reduced distributions which are involved in the calculations of cross-relaxation rates between magnetic resonance lines of electrons and radicals. For each configuration (α) of radicals and electrons remaining after recombination we calculate the radial part of the dipolar energy transfer rate

$$W_{\star o}^{(\alpha)} = 1/N_o^{(\alpha)} \sum_{\star} \sum_{o} (1/R_{\star o}^{6})^{(\alpha)};\ R_{\star o} > Ro \qquad (9)$$

In Eq. (9) $N_o^{(\alpha)}$ denotes the number of radicals remaining after recombination and $R^{o}_{\star o}$ are the electron-radical distances. In order to sample a large lattice we perform a cumulative averaging of $W_{\star o}^{(\alpha)}$ over a large number n of configurations α

$$\langle W_{\star o}\rangle = \sum_{\alpha=1}^{n} N_o^{(\alpha)}\, W_{\star o}^{(\alpha)} / \sum_{\alpha=1}^{n} N_o^{(\alpha)} \qquad (10)$$

From our calculation we also obtain the relative number γ of recombination events

$$\gamma = \sum_{\alpha=1}^{n} \Delta N_{\star}^{(\alpha)} / \sum_{\alpha=1}^{n} (\Delta N_{\star}^{(\alpha)} + N_{\star}^{(\alpha)}) \qquad (11)$$

where $\Delta N_{\star}^{(\alpha)}$ is the number of recombined electrons and $N_{\star}^{(\alpha)}$ is the number of electrons remaining after recombination in the configuration α.

4. RESULTS

4.1. Effective concentration

The relative concentration, C_{eff}/c, of isolated species (inside a volume of radius Ro) is displayed in Fig. 1a (2D lattice) and Fig 1b (3D lattice) as a function of the initially occupied sites for various values of Ro. The concentration dependence of the effective coordination number Z(c) is shown in the inserts. The calculation performed here represent a cumulative average of 50 configurations of 100 species for each concentration c. For a given dimensionality and a given value of the parameter Ro, which depend on the system studied, we obtain the relative number of isolated species at any concentration.

4.2. Homogeneous broadening

The homogeneous linewidth Λ associated with a particular transition of isolated species coupled by dipolar interactions may be calculated. While in the limit of high dilution ($c<10^{-3}$) the concentration dependence of Λ is well described by Eqs. 4a and 5a, deviations from the simple

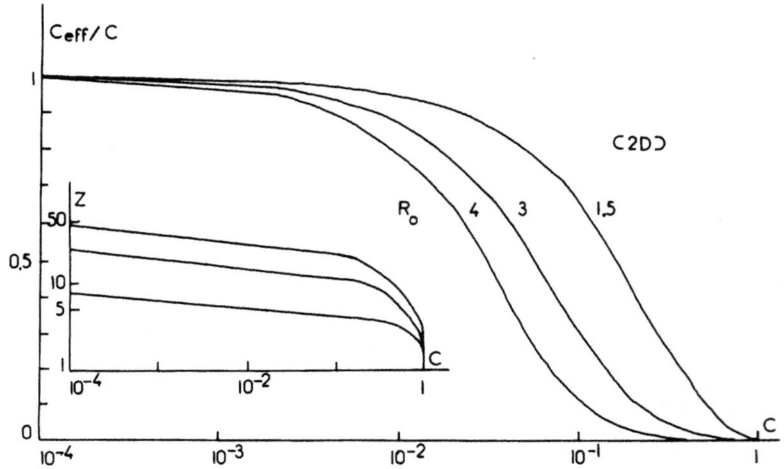

(a)

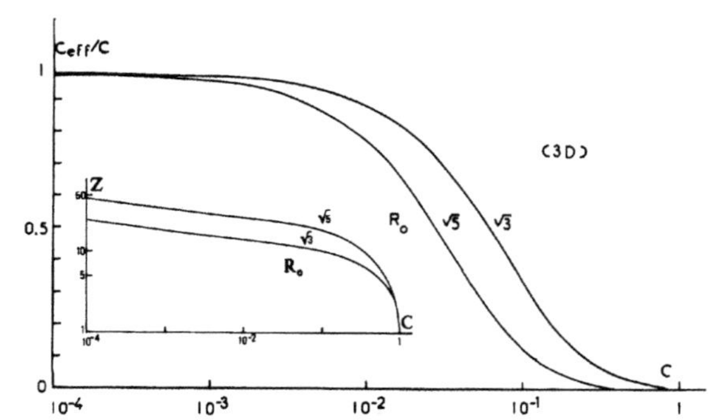

(b)

Figure 1. The ratio C_{eff}/c of the effective concentration of isolated species (inside a volume of radius Ro) to the sample concentration c for a 2D (a) and a 3D (b) lattice. The effective coordination numbers Z (c) are shown in inserts.

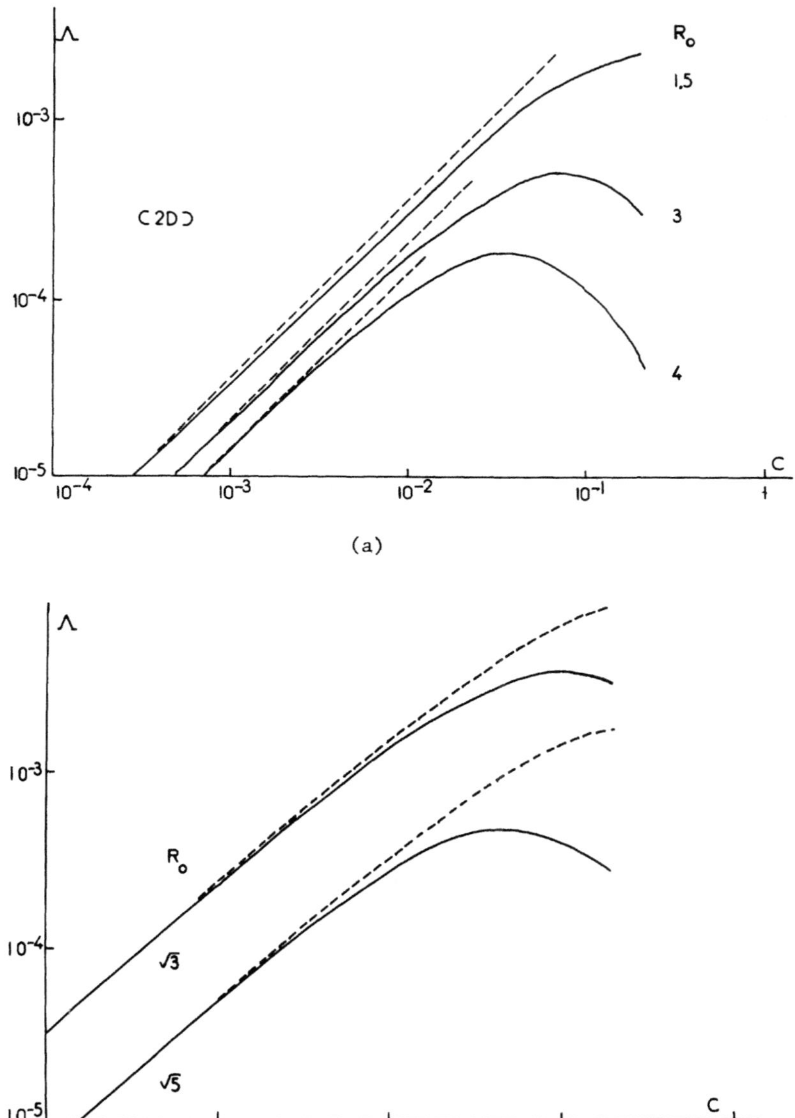

Figure 2. Concentration dependence of the dipolar broadened homogeneous linewidth $\Lambda = \langle 1/R^3 \rangle$ for various values of Ro in the cases of a 2D (a) and 3D (b) lattice.

relationships between Λ and c appear. Since Λ is an increasing function of c, one has $\Lambda\ (C_{eff}) < \Lambda\ (c)$, thus leading to a weaker dependence of the linewidth with c. This is illustrated in Fig. 2a (for a 2D lattice) and Fig. 2b (for a 3D lattice). Obviously the deviations become more important when one increases the concentration. Nevertheless , the calculated values obtained from our numerical experiments are well fitted by Eqs. 4a and 5a by simply replacing c by C_{eff} (Eq. 6) in these equations. We obtain for a 2D lattice

$$\Lambda = (2\ C_{eff}/Ro^3) \tag{12a}$$

or

$$\Lambda = (2\ c/Ro^3)(1 - c)^{Z(c)} \tag{12b}$$

and for a 3D lattice

$$\Lambda = (C_{eff}/Ro^3)(-\gamma_E - \ell n\ C_{eff}) \tag{13a}$$

or

$$\Lambda = [c(1-c)^{Z(c)}/Ro^3][-\gamma_E - \ell n\ c - Z(c)\ell n(1-c)] \tag{13b}$$

So, by means of the effective concentration law evaluated above (where Z(c) are given in inserts of Figs. 1a and 1b), we may recover the concentration behavior of Λ using the analytical expressions available at large dilution.

4.3. Cross relaxation rates

Magnetic energy transfer or cross-relaxation between paramagnetic species in solids strongly depends on the spatial distribution of the species involved. Most of the CW EPR and ELDOR methods which suppose a single rate for characterizing cross-relaxation involve some a priori relations between the measured cross-relaxation rate $<W_{\star o}>$ and an assumed unique average dipolar correlation distance $\overline{d_{\star o}}$. In what follows we shall consider only the 3D lattice case for comparison to experimental data. At large dilution the average time $<T_{\star o}>$ for cross-relaxation at high magnetic field between two paramagnetic spin species, due to dipole-dipole interaction is proportional to $<W_{\star o}>^{-1}$ or $\overline{d^6_{\star o}}$. According to Eq. 5b we obtain

$$<T_{\star o}> \propto Ro^6\ c^{-1} \tag{14}$$

In order to check the existence of an unique dipolar correlation distance at higher concentrations we have performed numerical calculations as described in section 3 over a large number n of configurations

$$\overline{d_{\star o}} \equiv [\ell im_{n>>1} <W_{\star o}>]^{-1/6} \tag{15}$$

Convergent values for $\overline{d_{\star o}}$ have been found provided that the recombination process is efficient above the first neighbors (Ro>1). Indeed the recombination process, which eliminates the closest pairs, leads to homogeneous distributions of the remaining pairs. The convergence behavior, as a function of the number n of configurations is illustrated in the inserts of Fig. 3 for Ro <1 and Ro> $\sqrt{5}$ lattice units at $C_o = C_\star$ = c = 10^{-3}. At high dilution (c<10^{-3}) the c^{-1} dependence of Eq. 5b is well verified as shown in Fig. 3 for any value of Ro. At higher concentration the theoretical relation for <T★o> is no longer valid. The decrease of $\overline{d^6_{\star o}}$ with c is weaker and one even notes a slight increase with c when the recombination becomes efficient (Ro>$\sqrt{5}$). As previously emphasized in the case of the homogeneous broadening, an approximate relationship between <T★o> and c can be obtained by using in Eq. 14 the effective concentration C_{eff} of the spin species ★ remaining after recombination instead of concentration c

$$<T_{\star o}> \alpha \ R_o^6 \ c^{-1}[1 - \gamma(c)]^{-1} \qquad (16)$$

where γ (c) is the relative number of recombination events defined in Eq. (11) whose the concentration dependence is shown in Fig. 4. We have obviously γ(c)=o for Ro<1. In the presence of recombination, the concentration dependence of <T★o> evolves from $c^{-0.7}$ to almost no variation with c for Ro>$\sqrt{5}$ lattice units at high concentration. This dependence can be compared to ELDOR measurements of cross-relaxation times between trapped electrons ans radicals diluted in disordered matrices (3). The observed c dependence of <T★o> is about $c^{-0.4}$ to $c^{-0.6}$ in a concentration range standing around 10^{-3}. So the calculated concentration dependence shown in Fig. 3 and described by Eq. (16) are in fair agreement with experiment.

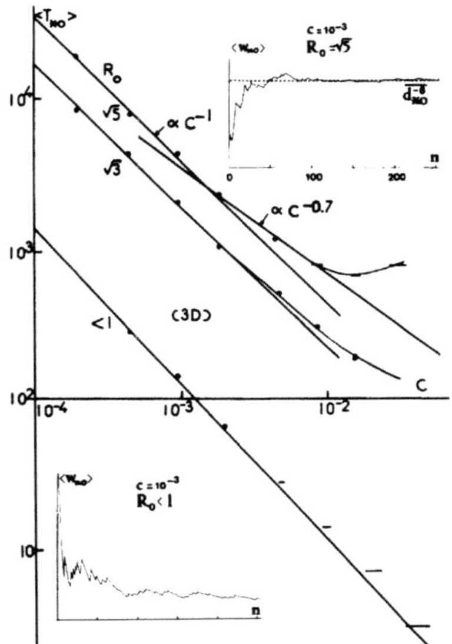

Figure 3. Average dipolar transfer time <T★o> in a 3D lattice as a function of the sample concentration and for various values of the recombination radius. The convergence behavior of <W★o>(=<T★o>$^{-1}$) as a function of the number n of configurations is illustrated in the inserts.

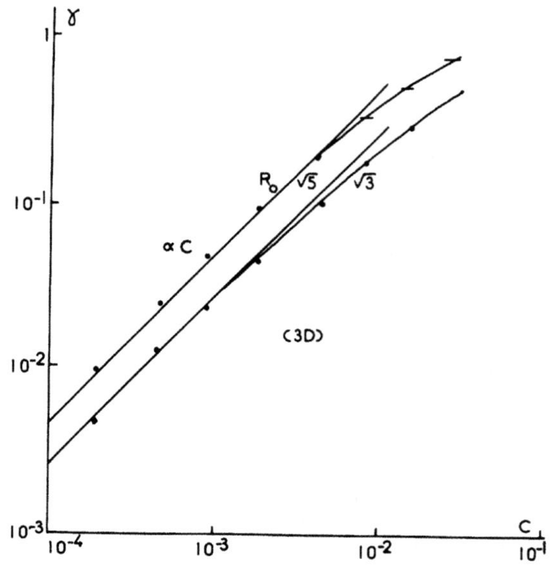

Figure 4. The relative number γ of recombined electrons to the total number of electrons before recombination for various values of Ro.

5. CONCLUSION

We have provided numerical calculations for the distributions of interacting species in 2D and 3D lattices. A procedure eliminating the species which do not participate to a given process led to the definition of effective reduced distributions over which the statistical averagings were performed. The ratio of this effective concentration to that of the sample was evaluated over the full range of concentrations with an unique parameter Ro ; Ro defines a volume inside which all the species are eliminated. The theoretical pair distributions and their averaged related quantities, available in the limit of the large dilution, were then extended to a larger domain of concentrations provided that the sample concentration was replaced by the effective concentration. Thus, deviations from the linear behavior were expected to occur. Indeed these modified pair distributions, in perfect agreement with the numerical calculations, predicted the non linear concentration behaviors experimentally observed.

REFERENCES

1. R. Brown, J.-P. Lemaistre, J. Megel, Ph. Pee, F. Dupuy and Ph. Kottis J. Chem. Phys. 76 (1982) 5719.
2. Ph. Pee, Y. Rebiere, F. Dupuy, R. Brown, Ph. Kottis and J.-P. Lemaistre, J. Phys. Chem. 88 (1984) 959.
3. D.P. Lin, D.F. Cheng, F.Q.H. Ngo and L. Kevan, J. Chem. Phys. 65 (1976) 3994.
4. J.-P. Korb and J. Maruani, J. Chem. Phys. 74 (1981) 1504.
5. M.K. Bowman and J.R. Norris, J. Phys. Chem. 86 (1982) 3385.
6. T. Endo and T. Muramoto, Phys. Rev. B29 (1984) 6043.
7. C. Lapersonne-Meyer, Thesis, Orsay (1978).
8. S. Chandrasekhar, Rev. Mod. Phys. 15 (1943) 1.
9. J.B.W. Morsink, B. Kruizinga and D.A. Wiersma, Chem. Phys. Letters 76 (1980) 218.
10. J.-P. Lemaistre and J.-P. Korb, Chem. Phys. Lett. (to be published).

THE EFFECT OF SPECTRAL DIFFUSION ON THE SATURATION TRANSIENT REGIME

R.Boscaino, F.M.Gelardi and R.N.Mantegna
University of Palermo-Institute of Physics
Via Archirafi 36
I-90123 PALERMO
Italy

ABSTRACT. Saturation kinetics of inhomogeneous resonance lines has been investigated in very dilute ruby samples at T=4.2°K by the saturation transient technique. Experimental evidence of the effectiveness of intraline spectral diffusion is reported. Satisfactory agreement is found between the experimental results and a theoretical model in which spectral diffusion is ascribed to the time fluctuations of the hyperfine field. Spectral diffusion times of the order of 10^{-5} sec are determined.

1. INTRODUCTION

1.1 The isolated packet model

The electron spin resonance (ESR) lines in diluted solid samples (paramagnetic impurities randomly diluted in a host lattice) are broadened both by homogeneous and inhomogeneous interactions (1).

A simple model to describe the properties of these lines and in particular their saturation properties is the "isolated packet model" (IPM) (2). In this model the inhomogeneous resonance line is considered to result from a continuous distribution $h(\omega'-\omega_0)$ of narrow homogeneous lines, each one resonating at some frequency offset $(\omega'-\omega_0)$ from the center frequency ω_0, all characterized by the same line shape.

The response of the overall spin system to an external excitation is given by the weighted superposition of the responses of the individual packets, i.e. by the convolution of the response of the generic packet with the frequency distribution function $h(\omega'-\omega_0)$. On the other hand the response of the individual packet is easily

R. Daudel et al. (eds.), Structure and Dynamics of Molecular Systems, 149–169.

calculated,as the packets are assumed to interact with the external radiation and with the thermal bath independently of each other and energy transfer processes between different packets are not considered in this model..

The saturation properties (steady state, saturation recovery, hole burning, saturation transient) of the inhomogeneous ESR lines, as obtained in the framework of this model have been extensively investigated in literature. Here we briefly recall some particular results, in view of their use in the following Sections.

1.1.1. The saturated steady state. The dynamic susceptibility of the whole system, saturated by a continuous resonant radiation,is given by:

$$\chi(\delta,s) = T_2^{-1}\int_{-\infty}^{+\infty} h(\varepsilon)\chi_p(\varepsilon+\delta,s)d\varepsilon \qquad (1)$$

with the usual meaning of the symbols: T_2 is the homogeneous spin-spin relaxation time; $\varepsilon=(\omega'-\omega_0)T_2$ is the spectral position of the generic packet and $\delta=(\omega_0-\omega)T_2$ is the detuning of the input radiation; $s^2= 2WT_1$ is the saturation parameter of the individual packet (W is the radiation-induced transition rate and T_1 is the spin-lattice ralaxation time); $h(\varepsilon)$ is the frequency distribution function (normalized to unit area) and $\chi_p(\varepsilon+\delta,s)$ is the steady state (saturated) response of the generic packet:

$$\chi_p(\varepsilon+\delta,s) = \tfrac{1}{2}\chi_0\omega T_2\left[g'(\varepsilon+\delta)-ig(\varepsilon+\delta)\right]/\left[1+s^2g(\varepsilon+\delta)\right] \qquad (2)$$

where $g(\varepsilon)$ is the spin packet line shape (normalized to its maximum) and $g'(\varepsilon)$ is the corresponding dispersion function.

1.1.2. The saturation transient (ST) regime. The ST regime (3,4) is the response of the spin system to an intense resonant radiation abruptly turned on (step-modulated) (Fig.1). Unlike the saturation recovery regime, the ST is the time evolution of the spin system from its initial state of thermal equilibrium toward a new stationary state, saturated by the incoming radiation.

The ST regime is peculiar of solid samples where $T_1 \gg T_2$ as its occurrence requires the following conditions to be satisfied:

$$T_2 \ll T_1/s^2 \ll T_1$$

T_1/s^2 is the typical time scale of the ST; the former condition ensures the absence of coherent transient effects, the latter is the high-power condition $s^2 \gtrsim 1$.

During the ST regime, $\chi(\delta,s,t)$ evolves from its unperturbed value

$\chi(\delta,0,0)$ (Eqs.(1,2) with s=0) toward its saturated steady state value $\chi(\delta,s,\infty)$ (Eqs.(1,2)),its time dependence being determined by the internal structure of the line. In the IPM and for a resonant

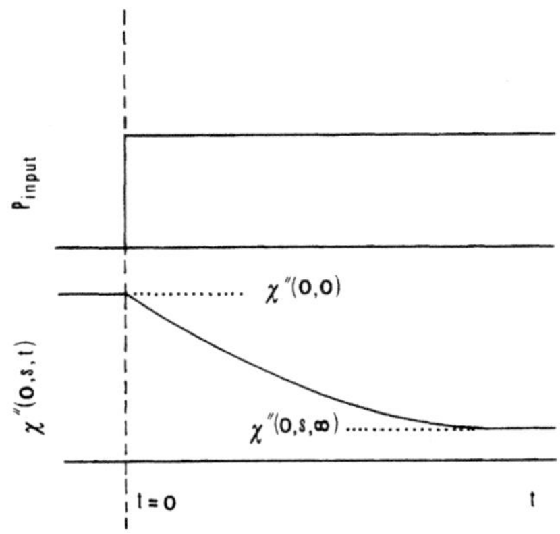

Figure 1. Schematic time dependence of the response of the spin system to an external step-modulated radiation $P_{inp}(t)$ for the particular case δ=0 (resonant excitation).

excitation (δ=0) of the line ($\chi(t)=-i\chi''(t)$), $\chi''(t)$ is given by:

$$\chi''(t)-\chi''(\infty) = \tfrac{1}{2}\chi_0\omega T_2\frac{as^2}{\sqrt{2\pi}}\int_{-\infty}^{+\infty}\frac{g}{1+s^2g}\exp\left[-\frac{a^2\varepsilon^2}{2}-\frac{t}{T_1}(1+s^2g)\right]d\varepsilon \quad (3)$$

where a Gaussian envelope

$$h(\omega'-\omega_0) = \frac{1}{\sqrt{2\pi}\sigma}\exp\left[-(\omega'-\omega_0)^2/2\sigma^2\right]$$

has been assumed and $a=1/\sigma T_2$ is the homogeneity parameter of the line. According to Eq.(3), $\chi''(t)$ in the IPM is the superposition of a continuous distribution of exponential decays, each one with different amplitude and decay time and each one corresponding to the ST regime of a single packet.

1.2 Failures of the isolated packet model

The isolated packet model is obviously an oversimplified picture of an inhomogeneous resonance line. Its main shortcomings are listed here.

1.2.1. The spin packet concept. In the original version by Castner (2) the packet lines are assumed to be broadened by the spin-spin interactions among the spins of the same packet (strong intrapacket interactions). This has raised some criticism in literature concerning the simultaneous neglect of the interpacket interactions and the criteria for a spin to be a member of a packet (5,6).

1.2.2. The variation of T_{SS}. The IPM does not take into account the possible variation of the temperature T_{SS} of the dipolar spin-spin energy reservoir which is known to play a relevant role in off-resonance saturation of magnetic resonance lines in solid samples (7).

1.2.3. The spectral diffusion. The IPM does not take into account the diffusive propagation of the spin excitations over the whole line; this may be caused by the time fluctuations of the local inhomogeneous field that tend to randomize the spectral position $\omega'(t)$ of the spins (5,8).

As regards the first point, we refer to a previous paper (9), where we suggested that this difficulty may be overcome by properly defining the homogeneous line shape $g(\varepsilon)$, at least in the case of randomly diluted systems and in the limit of low concentration. Moreover, in the present paper we will consider only resonant saturation of symmetric lines, so that the effects related to dipolar reservoir heating or cooling will be disregarded. So, in the following we will be concerned only with the effects of Spectral Diffusion.

2. SPECTRAL DIFFUSION

2.1 The model

In this Section we wish to examine the effect of Spectral Diffusion (SD) on the saturation dynamics of a randomly diluted spin system. In particular we consider a situation where SD arises from random changes of the spectral positions of the spins caused by the time fluctuations of the local inhomogeneous field. Cross-Relaxation processes induced by the spin-spin dipolar interaction (6,10,11) are not considered here as possible sources of SD.

We use a simplified model based on the following hypotheses:

a) the time fluctuations of the local field originate in completely uncorrelated events and yield random abrupt changes of the spin frequencies.
b) the scattering probability of a spin excitation from its initial position ε' to the final one ε'' is independent of the frequency distance $(\varepsilon''-\varepsilon')$ and is limited only by the frequency distribution function $h(\varepsilon)$.
c) the rate ω_d of the frequency changing events lies within the range:

$$T_2 \ll \omega_d^{-1} \ll T_1$$

These conditions deserve a few comments. Outside this range, for $\omega_d T_1 \ll 1$, the local field is essentially static and the validity of the IPM is restored; the opposite limit, $\omega_d T_2 \gg 1$, where the fluctuations of the local field determine the homogeneous line shape has been extensively investigated in literature in connection with line shape theories (12) and with spin echo decay experiments and theories (13). The peculiarity of the range of ω_d considered here is that the spin packet concept keeps its validity and usefulness and the investigation of the saturation properties of the resonance line is a suitable method to evidence the effectiveness of SD.

Our model is different from the Wolf's model (8) because of hypothesis c). A similar model was also used by Kofman et al (14) to describe the saturation kinetics of a Doppler broadened optical line.

Under the above hypotheses the rate equation for the generic packet of an inhomogeneous line excited by an external radiation tuned to the center of the line ($\delta=0$) is given by:

$$n(\varepsilon,t) = -\frac{n(\varepsilon,t)}{T_1}(1+s^2g) + \frac{n_0}{T_1} - \omega_d n(\varepsilon,t) + \\ +\omega_d\int_{-\infty}^{+\infty} h(\varepsilon')n(\varepsilon',t)d\varepsilon' \qquad (4)$$

The last two terms of right hand side of Eq.(4) take into account the effects of SD. The former describes the scattering of the spin excitations away from the considered packet, the latter their back-scattering from everywhere within the line profile toward the packet under consideration. According to Eq.(4), the effect of SD is two-fold, as it results from the interplay between these two terms: SD may be considered equivalent to an additional relaxation mechanism if the former term overcomes the latter, whereas it is expected to act as a saturating excitation in the opposite case. This may be pointed out by examining in some detail the steady state solution of Eq.(4).

2.2 The saturated steady state

The steady state solution of Eq.(4) is:

$$n(\varepsilon,\infty) = \frac{n_0}{1+S}\,\frac{1+\omega_d T_1}{1+\omega_d T_1+s^2 g} \qquad (5)$$

with

$$S = \sqrt{\pi/2}\,a\omega_d T_1 s^2 \mathrm{erfc}(y)\exp(y^2)(1+\omega_d T_1)^{-1}\left[1+s^2(1+\omega_d T_1)^{-1}\right]^{-\frac{1}{2}}$$

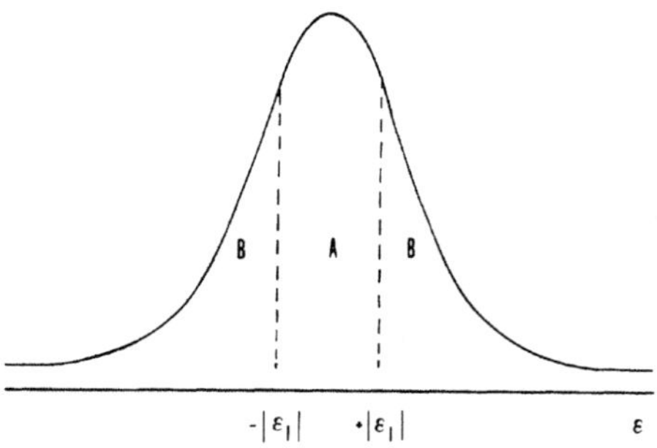

Figure 2. SD causes a net transfer of the spin excitations from the spin packets located within the spectral region A to B.

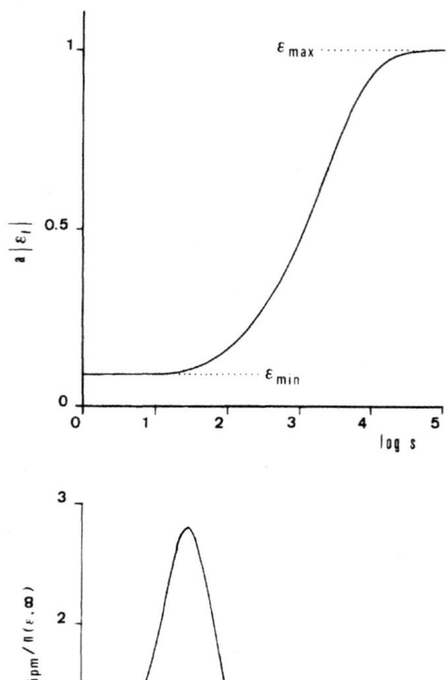

Figure 3. Typical power dependence of the boundary between zones A and B of Fig.2, as obtained by the theoretical model of Sect.2. The spectral boundary is given in units of the halfwidth of the whole line. The curve refers to a=0.01, $\omega_d T_1=10^3$.

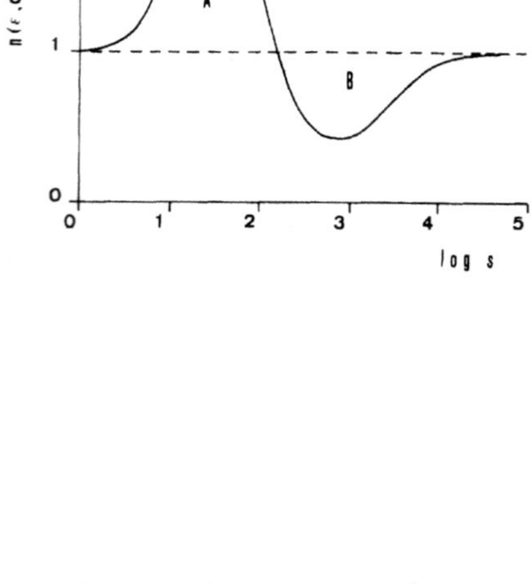

Figure 4. In a resonance line with a=0.01 and $\omega_d T_1=10^3$ (the same of Fig.3) the spin packet located at ε=20 belongs either to the zone A or B of Fig.2 depending on the value of the saturation parameter.

$$y = \left[1+s^2(1+\omega_d T_1)^{-1}\right]^{\frac{1}{2}} a/\sqrt{2}$$

It is of interest to compare the solution $n(\varepsilon,\infty)$ in the presence of SD (Eq.(5)) with the corrisponding one derived by using the IPM:

$$\left[n(\varepsilon,\infty)\right]_{ipm} = \frac{n_0}{(1+s^2 g)}$$

The comparison shows that the role played by SD in the saturation behaviour of a single packet is not the same for all the packets. In fact,two different spectral regions may be distinguished in the overall line (Fig.2): for the packets within the central part (A) $n(\varepsilon,\infty)$ is larger than $\left[n(\varepsilon,\infty)\right]_{ipm}$,so SD acts as an energy relaxation mechanism, whereas the opposite is true for the packets in the wings, for which SD is the main saturation source. The limit ε_1 between these two zones depends on a and on $\omega_d T_1$ and, for a given line, on the saturation parameter s^2, as shown in Fig.3 for a particular case (a=0.01 and $\omega_d T_1 = 10^3$). Note that for the packets with ε such that $\varepsilon_{min}<\varepsilon<\varepsilon_{max}$ SD acts either as a saturation source or as a relaxation channel depending on s^2,as shown in Fig.4.

The saturated steady state of the overall line is described by:

$$\chi''(0,s) = \chi''(0,0)\frac{a\ \mathrm{erfc}(y)\ \exp(y^2)}{\sqrt{2}(1+S)y\ \mathrm{erfc}(a/\sqrt{2})\ \exp(a^2/2)} \tag{6}$$

This solution is plotted in Fig.(5) as $s\chi''(0,s)/\chi''(0,0)$ versus s. This

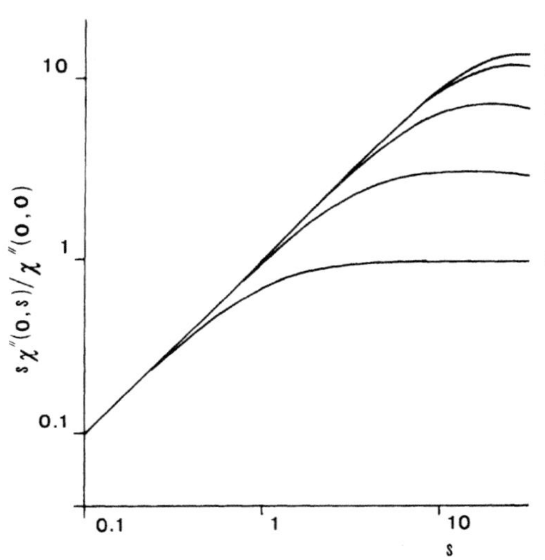

Figure 5. Normalized power dependence of $\chi''(0,s)$ for an inhomogeneous line with a=0.001 at various values of $\omega_d T_1$: 0 (IPM limit), 10, 10^2, 10^3, 10^4 from a to e, respectively.

kind of plot (so-called Castner's plot) is currently used in ESR work to analyze experimental data obtained by the c.w. progressive saturation method. It is worth noting that , apart from scale factors, the sequence of curves in Fig.5 closely resembles the sequence of curves obtained by using the IPM and on scanning the homogeneity parameter a. This circumstance indicates that the c.w. progressive saturation method is unsuitable for pointing out the effectiveness of SD mechanisms within an inhomogeneous line,since the same experimental data can be explained by the IPM by hypothesizing a larger value of a. This shortcoming was already noticed by other authors (5,8).

In the next Section we show that the effectiveness of SD mechanisms can be revealed unambiguously by saturation transient experiments.

2.3 The saturation transient

We report here approximate analytical solution of $\chi''(0,s,t)$, as derived from the spin packet rate equation of Eq.(4). We recall that $\chi''(0,s,t)$ is the response of the spin system, initially at thermal equilibrium,to a step-modulated radiation tuned to the line center. The approximate solutions reported below were obtained by Laplace transforming Eq.(4) and by approximatively solving it in order to get the time evolution $\chi''_p(\varepsilon,s,t)$ of the generic packet, at short and at long times separately. $\chi''(0,s,t)$ was then obtained by integrating $\chi''_p(\varepsilon,s,t)$ over the frequency distribution function $h(\varepsilon)$. The detailed procedure (15) will be published elsewhere; here we limit ourselves to report the final expressions.

According to our approximate solutions, at short times ($t \ll \omega_d^{-1}$) and in the high power limit ($s^2 \gg \omega_d T_1$) the time evolution of the spin system during the ST is just the same as predicted by the IPM, namely a non-exponential time decay (Eq.(3)). This may be qualitatively understood by considering that at short times after the onset of the input power, the main contribution to $\chi''(0,s,t)$ comes from the very near resonant packets, on which SD acts as a relaxation mechanism.

The main result of our calculations is that at longer times $\chi''(0,s,t)$ is predicted to deviate from its initial behaviour and to assume, at times $t \gg \omega_d^{-1}$, the form of a single exponential decay:

$$\chi''(0,s,t) = A_e \exp(-t/\tau_e) + \chi''(0,s,\infty) \tag{7}$$

with $\chi''(0,s,\infty)$ given by Eq.(6) and

$$A_e = \chi''(0,0)\frac{\operatorname{erfc}(y)\exp(y^2)\ a\ S}{\sqrt{2}(1+S)\ y\ \operatorname{erfc}(a/\sqrt{2})\exp(a^2/2)} \tag{8}$$

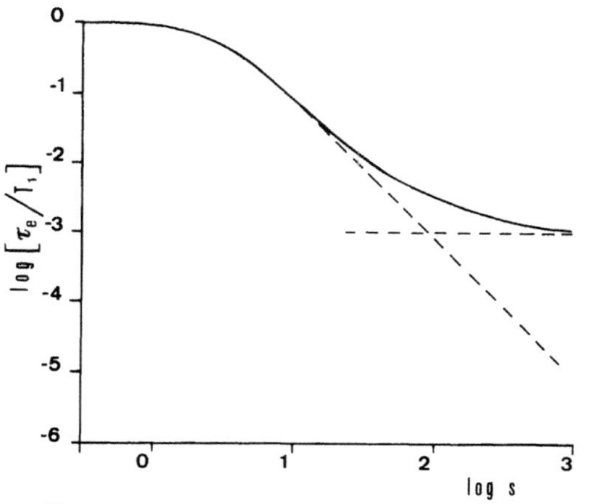

Figure 6. Theoretical power dependence of τ_e/T_1, as derived from Eq.(9). The curve refers to a=0.1 and $\omega_d T_1 = 10^3$. Dashed lines are the asymptotic approximations: $\tau_e = \omega_d^{-1}$ (for $s \to \infty$) and $\tau_e/T_1 = [1+(\pi/2)^{\frac{1}{2}} as^2]^{-1}$ (for $s \to 0$).

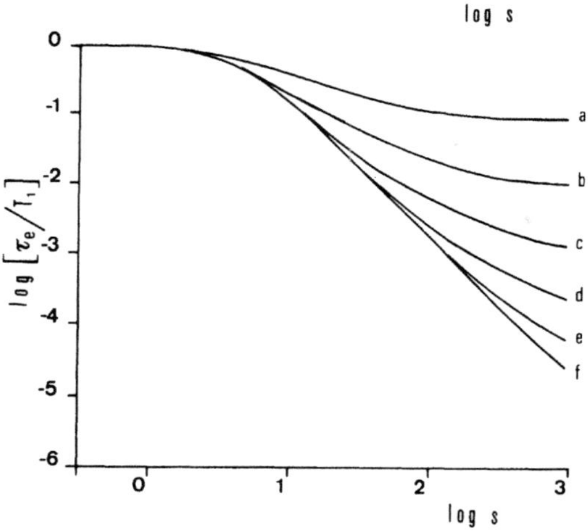

Figure 7. Power dependence of τ_e/T_1, as derived from Eq.(9) for a=0.05 and at various values of $\omega_d T_1$: $10, 10^2, 10^3, 10^4, 10^5, 10^6$ from a to f, respectively.

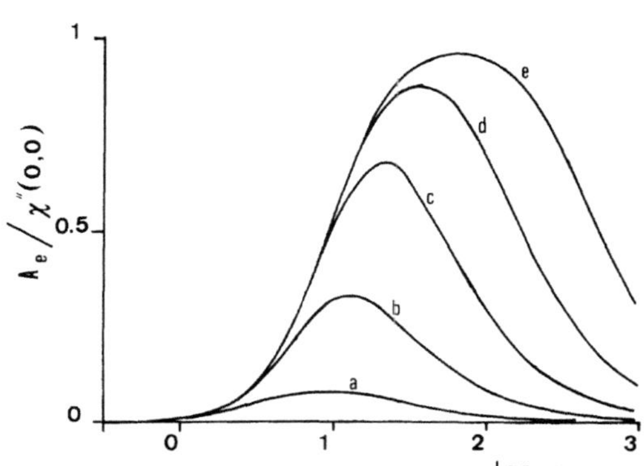

Figure 8. Theoretical power dependence of the normalized amplitude A_e as given by Eq.(8). The curves refer to a=0.01 and $\omega_d T_1 = 10, 10^2, 10^3, 10^4, 10^5$ from a to e, respectively.

$$\frac{1}{\tau_e} = \frac{1}{T_1}\left|1 + \frac{(\pi/2)^{\frac{1}{2}} a\omega_d T_1 s^2 \text{ erfc}(y) \exp(y^2)}{|(1+\omega_d T_1)(1+\omega_d T_1+s^2)|^{\frac{1}{2}}}\right| \tag{9}$$

A qualitative reason for the single exponential time behaviour is that at long times contributions to $\chi''(0,s,t)$ come mainly from the distant packets: these are driven toward the saturation by the SD rather than by the input radiation. As a consequence of the assumed independence of ω_d of the spectral distance, the rate at which the distant packets evolve toward the saturated state is the same for all the packets, thus becoming the rate of evolution of the whole system.

The characteristic time τ_e of the exponential decay is expected to decrease on increasing the input power, starting from $\tau_e=T_1/(1+\sqrt{\pi/2}s^2a)$ for $s^2<<\omega_d T_1$ and tending to $\tau_e=\omega_d^{-1}$ for $s^2>>\omega_d T_1/a^2$. A typical power dependence of τ_e is reported in Fig.6: the curve refers in particular to $a=0.1$ and $\omega_d T_1=10^3$. The extent to which τ_e is affected by the SD rate is shown in Fig.7, for $a=0.05$; as shown, τ_e decreases on increasing $\omega_d T_1$.

As regards the amplitude of the single exponential decay at long times, its dependence on s^2 and on $\omega_d T_1$ is shown in Fig.8 for $a=0.01$. Note that, in the high power limit $(s^2>>\omega_d T_1/a^2)$, $A_e/\chi''(0,0)$ tends to zero; in other words, the single exponential decay tends to disappear and the validity of the IPM is restored both for the ST regime and for the steady state.

To summarize the theoretical results of this Section, the effectiveness of SD is expected to affect strongly the ST curves $\chi''(0,s,t)$, causing them to depart from the time behaviour predicted by the IPM at some time after the onset of the input power and to assume a single exponential decay form. The effect is more and more evident at lower and lower input power levels. This peculiar behaviour, when observed in the experimental ST curves, can be taken in turn as evidence of the effectiveness of SD mechanisms and we will show in the next Section that the value of ω_d can be inferred from the experimental data by comparing experimental and theoretical results of the power dependence of τ_e and A_e.

3. EXPERIMENTAL RESULTS

3.1 The samples

The experimental results reported here were obtained in samples of dilute ruby, where the Cr^{3+} ion is the paramagnetic impurity in the host lattice of Al_2O_3. The ground state of the Cr^{3+} ion has $S=3/2$ and, owing to the axial symmetry of the lattice, the energy level diagram depends on the angle θ between the static field $\vec{H}$ and the $\vec{c}$ crystal axis. We report here the measurements of the ST curves $\chi''(0,s,t)$ taken at the resonance (1-2 h.f., $\theta=0°$) and at the resonance (3-4, $\theta=90°$), which fall at 6.2 kG and at 3.9 kG, respectively, at our working frequency $\nu=6$ GHz (energy levels are labelled as in ref.16). The choice of these lines is motivated by the fact that they are well isolated from other resonance lines and are distant from any fundamental or harmonic interline Cross Relaxation (CR) resonance condition.

3.2 The resonance lines

The Cr^{3+} ions substituting Al^{3+} ions in the lattice of Al_2O_3 are surrounded by ^{27}Al nuclei of the host lattice, to which they are coupled by the superhyperfine interaction. This causes inhomogeneous broadening with a frequency spreading of 20 MHz, corresponding to an inhomogeneous halfwidth of 6.7 G of the ESR absorption spectrum or to a $T_2^*=\sigma^{-1}=8\times10^{-9}$ sec. The experimental results reported here were taken in three samples (N_o.1,2 and 3) of diluted ruby with nominal concentrations (atomic by weight): c=0.065%, c=0.025% and c=0.006%, respectively.

The homogeneous broadening is caused by the dipolar interaction between electronic spins and depends on the concentration. The values of the homogeneous dipolar width in our samples, estimated on the basis of literature data (17) are reported in Table I, together with the values of T_2 and of the homogeneity parameter a.

All the experiments were carried out at T=4.2°K, where T_1 is of the order of 0.1 sec., the exact value depending on the resonance line, on the crystal growth method and on the quantity of foreign impurities. In view of the use that will be made in the following analysis, T_1 was measured for each sample by the saturation recovery method in the same experimental conditions as the ST measurements.

The experimental method for measuring $\chi''(0,s,t)$ during the ST

TABLE I

Sample N_o	c (%)	ΔH_d (G)	T_2 (10^{-8}s)	a	T_1 (msec) 3.9 kG $\theta=90°$	T_1 (msec) 6.2 kG $\theta=0°$
1	0.065	0.95	5.7	0.15	70	42
2	0.025	0.35	16	0.05	330	150
3	0.006	0.085	67	0.013	95	110

regime was described previously (4).

3.3 Experimental results

3.3.1. The sample N_o.1. A typical experimental curve of $\chi''(0,s,t)$ (hereafter referred to as $\chi''(t)$) obtained in sample N_o.1 is reported in Fig.9. The curve is taken from ref.(4), where it was shown that experimental curves in this sample agree with the time behaviour predicted by the IPM, in the sense that they are well described as superposition of a continuous distribution of exponential terms, each one

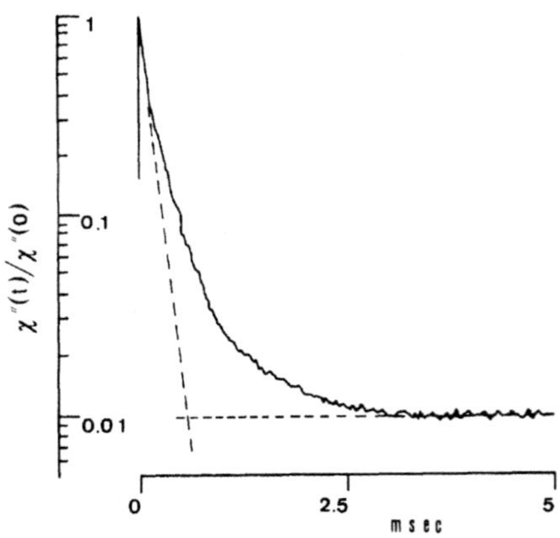

Figure 9. Experimental ST curve at the resonance 1-2 h.f. $\theta=0°$ in sample N_o.1 for $s^2=350$. Broken lines indicate the initial slope and the final value (from ref.(4)).

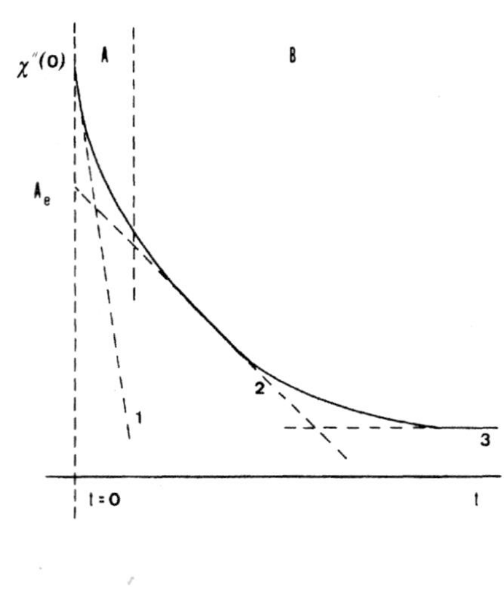

Figure 10. Outline of a typical experimental curve $\chi''(t)$ during ST in very dilute samples. Logarithmic scale with arbitrary units is used for $\chi''(t)$. Broken lines: 1 indicates the initial slope; 2 is a plot of the exponential function $A_e\exp(-t/\tau_e)$; 3 is the final value.

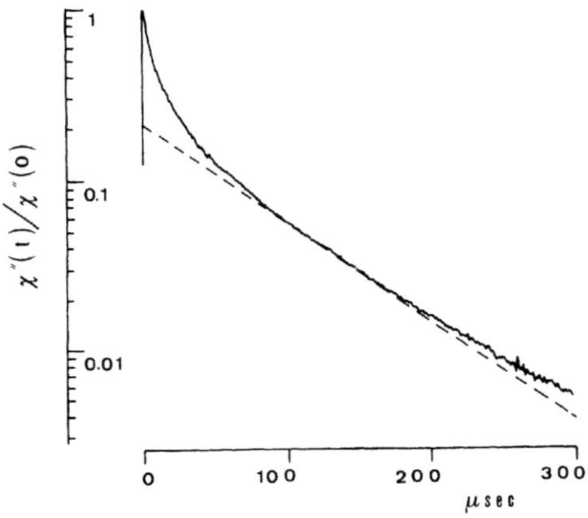

Figure 11. Experimental ST curve at the resonance 1-2 h.f. θ=0 in sample N$_o$.2 for s^2=1.2×10^5. The curve $A_e\exp(-t/\tau_e)$ with A_e and τ_e determined by best fitting is also shown (broken line).

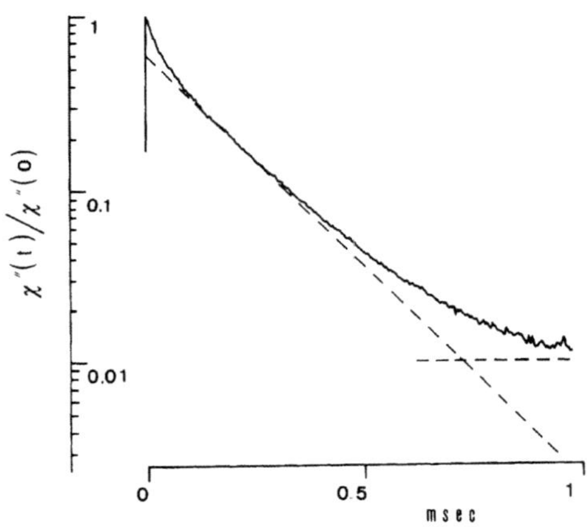

Figure 12. The same as Fig.11 but at a lower power level (s^2=2.5×10^4). The curve $A_e\exp(-t/\tau_e)$ and the final value are shown by the broken lines.

with its proper decay time and amplitude. The agreement with the IPM was shown to be also quantitative as regards the values of the initial slope and the whole amplitude excursion. Essentially similar results were obtained in other resonances of this sample as well as in more concentrated samples. We conclude that SD does not affect the ST curves of dilute ruby samples for c⩾0.06%.

3.3.2. The diluted samples N$_o$.2 and N$_o$.3. The ST behaviour of very diluted samples N$_o$.2 and N$_o$.3 is quite different from sample N$_o$.1. A typical experimental curve of χ''(t) in these samples is outlined in

Fig.10. Two different regimes of time evolution may be singled out in this curve: at short times after the onset of the input power (A), χ"(t) follows a non-exponential decay law resembling closely the time behaviour predicted by the IPM (line 1 indicates the initial slope); at longer times (B) the experimental curve departs from its initial behaviour and assumes a single exponential decay form which is maintained down to its stationary value.

The time behaviour outlined above is illustrated in Figs.11-14, where we report the experimental curves taken at both resonances and at different power levels.

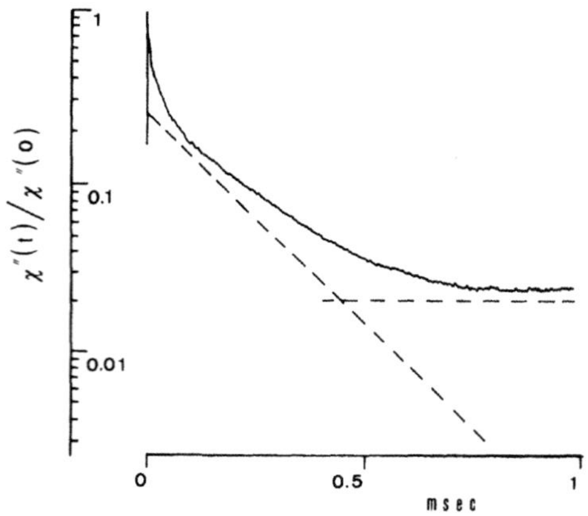

Figure 13. Experimental ST curve at the resonance 3-4 θ=90° in sample N_o.3 for $s^2=1.5\times10^5$ (broken lines as in Fig.12).

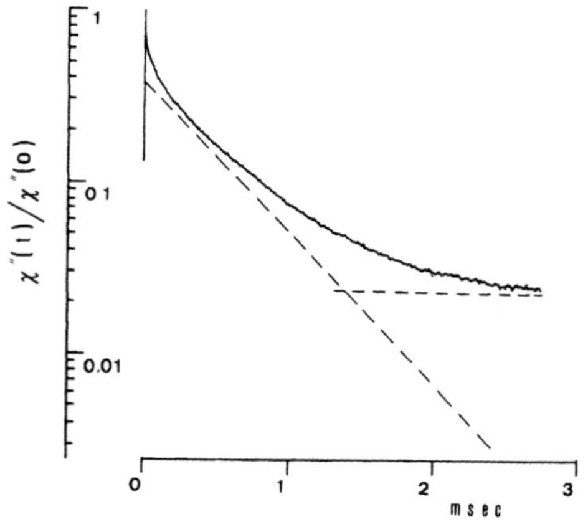

Figure 14. The same as Fig.13,but for $s^2=3.4\times10^4$.

The extent to which the experimental curves follow a single exponential law at long times is shown in Fig.15, where the experimental curve of Fig.12 is redrawn after subtracting the final value.

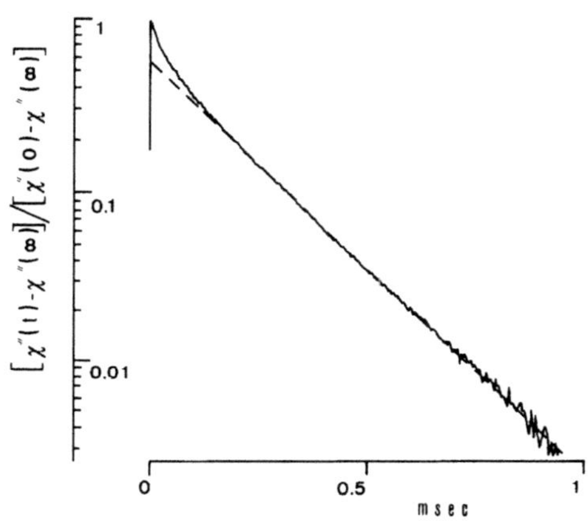

Figure 15. The same as Fig.12, but redrawn after subtracting the final value.

The comparison between the curves in Fig.11 and Fig.12 and between Fig.13 and Fig.14 indicates that the relative relevance of the two time behaviours depends on the input power level. On decreasing s the single exponential behaviour tends to overcome the isolated packet behaviour, which is confined to a relatively shorter time span and to a smaller amplitude excursion. At still lower power levels, the whole ST curves tend to a single exponential but for a very small peak near t=0.

The characteristic time τ_e and the amplitude A_e of the exponential part of $\chi''(t)$ at long times were measured in the ST curves as a function of the input power and arranged in plots of (τ_e/T_1) vs s and of $A_e/\chi''(0,0)$ vs s to be used for a quantitative comparison with theoretical results of Figs.6-8. This will be done in Sect.4.2.

A final comment of this Section concerns the values of s indicated for each experimental curve in Figs.11-14. In view of the relevance of its value for the analysis to be made below, the microwave field amplitude at the sample position was measured at each input power setting by revealing the nutational period of a long-T_2 sample (E_1' centers in silica) placed in the cavity for power calibration purposes together with the ruby sample. s was then determined by using the values of the microwave field amplitude, of T_1 and of T_2 (TableI).

4. COMPARISON WITH THEORETICAL RESULTS

4.1 Remarks on the suitability of the model

Firstly we wish to comment on the hypotheses that are on the basis of the model used in Sect.2.1 and on their suitability to the experimental situation considered here.

The failure of the IPM to account for the ST behaviour of dilute samples ($c \leqslant 0.05\%$) suggests by itself that SD mechanisms are effective in these samples both at $\theta=0°$ and at $\theta=90°$.

The experimental observation that the departure from the IPM behaviour is more evident in dilute rather than in concentrated samples rules out the possibility of relating the observed SD to the CR processes induced by spin-spin dipolar interaction. This is also supported by the following consideration : were the interpacket CR processes responsible of the observed SD effects,they should yield values of ω_d^{-1} of the same order as T_2; in variance, we show below that our experimental curves are compatible with values of ω_d^{-1} of the order of 10^{-5} sec much longer than T_2.By the way we note that this gives further support to the suggestion put forward in a previous paper (9) that, at least in randomly diluted samples at very low c, the effect of CR processes can be taken into account by properly defining the homogeneous line shape $g(\varepsilon)$, so that their effectiveness does not imply by itself that the isolated packet hypothesis is no longer valid.

On the basis of the above arguments, it seems reasonable to ascribe the SD effects experimentally observed in the ST curves of very diluted ruby samples to the time fluctuations of the hyperfine field created at each electronic spin site by its neighbouring nuclear spins. In this picture the sources of SD are the flipping motions of the nuclear spins. In this regard it is worth noting that Monte Carlo calculations carried out recently (18) for a similar microscopic situation have shown that nuclear spin correlation are not relevant for this kind of SD. So, hypothesis a) of Sect.2.1 seems to suit the system considered here.

As regards the frequency independence of the scattering probability function (hypothesis b) of Sect.2.1), we note that this simplifying approximation may be not completely true in the system considered here, where more than one nuclear spin flipping may be required to cover a spectral distance of the order of a linewidth.

Finally, we recall that the solutions of $\chi''(0,s,t)$ reported in Sect.2.2 were derived by hypothesizing Lorentzian packets, for calculation convenience. This hypothesis is not fullfilled in our samples, where the spin packet lines are known to be described by

Gaussian rather than by Lorentzian shapes (19,20). However, we remark that the particular shape of $g(\varepsilon)$ should be of poor relevance to the SD mechanism considered here, which by no means involves overlapping of adjacent packets.

4.2 The comparison with theoretical results

The theoretical model of Sect.2 is able to account for the occurrence of two different regimes of time evolution (at short and at long times, respectively) of the ST curves $\chi''(0,s,t)$ experimentally observed in our dilute samples, in particular as regards the single exponential behaviour at long times after the onset of the input power.

In agreement with the solutions and the qualitative considerations reported in Sect.2.2, the relative relevance of the single exponential decay was experimentally observed to lower on increasing the input power level. In the present Section we wish to carry out a more quantitative comparison between the theoretical and experimental results of the power dependence of τ_e and $A_e/\chi''(0,0)$.

In Fig.16 and in Fig.17 we report the experimental values of $A_e/\chi''(0,0)$ and of τ_e/T_1, respectively, as obtained from the ST curves measured in sample N$_o$.2 for a parallel orientation of the static field (1-2,θ=0°) at various input power levels. The full lines in these Figures are the theoretical curves, as obtained from Eq.(8) and from Eq.(9), respectively, that best fit the experimental points. The curves shown in the Figures refer to a=0.08 and $\omega_d T_1=6.5\times10^3$. The value of $\omega_d T_1$ was obtained by fitting the data of the amplitude, as this quantity is nearly independent of a in the range of s where the

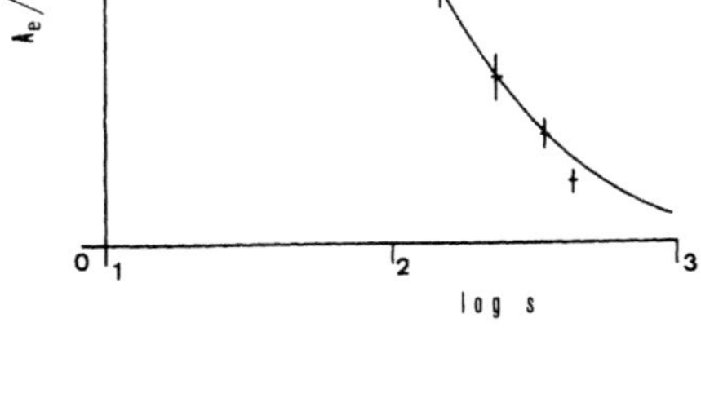

Figure 16. Normalized amplitude of the single exponential decay of the ST curve in sample N$_o$.2 at the resonance 1-2 h.f. θ=0°. The full line is the theoretical curve (a=0.08,$\omega_d T_1=6.5\times10^3$) that best fits the experimental points.

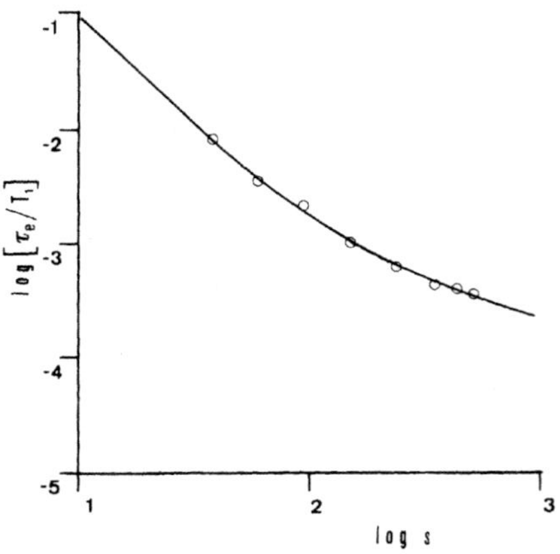

Figure 17. τ_e, as measured in sample N$_\circ$.2 at the resonance 1-2 h.f.,θ=0°, versus s. The full line is the theoretical curve (a=0.08, $\omega_d T_1$=6.5×10^3) that best fits the experimental points.

experimental points fall; the obtained value of $\omega_d T_1$ was then used to determine a by fitting the experimental data of τ_e.

The obtained value of a (a=0.08) is larger than the expected one (a=0.05) by a factor 1.5. We note that the latter was determined from literature data using the nominal concentration, which may be lower than the actual one. However, the disagreement might be related to the approximation made above on hypothesizing a frequency independent scattering probability of the spin excitations within the line.

From the value of $\omega_d T_1$ determined by the fitting procedure described above and using the measured value of T_1 (Table I) we determined the experimental value of $T_d=\omega_d^{-1}$=(23±5) μsec.Note that this value fully satisfies the condition c) of our model (Sect.2.1).

A similar procedure was followed for the other resonance and for the other sample. The obtained values of T_d are listed in Table II. In spite of the relatively large uncertainty (∿25%) of our determination of T_d, a trend of T_d may be singled out in Table II, namely to increase on increasing the concentration and, perhaps,to get

TABLE II

Sample N$_\circ$	Resonance	T_d (μsec)
2	1-2 h.f.,θ=0°	23±5
2	3-4,θ=90°	40±5
3	1-2 h.f.,θ=0°	14±3
3	3-4,θ=90°	10±3

in sample N$_o$.1 so long values that SD effects are no longer detectable. Contrarywise, T_d is expected to be independent on c,at least in the limit of low c, if the SD sources are the spin flipping motions of the nuclei surrounding each electronic spin. Further improvements of the experimental apparatus and procedures are in progress to reduce the uncertainty of our evaluation of T_d and to get more accurate measurements of its dependence on c. Here we wish to recall that other effects in dilute ruby samples are known to have an anomalous dependence on the concentration of Cr^{3+} ions, e.g. the stirring effect (21) and the (nuclear) spin-lattice relaxation (22), which has not yet found a satisfactory explanation.

5. CONCLUSIONS

The experimental results on the ST regime reported here show that SD mechanisms are effective in dilute ruby samples.Our experimental results can be explained by a model in which the SD is ascribed to the flipping motions of the nuclear spins surrounding the electronic spins. The comparison between experimental and theoretical results supports this picture even if some additional refinements of the model seem to be required to get a complete agreement.

In conclusion, we wish to point out a problematic aspect of the model used here. Consider an electronic spin surrounded by the nuclear spins of the host lattice. The ESR line is broadened by the local field created by the neighbouring nuclei at the electronic spin sites. On the other hand electronic spins create at the nuclear sites a local field by the same interaction, which tends to detune the nearest nuclei from the resonance frequency of bulk nuclei, thus forming around each electronic spin a so-called "frozen core" of nuclear spins uncapable of spin flipping with each other and with bulk nuclei (18). This circumstance seems to counteract the effectiveness of the SD mechanism considered here. In this regard we recall that in a recent paper T.Endo et al (23) reported on some optical experiments in dilute ruby samples that yield conclusions seemingly contrasting with the ones drawn in the present paper. In their experiments a hole is burnt in the optical line of Cr^{3+} ion and its spectral shape is recorded at various times during the recovery toward the equilibrium state. These authors report that the hole burnt in the optical line relaxes to zero without changing its shape in time. They ascribe the absence of spreading effect to the uneffectiveness of SD caused by the existence of a frozen core of ^{27}Al nuclei around each Cr^{3+} ion.

This result is seemingly in disagreement with those obtained from ST experiments. However, we wish to point out that the effectiveness of SD in our experimental conditions not necessarily contradicts the frozen core hypothesis. In fact, even if the frozen core exists at equilibrium, the local field at the nuclear spin sites may average to zero if the electronic spins are flipped up and down by the intense resonant radiation at a rate high enough (21,24). As a consequence, the nearest nuclei recover their natural frequencies, resonant flip-flops of neighbouring nuclei with bulk nuclei are restored and the SD mechanism mentioned above becomes effective.

Several facts support this hypothesis:firstly, the value of T_d inferred from our ST curves is of the same order as the spin-spin relaxation time of the ^{27}Al bulk nuclei; secondly, some ENDOR measurements carried out under heavy saturation condition of the electronic spin system yield resonance values equal to the bulk nuclei frequencies rather than the neighbouring ones (21); finally, measurements of the T_{ss} variations under heavy saturation indicate that up to 30% of the ^{27}Al nuclei are in good thermal contact with the electronic spin dipolar reservoir (25). On the other hand, the saturation induced disappearance of the frozen core is expected not to affect the optical hole burning experiments (23), as there the effectiveness of SD is tested during the saturation recovery, namely when the saturating radiation is turned off.

6. ACKNOWLEDGEMENTS

We wish to thank Mr G.Duro, Mr G.Lapis and Mr V.Greco for their continuous technical assistance in microwave and cryogenic work. Finantial support was provided by the Gruppo Nazionale Struttura della Materia del C.N.R.,by Ministero Pubblica Istruzione (national and local funds) and by the Comitato Regionale Ricerche.

REFERENCES

1. A.M.Portis - Phys. Rev. **91**, 107 (1953)
2. T.G.Castner - Phys.Rev. **115**, 1506 (1959)
3. P.W.Atkins,K.A.McLauchlan and P.W.Percival - Mol.Phys. **25**,281 (1973)
4. R.Boscaino and F.M.Gelardi - J.Phys.C**13**, 3737 (1980)
5. S.Clough and C.A.Scott - J.Phys.C**1**, 919 (1968)
6. P.R.Cullis - J.Magn.Res. **21**, 397 (1976)
7. B.N.Provotorov - Sov. Phys.JETP **14**,1126 (1962)
8. E.L.Wolf - Phys.Rev. **142**, 555 (1966)
9. R.Boscaino and F.M.Gelardi - J.Magn.Res. **49**, 476 (1982)
10. L.L.Buishvili,M.D.Zviadadze and G.R.Khutsishvili - Sov.Phys.JETP **27** 469 (1968)
11. J.P.Korb and J.Maruani - Phys.Rev. B**23**, 971;5700 (1981)

12. P.W.Anderson and P.R.Weiss - Rev.Mod.Phys. 25, 269 (1953)
13. W.B.Mims - Phys.Rev. 168, 370 (1968)
14. A.G.Kofman and A.I.Burshtein - Sov.Phys.JETP 49, 1019 (1979)
15. R.N.Mantegna - Thesis, Institute of Physics,University of Palermo, Palermo (Italy) (1984) (in Italian)
16. E.O.Schulz du Bois - Bell Syst.Tech.J. 38,271 (1959)
17. J.P.Gayda and C.Blanchard - J.Physique 30, 827 (1969)
18. R.G.De Voe, A.Wokaun, S.C.Rand, and R.G.Brewer - Phys.Rev. B23, 3125 (1981)
19. R.Boscaino and F.M.Gelardi - J.Phys. C15, 6245 (1982)
20. R.Boscaino and J.P.Korb - J.Magn.Res. 57, 127 (1983)
21. R.Boscaino,M.Brai and I.Ciccarello - Phys.Rev. B13, 2798 (1976)
22. H.H.Niebuhr,E.E.Hundt and E.Brun - Helv.Phys.Acta 43, 777 (1970)
23. T.Endo,T.Muramoto and T.Hashi - Phys.Lett. 99A, 128 (1983)
24. L.Q.Lambert - Phys.Rev. B7,1834 (1973)
25. V.A.Atsarkin, O.A.Ryabushkin and V.A.Skidanov - Sov.Phys.JETP 45 584 (1977)

SIMULATION OF THE ELECTRON SPIN RESONANCE SPECTRA OF SPIN-LABELED FATTY ACIDS IN PROTEIN-CONTAINING MODEL MEMBRANES

M. Seigneuret, J. Davoust and P.F. Devaux
Institut de Biologie Physico-chimique
13, rue Pierre et Marie Curie, F-75005 Paris

ABSTRACT. In this review we show that Electron Spin Resonance (ESR) can be used to investigate the time scale and specificity of lipid-lipid and lipid-protein contacts in membranes. Details are given of the simulation of the spectra obtained with labeled lipids either covalently attached to the boundary of rhodopsin or freely diffusing in rhodopsin containing vesicles. It is shown that the time scale of conventional ESR is idealy suited to the determination of the collision frequencies between lipids and proteins and/or lipids-lipids in membranes. In addition we show that the use of a mixture of ^{14}N and ^{15}N nitroxides allows one to detect unambiguously the collision frequency between unlike molecules such as two different phospholipids or a phospholipid and a labeled protein.

1. INTRODUCTION

Spin labeling macromolecules with nitroxide radicals allows one to study their dynamical properties by electron spin resonance (ESR). The basic principles of the spin labeling technique can be found in several books and reviews (1 - 5). The technique has been widely used to investigate the mobility of spin labeled fatty acid or phospholipids diluted in model membranes or in biological membranes. The ESR spectra obtained with nitroxide probes covalently linked to the acyl chains of the lipid molecules bring informations on the internal mobility of the acyl chains and thus report on the "local viscosity". The details of the computer simulation of the spectra obtained with a spin labeled fatty acid incorporated, at low concentration, in a homogenous phospholipid bilayer can be found, for example, in the article by Gaffney and Mc Connell (6). The basic assumption is that for each orientation of the membrane, each line (m = -1, 0, +1) gives a lorentzian line, which can be considered as the solution of a set of classical Bloch equations. If the probe experiences two physical environments with a rapid exchange frequency, the Bloch equations must be modified to include this heterogeneity. One adjustable parameter obtained from the simulation is the exchange frequency. The heterogenous environment of the probe can be due to the presence, within the lipid

R. Daudel et al. (eds.), Structure and Dynamics of Molecular Systems, 171–185.

bilayer, of rigid intrinsic proteins, imposing a local order to the chains. If the concentration of spin labels per phospholipids is high, a probe can encounter another probe and exchange its electron spin. This spin diffusion mechanism allows one, again through the solution of the modified Bloch equations, to inferr an exchange frequency which, in turns, is related to a molecular collision frequency. The advantage of using hetero-isotopic mixtures of spin labels will be demonstrated.

2. ESR SPECTRA OF SPIN-LABELED LIPID CHAINS COVALENTLY LINKED TO AN INTRINSIC PROTEIN. A PHYSICAL EXCHANGE MODEL

2.1. ESR spectra of a spin labeled fatty acid covalently linked to the intrinsic protein rhodopsin

We have developped a spin labeling approach which permits us to visualize selectively boundary lipids while minimizing the contribution of the bulk lipids. For this purpose a spin labeled fatty acyl chain is covalently bound to an intrinsic protein thus reports upon the behavior of a phosholipid chain at the protein boundary. We made use of the following spin label which bears a meleimide reactive group, allowing us to anchor the label chain to a protein sulfhydryl group (7).

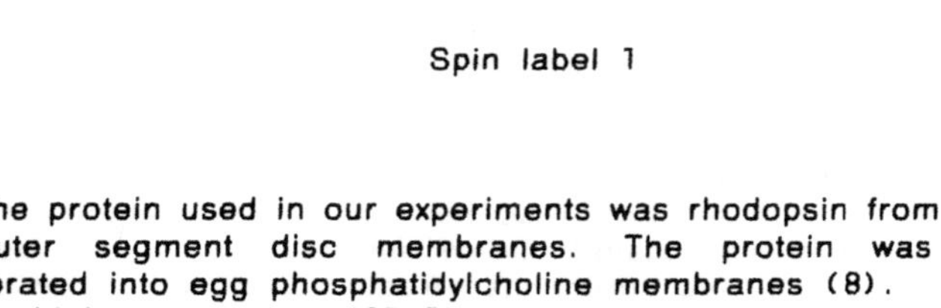

Spin label 1

The protein used in our experiments was rhodopsin from bovine retinal rod outer segment disc membranes. The protein was labeled and incorporated into egg phosphatidylcholine membranes (8).

At high temperature (37°C), a single narrow component is observed on the ESR spectrum of the protein bound label (Fig. 1, top left spectrum). At low temperature, the spectrum looks heteregenous with the superposition of a broad and a narrow component (see arrow in figure 1, bottom left spectrum).

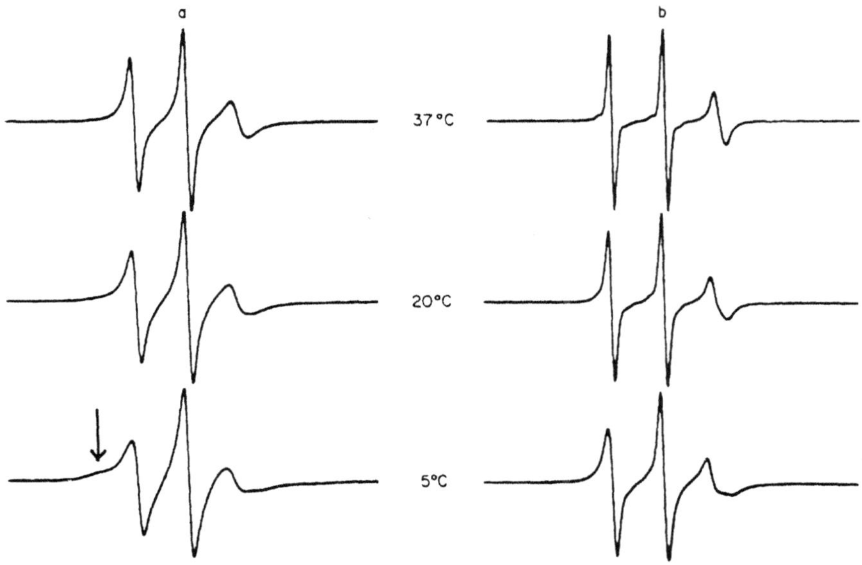

Fig. 1. a) ESR spectra of spin label 1 attached to rhodopsin reconstituted in egg phosphatidylcholine membranes (lipid to protein mole ratio 500/1); b) incorporated in pure egg phosphatidylcholine membranes. Taken from (8).

2.2. Principle of the physical exchange model

To account for the spectra of fig. 1, we derived a two state physical exchange model (9) (In reference 9, we used the term "chemical exchange model". However, as pointed out to us by A. Lee, a "physical exchange model" is a more appropriate description). The label is supposed to probe two environments corresponding to two states A and B. State A is the label in contact with the protein surface and has a high order parameter (broad component) while state B is the label removed from the protein surface and of mobility comparable to bulk bilayer lipid chains (narrow component). The label can exchange between A and B (i.e. go on and off the protein) at a rate which is determined by the collisions with freely diffusing lipids and thus which increases with temperature. The limit between slow and rapid exchange corresponds to a rate comparable to the frequency difference between ESR lines associated with the two states.

Additionally, two refinements of the exchange model are made. In model I, the label in both states A and B has an average orientation which is normal to the membrane (smooth protein). In model II, state B still has a normal average orientation while state A corresponds to an isotropic distribution of orientations (Fig. 2).

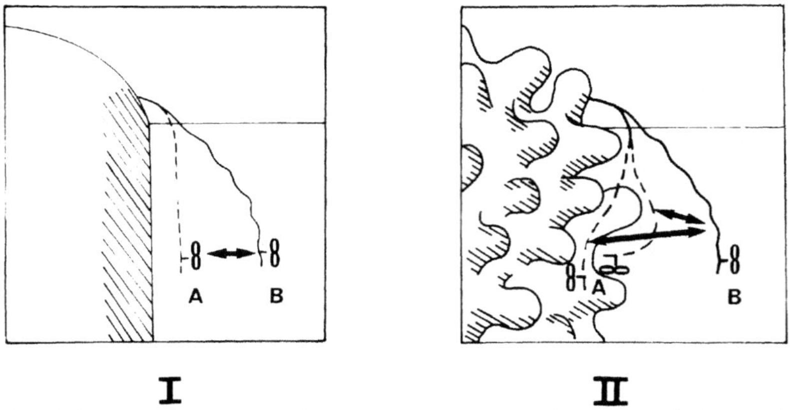

Fig. 2. Schematic representation of the average orientation of the protein bound spin label in state A and state B according to models I and II. Taken from (9).

The ESR spectra associated with state A and B are taken respectively as that of spin label 1 trapped inside protein aggregates (see (9) for details) and that of spin label 1 incorporated into pure egg phosphatidylcholine membranes at the corresponding temperature (Fig. 1b).

2.3. Spectral simulation

The first step is the simulation of the ESR spectra associated respectively with the pure states A and B, using two effective time-independent spin hamiltonians of axial symmetry. For an isotropic membrane suspension, each orientation θ of the director axis of the label corresponds to three resonance lines ($m = \pm 1, 0$) of the motionally averaged values of the T (hyperfine) and g tensor as follows :

$$H_m(\theta) = H_o + \Delta H(2/3 - \sin^2 \theta) - m(T_{/\!/}'^2 \cos^2 \theta + T_\perp'^2 \sin^2 \theta)^{1/2}$$

H_o is the mean field position of the central line and ΔH the field shift due to the g factor anisotropy.

Each resonance corresponds to a complex lorentzian lineshape with linewidth $(T_{2m})^{-1}$:

$$L_m[H - H_m(\theta)] = [(T_{2m})^{-1} + i\ \gamma/2\pi\ (H - H_m(\theta))]^{-1}$$

The ESR absorption spectrum is obtained by isotropic summation of the real parts of the lorentzians :

$$A(H) = \sum_{m=-1,0,+1} \gamma H_1 M_o \int_0^{\pi/2} \text{Real}\ [L_m(H - H_m(\theta))]\ \sin\theta\ d\theta$$

For both states A and B, each average orientation of the label θ and each nuclear spin value m is associated with one set of Bloch equations. Allowing chemical exchange between A and B means coupling between both systems using the appropriate exchange rates. τ_{AB}^{-1} is the probability per unit time that an electron spin corresponding to state A undergoes an exchange to state B, while τ_{BA}^{-1} is associated with the opposite mechanism. The average exchange rate τ_{ex} is defined by the following relation (which also express the conservation of the total longitudinal magnetization of each state M_{OA} and M_{OB}) :

$$\nu_{ex} = \frac{M_{OB}\ \tau_{BA}^{-1}}{M_{OA} + M_{OB}} = \frac{M_{OA}\ \tau_{AB}^{-1}}{P_{OA} + M_{OB}} = R\ \tau_{AB}^{-1}$$

R being the relative weight of state A. The exact coupling of the Bloch equations depends upon the details of the exchange model.

Model I. The exchange occurs between state A and B having the same average orientation θ relative to the magnetic field. This means coupling of two lorentzian lines for each value of m and θ.

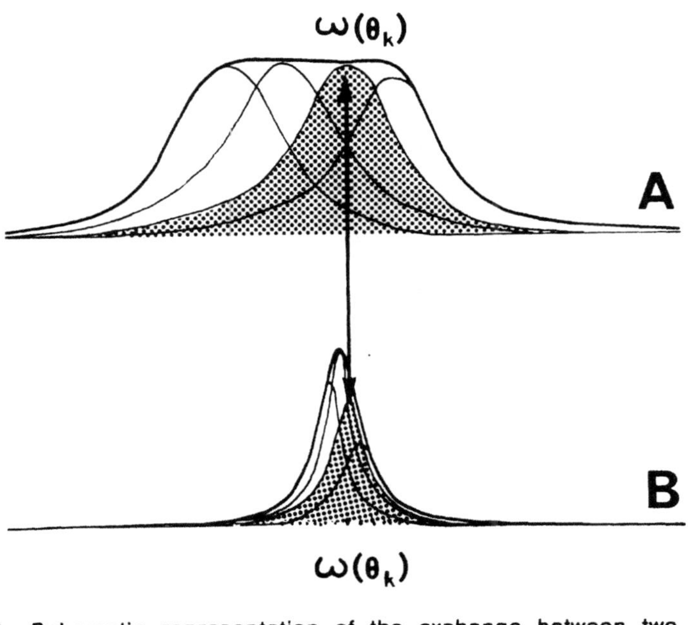

Fig. 3. Schematic representation of the exchange between two lorentzian lines of same nuclear spin corresponding resectively to state A (boundary) and state B (bulk) within the framework of model I. Taken from (9).

The corresponding modified Bloch equations can be written for the longitudinal and complex transversal magnetization of state A in frequency units ($\omega = \gamma/2\pi H$), at non saturating microwave frequency and in the slow passage limit :

$$M^{m}_{ZA} = M_{OA}/3$$

Let us write : $\hat{M}_A = u_A + iv_A$ and $\hat{M}_B = u_B + iv_B$

Then

$$\dot{\hat{M}}_A + [T_{2A}^{-1} + \tau_{AB}^{-1} - i(\omega-\omega_A(\theta))]\hat{M}_A = \tau_{BA}^{-1}\,\hat{M}_B + i\,\gamma\,H_1\,M^{m}_{ZA}.$$

The corresponding equations for state B are obtained by permutation of the undexes A and B. The coupled equations are solved for steady state conditions $\dot{\hat{M}}_A + \dot{\hat{M}}_B = 0$ and the absorption spectrum is obtained by summing the imaginary part of $\hat{M}^{m}_{A} + \hat{M}^{m}_{B}$ over all values of m and all orientations θ (9).

Model II. The calculation is basically the same, except that each orientation θ of state B can exchange with all possible orientations of state A. Thus each lorentzian line of state B is coupled to a complete manifold of lorentzian lines, to state A (fig. 4).

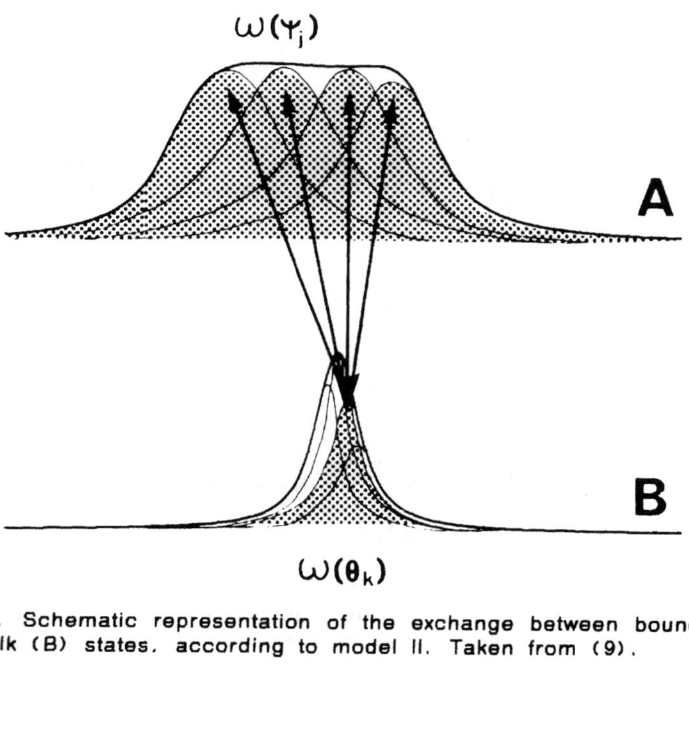

Fig. 4. Schematic representation of the exchange between boundary (A) and bulk (B) states, according to model II. Taken from (9).

The angular space of state A is divided into discrete angular zone, each of which is associated with a set of modified Bloch equations coupled to that of state B. Solving this system yields an expression for $M^{m}_{A} + M^{m}_{B}$ which has the same form as in model I except that the contribution of state A is now a sum of lorentzian lines. The absorption spectrum is calculated as preceedingly.

Both models allow one to adequately represent what is observed experimentally when the exchange rate ν_{ex} is increased (Fig. 5).

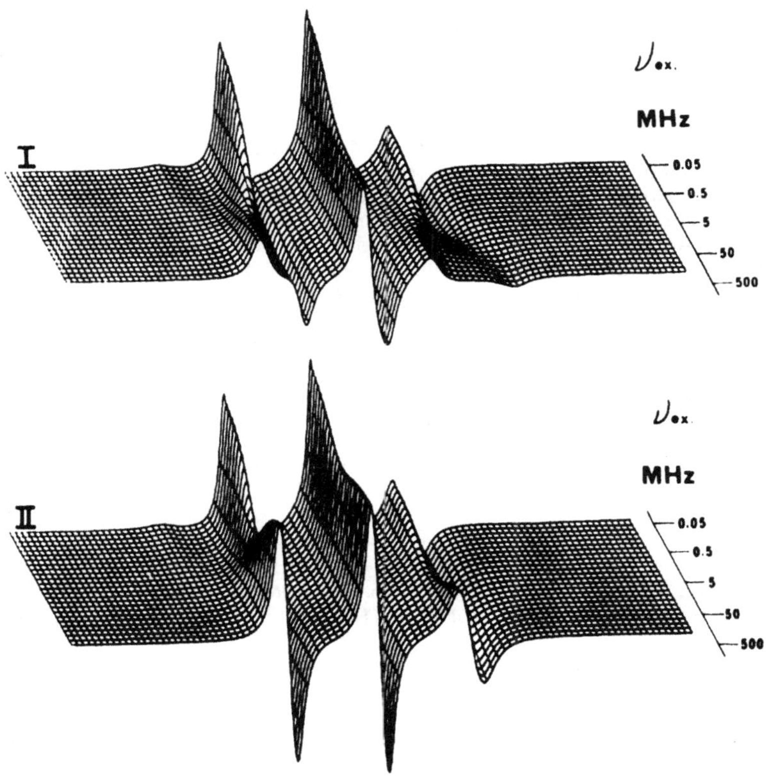

Fig. 5. Results of a series of simulations performed according to models I and II. The exchange frequency is varied while all other spectral parameters are kept constant. The relative weight of state A is R = 35%. One block in the horizontal scale correspond to 1 G.

The major effect of ν_{ex} is to broaden each peak of the narrow component. When the exchange frequency is such that $\nu_{ex} \geq (\gamma/2\pi)(T_{\parallel}'^{A} - T_{\parallel}'^{B})$, the broad component is no longer resolved. Note that, in the limit of fast exchange, the two models yield very different lineshapes. Indeed, under such conditions, model I leads to a situation which is similar to that of a single environment with an order parameter intermediate between those of A and B. While model II allows the exchange with an isotropic distribution of orientations and averages the spectral anisotropy of A and B to zero.

2.4. Experimental exchange rates of lipid chains at the protein boundary

Both models were found to adequately simulate the experimental spectra of fig. 1. Model I gives slightly better results at high temperature. Thus each experimental spectrum can be accounted for by several couples of parameters τ_{AB}^{-1} and R. However, the average on-off exchange frequency was found to increase with temperature. Values were found comparable to the lateral hopping frequencies of freely diffusing phospholipids ($\simeq 10^7 sec^{-1}$, see next section). This confirms that the exchange process is mediated through collisions between the covalent label and the chains of freely diffusing lipids. Additionally, the fraction of the immobilized environment A decreased with increasing temperature. It is possible that the increased disorder of the surrounding lipid chains allows the covalent label to explore a larger spatial domain in the bilayer at high temperature.

In conclusion, a spin labeled lipid chain bound to rhodopsin is able to probe two lipid environments : the motionally restricted boundary lipids and the bulk lipids. The fast exchange observed between both states at physiological temperature suggests that the lipid chains at the boundary of the protein exchange rapidly with the bulk lipid chains.

Recently Lee has extended this calculation to the simulation of freely diffusing spin labeled lipids in model membranes containing Ca^{2+}ATPase (10).

3. MEASUREMENT OF LIPID-INTRINSIC PROTEIN COLLISION RATES BY THE USE OF SPIN-SPIN INTERACTIONS BETRWEEN ^{14}N AND ^{15}N SPIN LABELS

3.1. Spin-spin interactions in membranes. Advantages of an hetero-isotopic labeling

Besides their use as rotational mobility markers, lipid spin labels can also be used to investigate lipid collision rates and lateral diffusion constants. This requires that the local spin label concentration be high enough in the membrane for spin-spin interaction to occur between nitroxides. One of these interactions, the Heisenberg spin exchange is mediated by collisions between labels. Quantitation of spin exchange rates from spectral simulations allows one to calculate lipid collision and diffusion rates (11, 12).

The method was formerly limited to collisions between identical spin labels. We have recently shown that, by using spin labels bearing nitroxides with different nitrogen isotopes (^{14}N and ^{15}N), collision frequencies between chemically distinct molecules can be selectively monitored from hetero-isotopic spin-spin interactions. The method permits the study of collision rates between different lipids or between a lipid and a protein (13, 14, 15).

3.2. Theory and spectral simulations

^{15}N nitroxides give rise to two ESR lines which intercalate between the three lines of ^{14}N nitroxides, allowing one to resolve both isotopes when mixed in the same membrane. If the local concentration of at least one of the label is high enough (i.e. more than 1% relative to total lipid concentration), spin-spin interactions can be detected. For lipid labels having a rapid quasi-isotropic rotation, dipole-dipole interactions are largely averaged out and spectral modifications are due to Heisenberg spin exchange. The exchange rates can be determined from spectral simulation.

Such procedure requires first a correct representation of the spectra obtained with either ^{14}N and ^{15}N labels alone, i.e. without interaction. The method has been described in the preceeding section for ^{14}N labels and applies also to ^{15}N labels except that only two lines are considered.

For a mixture of ^{15}N and ^{14}N labels, the presence of spin exchange means coupling between the sets of Bloch equations of each nuclear spin of each isotope. If one defines the spin exchange rates τ_{ij}^{-1} as the average probability per unit time for a given spin label ^{i}N to exchange its electron spin with any of the spin labels ^{j}N, one can introduce four exchange rates : $\tau^{-1}_{14,14}$, $\tau^{-1}_{15,15}$, $\tau^{-1}_{15,14}$ and $\tau^{-1}_{14,15}$. To give rise to a spin exchange process affecting the lineshape, collision must occur between nitroxide having distinct electron and nuclear spins. The modificed Bloch equation of the complex transversal magnetization corresponding to one nuclear spin m of the ^{14}N label can thus be written for each orientation θ of the membrane normal (n′ representing the nuclear spin states of the ^{15}N label) :

$$M^{m}_{14} + [(T^{m}_{2,14})^{-1} + \tau^{-1}_{14,14} + \tau^{-1}_{14,15} - i(\omega - \omega^{m}_{14}(\theta))]\, M^{m}_{14}$$
$$= 1/3\, [{}_{m'}\, \tau^{-1}_{14,14}\, M^{m'}_{14} + {}_{n'}\, \tau^{-1}_{15,14}\, M^{n'}_{15} + i\, \gamma\, H_1\, M_{0,14}]$$

Here, we have directly introduced the non saturation condition : $M^{m}_{z,14} = M_{0,14}/3$ (all notations follow from the preceeding section). A similar equation can be written for the ^{15}N label. Resolution and calculation of the absorption spectrum is obtained as in the preceeding section.

The effect of increasing ^{14}N-^{15}N spin exchange rate is illustrated in the set of simulated spectra shown in fig. 6.

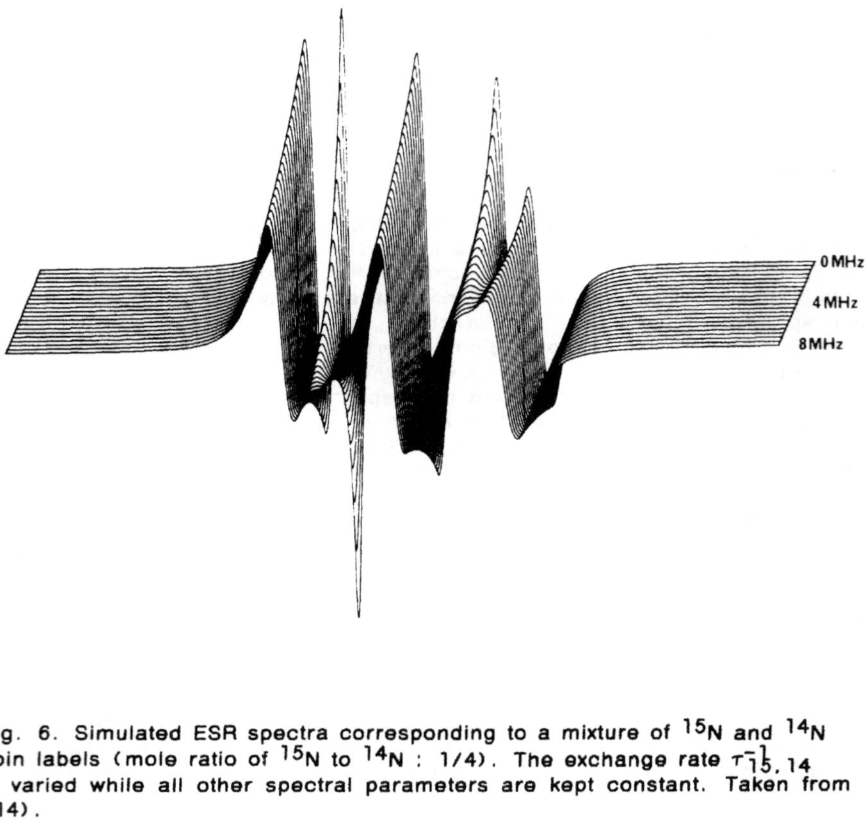

Fig. 6. Simulated ESR spectra corresponding to a mixture of ^{15}N and ^{14}N spin labels (mole ratio of ^{15}N to ^{14}N : 1/4). The exchange rate $\tau^{-1}_{15,14}$ is varied while all other spectral parameters are kept constant. Taken from (14).

Increasing $\tau^{-1}_{15,14}$ has a broadening effect upon all ESR lines. However this effect is particularly spectacular on the ^{15}N lines in figure 6, because the ^{15}N spin labels are more dilute than the ^{14}N spin labels in this example. The line broadening, which can be calculated from line height variations assuming lorentzian lines, is, to a first approximation, proportional to $\tau^{-1}_{15,14}$.

3.3. Measurement of lipid-lipid collision rates. ESR approach

In these experiments performed in pure egg phosphatidylcholine membranes, a ^{15}N spin labeled fatty acid, was allowed to interact with two ^{14}N spin labeled phospholipids bearing nitroxide groups situated at very different depths in the bilayer, (see fig. 7 a and b).

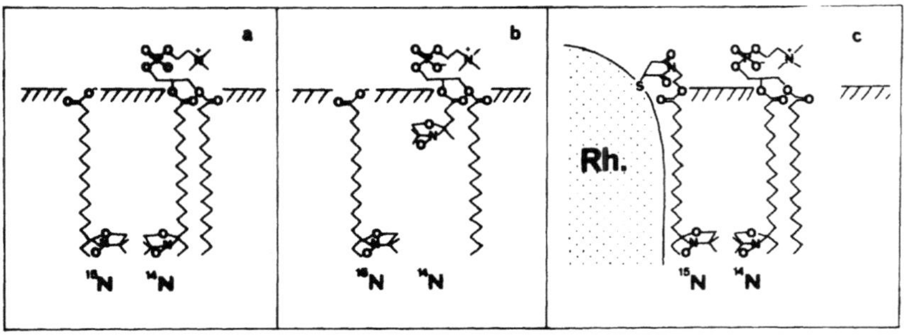

Fig. 7. Schematic representation of double labeling experiments designed to study lipid–lipid or protein–lipid collision rates. The labeled lipids are diluted with unlabeled lipids, not shown in these diagrams.

Spectral simulation were possible because all probes in figure 7 a and b experience a quasi-isotropic motion at T > 35oC, so that spin exchange is the dominant interaction mechanism. The values of homo-isotopic ($\tau^{-1}_{14,14}$) and hetero-isotopic ($\tau^{-1}_{14,14}$) spin exchange rates corresponding to the two couples of labels are shown in fig. 7 a and b.

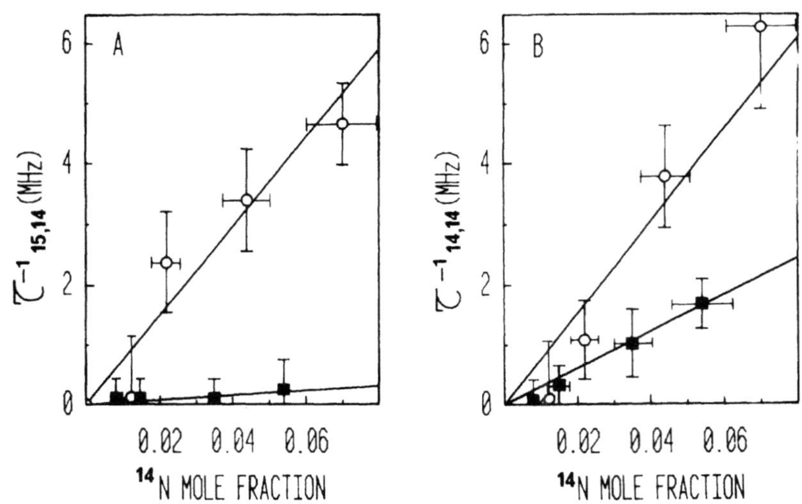

Fig. 8. ^{15}N–^{14}N (A) and ^{14}N–^{14}N (B) exchange rates vs. ^{14}N mole fraction for mixtures of ^{15}N-fatty acids (1% relative to total lipid) with ^{14}N-phospholipids diluted in egg phosphatidylcholine. The ^{14}N mole fraction is calculated versus unlabeled lipids. Open circles correspond to the experiment described in figure 7 a; closed squares correspond to the experiment described in figure 7 b.

If one pay attention first to $\tau^{-1}_{15,14}$ values (fig. 8 A), the difference between the physical situation of figure 7 a and 7 b appears strickingly. In fact the result indicates clearly the absence of hetero-probe collisions in case 7 b (which corresponds in figure 8 A to the closed squares). $\tau^{-1}_{14,14}$ (figure 8 B) indicates the values of homo-isotopic spin exchange and is proportional to the collision rates between like molecules. Figure 8 B tells us that both ^{14}N phospholipids in figure 7 a and 7 b collide with like molecules. However, the difference in slope of the two curves in figure 8 B is remarkable. One interpretation would be that the long chain phospholipid (see figure 7 a) and the short chain phospholipid (see figure 7 b) have different lateral diffusion constants. Indeed, the frequency of lipid collision is proportional to the rate of lateral diffusion (11, 12, 14). However it is generally found that the rate of lateral diffusion of membrane components depends very little on their actual size. Thus a more likely explanation for the observed diference in $\tau^{-1}_{14,14}$ values (figure 8 B) would be a difference in cross section of the two types of phospholipids. When the probe is near the methyl terminal (figure 7 a), it sweeps rapidely a large area even when the phospholipid head group is immobile. This is due to the flexibility of the long hydrocarbon chain. On the contrary with the short chain phospholipids (figure 7 b) the probe covers a more limited area. Thus spin exchange in the latter case requires direct contact between spin labeled phospholipids, whereas in the former case spin exchange may be possible between second neighbours. This discussion shows that it is not straightforward to deduce the actual diffusion constant of phospholipids from the measurement of the spin exchange frequency. The reader is reported to the discussion section of reference 14.

3.4. Measurement of lipid-lipid collision rates. ELDOR approach

Hyde and collaborators have recently described the use of stationary electron-electron double resonance (ELDOR) to measure spin exchange between lipid spin labels (16, 17). The ELDOR process makes use of two distinct microwave frequencies, one (pump) saturating the spin system at one resonance position and the other (probe) measuring the effect of saturation at another position. Since spin exchange induces transfer of saturation from one nuclear spin state to another, it can be detected using ELDOR by pumping one ESR line and probing another. Advantages lie in the fact that spin exchange frequencies about ten times lower can be measured, allowing one to use much lower spin label mole ratio or to measure slow diffusion processes. Drawbacks are the necessity of a sophosticated ESR set-up (ELDOR microwave bridge, bimodal cavity) and of much more concentrated samples (100 to 1 000 times higher than for classical ESR). Additionally, calculation of exact spin exchange rates from ELDOR data requires knowledge of the electron spin-lattice relaxation time, which must be determined from a saturation recovery experiment (16).

When studying homo-isotopic ^{14}N-^{14}N spin exchange, an additional limitation is the contribution of the ^{14}N nuclear spin lattice relaxation which is also an ELDOR active process and precludes the measurement of very low exchange rates (16). Hyde and collaborators obviated this difficulty by using mixtures of ^{15}N and ^{14}N spin labels (17). Hetero-isotopic spin exchange was measured by pumping the ^{15}N m=-1/2 line and probing the ^{14}N m=0 line. Data are expressed as the ELDOR reduction rate which is defined as R = [(signal with pump off) - (signal with pump on)]/(signal with pump off). The value of R extrapolated at infinite pumping power, R_∞, is directly related to the spin exchange rate.

Lateral diffusion coefficients obtained by ELDOR from spin exchange rates between ^{15}N-^{14}N-fatty acids agree quite well with our values obtained directly from spectral simulations.

3.5. Measurements of lipid-intrinsic proteins collision rates (15)

We have also applied hetero-isotopic spin exchange to the study of lipid collision rates with a protein boundary. For this purpose, rhodopsin was covalently labeled with spin label 1 (see preceeding section) bearing a ^{15}N nitroxide and allowed to interact with a freely diffusing ^{14}N spin labeled phospholipid (either phosphatidylcholine, phosphatidylserine or phosphatidylethanolamine) in a reconstituted system (fig. 7C). Simulation of the spectra yielded $\tau^{-1}_{15,14}$ and $\tau^{-1}_{14,14}$ values (fig. 9).

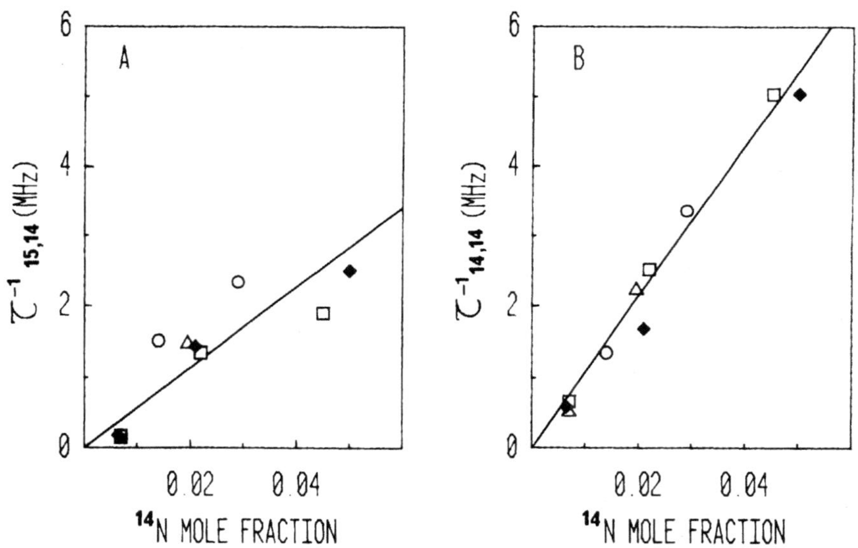

Fig. 9. ^{15}N-^{14}N (A) and ^{14}N-^{14}N (B) exchange frequencies vs. ^{14}N label mole fraction for rhodopsin bound with ^{15}N-spin label 1 and diluted with ^{14}N-PC (open circles), ^{14}N-PE (open triangles) and ^{14}N-PS (open squares) in reconstituted systems (lipid to protein mole ratio 80/1) (taken from ref. 15).

$\tau^{-1}_{15,14}$ values obtained with the three different phospholipids were similar. This implies an absence of polar head group selectivity for the interaction of phospholipids with rhodopsin. These $\tau^{-1}_{15,14}$ values are about two times smaller than the corresponding $\tau^{-1}_{14,14}$ values which represent collisions between free lipid chains (see figure 8). This reduction is probably due to the presence of the nearby protein surface which reduce the number of nearest neighbor lipid chains sensed by the ^{15}N label as well as to the absence of lateral diffusion of this label.

These results clearly indicate that the collision rates of lipid chains at the boundary of the intrinsic protein rhodopsin is of the same order of magnitude as in the bulk lipid phase.

4. CONCLUSIONS

The spin labeling technique can be used in many different ways to investigate the motion of fatty acids and phospholipids in membranes. Here we have shown that including exchange mechanisms in the Bloch equations allowed us to simulate situations corresponding to phospholipids hopping between different environments. The time scale for the exchange mechanism to be efficient on the spectral line shape is precisely that of phospholipid diffusion rates in membranes ($\simeq 10^7 sec^{-1}$). Thus, conventional ESR of spin labeled phospholipid provides a useful mean of investigating lipid diffusion and lipid-protein interactions. The absolute sensitivity of ESR is another advantage of the method. The drawbacks are the followoing :

i) the obligation of lipid modification by attachment of a nitroxide probe;
ii) the incorporation of a synthetic lipid in a membrane containing membrane;
iii) possible local steric perturbation due to the nitroxide ring.

In spite of these drawbacks the time scale and sensitivity of ESR makes it still very competitive with other spectroscopic techniques which are commonly used to investigate lipid mobility and lipid-protein interactions (see reviews 1 and 2).

ACKNOWLEDGEMENTS

This work was supported by grants from the "Ministère de la Recherche et de la Technologie", the "Centre National de la Recherche Scientifique" (UA 526), and the Université Paris VII.

Present addresses :
M.S. Centre d'Etudes Nucléaires de Saclay
Département de Biologie, Service de Biophysique
F - 91191 Gif sur Yvette Cedex
J.D. European Molecular Biology Laboratory
Postfach 10.2209, Meyerhofstrasse 1
D - 6 900 Heidelberg

REFERENCES

1. Devaux, P.F. (1983) in "Biological Magnetic Resonance" vol. 5, Berliner, L.J. and Reuben, J., eds (Plenum Press, New York), pp. 183-299.
2. Devaux, P.F. and Seigneuret, M. (1984) Biochim. Biophys. Acta, in press.
3. Berliner, L.J., ed. (1976) Spin Labeling - Theory and Applications Vol. 1 (Academic Press, New York).
4. Berliner, L.J., ed. (1979) Spin Labeling - Theory and Applications Vol. 2 (Academic Press, New York).
5. Marsh, D. (1981) in "Membrane Spectroscopy", Grell, E., ed. (Springer Verlag, Berlin) pp. 51-142.
6. Gaffney, B.J. and Mc Connell, H.M. (1974) J. Magn. Res. 16, 1-28.
7. Favre, E., Baroin, A., Bienvenüe, A. and Devaux, P.F. (1979) Biochemistry 18, 1156-
8. Davoust, J., Schoot, B.M. and Devaux, P.F. (1979) Proc. Natl. Acad. Sci. USA 76, 2755-2759.
10. Lee, A.G., XI International Conference on Magnetic Resonance in Biological Systems, Fort Aguada Beach Resort, Goa, India, sept. 1984, abstract IT 13.
11. Devaux, P.F., Scandella, C.J. and Mc Connell, H.M. (1973) J. Magn. Res. 9, 474-485.
12. Sackmann, E. and Traüble, H. (1972) J. Am. Chem. Soc. 94, 4492-4498.
13. Seigneuret, M., Davoust, J., Hervé, P. and Devaux, P.F. (1981) Biochimie 63, 867-870.
14. Davoust, J., Seigneuret, M., Hervé, P. and Devaux, P.F. (1983) Biochemistry 22, 3137-3145.
15. Davoust, J., Seigneuret, M., Hervé, P. and Devaux, P.F. (1983) Biochemistry 22, 3145-3151.
16. Popp, C. and Hyde, J.S. (1982) Proc. Natl. Acad. Sci. USA 79, 2559-2563.
17. Feix, J.R., Popp, C., Venkataramu, S.D., Beth, A.H., Park, J.H. and Hyde, J.S. (1984) Biochemistry 23, 2293-2299.

NUCLEAR MAGNETIC RESONANCE AND MOLECULAR MOTIONS IN LYOTROPIC LIQUID CRYSTALS

C. Chachaty, J.-P. Quaegebeur and B. Perly
Département de Physico-chimie
C.E.N. de Saclay
91191 Gif sur Yvette Cedex
France

ABSTRACT. Some specific applications of NMR to lyotropic liquid crystals are reported on examples taken from studies in progress on sodium alkyl-phosphate/water systems. The reorientation dynamics of interstitial water is obtained from the splittings and relaxation rates of ^{1}H, ^{2}H and ^{17}O. The molecular order parameters and alkyl chain conformation are derived from ^{2}H quadrupolar splittings and interpreted in term of intermolecular interaction confining the molecules in restricted volumes. The rotational diffusion coefficients of surfactant molecules are obtained from the temperature and frequency dependences of the ^{31}P longitudinal relaxation which depends on the chemical shift anisotropy and on the dipolar coupling with nearby protons.

1. INTRODUCTION

In concentrated aqueous solutions, most of surfactant molecules, generally formed of a polar head and of one or two hydrocarbon chains give rise to large size disk or rod shaped micellar aggregates. By lowering the temperature or the water content, these aggregates become organized in lyotropic lamellar, cubic, direct or reverse hexagonal liquid crystalline phases. The most frequently encountered lyotropic mesophases are the lamellar (L_α) and direct hexagonal (H_I) ones. They are respectively formed of alternate water layers and surfactant double layers or of long rods ordered in an hexagonal lattice.

These mesophases are characterized by their optical properties as well as by X ray diffraction. On the other hand the NMR is the most convenient method for investigating their properties at the molecular scale since most of the nuclei of non zero magnetic moment may be observed, even in low isotopic abundance like ^{13}C and ^{17}O. Some specific applications of NMR to lyotropic liquid crystals will be presented on examples taken from studies in progress on sodium mono and dialkylphosphates. After a short description of their phase diagrams which are given elsewhere [1,2], an outline of the methods used for the determination of the ordering and dynamics of water and surfactant molecules will be reported.

R. Daudel et al. (eds.), Structure and Dynamics of Molecular Systems, 187–201.

2. PHASE DIAGRAMS

The lyotropic mesophases are unambiguously identified by X ray diffraction which gives the main lattice parameters like the thicknesses of the polar and non polar layers of the L_α phases or the spacement of the rods in the H_I phase. However the NMR of water deuterons or of the ^{31}P of the polar heads provides a simple and fast method for delineating the phase diagram. Figure 1 shows for example that the lamellar → hexagonal transition may be evidenced by a sign change and a reduction of the ^{31}P chemical shift anisotropy (CSA) resulting from differences in the the averaging of the chemical shielding tensor $\underset{\sim}{\sigma}$. While in the L_α phase this tensor is averaged by the reorientation of surfactant molecules about the director (normal to the bilayers), it is averaged in the H_I phase both by the axial rotation of the rods and by the reorientation of surfactants perpendicular to their axis.

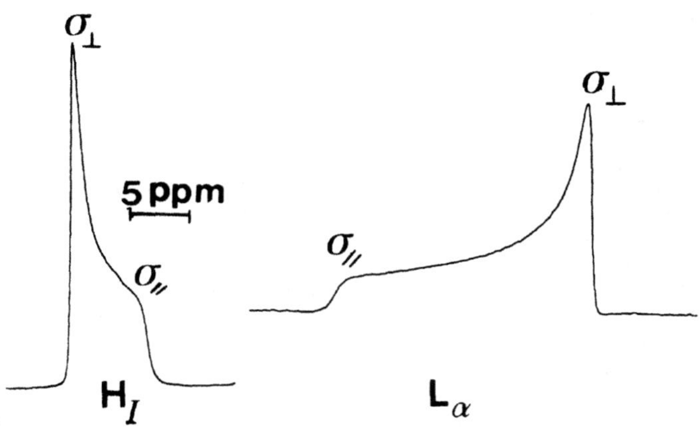

Figure 1. ^{31}P NMR spectra of randomly oriented hexagonal and lamellar phases of monobutyl phosphate/H_2O.

At room temperature and above, the dibutyl and dihexyl phosphates give only rise to the L_α phase. On the other hand, at increasing surfactant concentrations, the monobutyl and monohexyl phosphates form successively the H_I and L_α phases (figure 2). By cooling down the samples from the isotropic melt, the directors of the dialkyl phosphate L_α and the monoalkyl H_I phases become oriented perpendicular or parallel to the magnetic field B_0. This behaviour is typical of the DMII and CMI phases with negative or positive magnetic susceptibilities, formed respectively of giant disk or cylindrical rod shaped micelles [3].

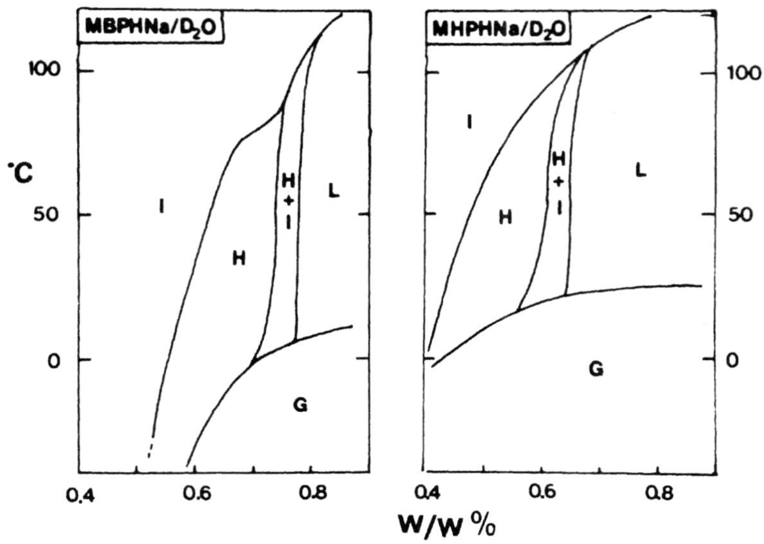

Figure 2. Phase diagrams of the sodium monobutyl and monohexyl hydrogeno-phosphate/H_2O systems. H = hexagonal, L = lamellar, I = isotropic, G = gel.

3. ORDER PARAMETERS

In uniaxial lyotropic liquid crystals, the magnetic tensors, such as the chemical shielding, quadrupolar or dipolar tensors of principal value T_{xx}, T_{yy} and T_{zz} are averaged by the overall molecular tumbling about the director. The principal values of the axially symmetric averaged tensor are:

$$T_{/\!/} = - 2T_{\perp} = 2/3 \; (T_{xx} \; S_{xx} + T_{yy} \; S_{yy} + S_{zz} \; T_{zz}) \qquad (1)$$

with $S_{ii} = 1/2 \; < 3l_{ii}^2 - 1 >$ where l_{ii} is a direction cosine of the director in the xyz frame. If two of the planes of this frame are symmetry elements of the molecules, the x, y and z axes are the principal directions of the ordering tensor:

$$S_{ij} = 1/2 <3 \; l_{ij}^2 - \delta_{ij}> \qquad (2)$$

In the case of a flexible molecule, to each conformation corresponds a different ordering tensor, the principal directions of which are those of the inertia tensor. Nevertheless, one may consider for simplicity an effective ordering tensor corresponding to a weighted average over all the molecular conformers. It is often assumed that

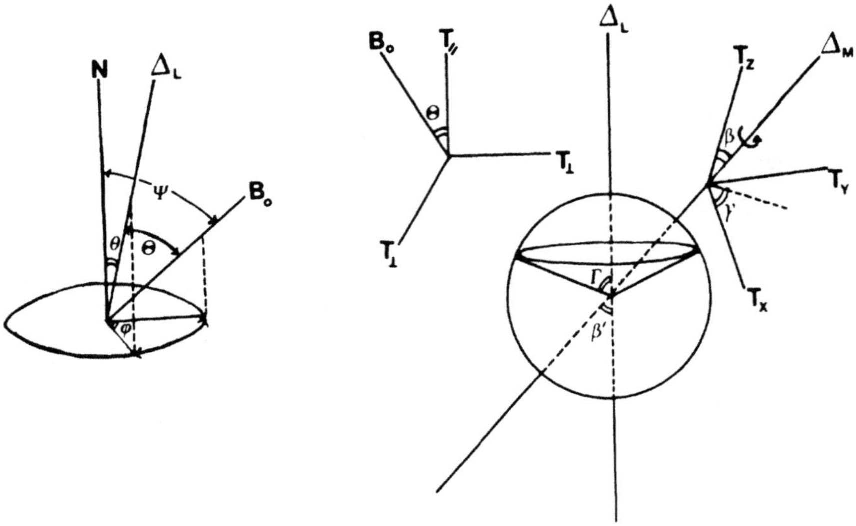

Figure 3. Definition of angles and axes. N is a reference axis of the sample e.g. the axis of NMR cylindrical tube or the normal to a rectangular cell.

an elongated molecule, or a molecule fixed to a binding site undergoes a quasi axial reorientation about a Δ_M axis which reorients itself about a local director Δ_L of a microdomain of the liquid crystal. The order parameters of equation 1 may then be written as :

$$S_{33} = 1/2\ P_2(\cos\beta')(3\ \cos^2\beta - 1)$$
$$S_{22} = 1/2\ P_2(\cos\beta')(3\ \sin^2\beta\ \sin^2\gamma - 1) \qquad (3)$$
$$S_{11} = 1/2\ P_2(\cos\beta')(3\ \cos^2\beta\ \sin^2\gamma - 1)$$

where $P_2\ (\cos\beta') = 1/2 < 3\ \cos^2\beta' - 1 >$, β' being the time dependent angle between Δ_M and Δ_L and β, γ the polar angles of Δ_M in the principal axes frame of the magnetic tensor. For an axial or quasi axial molecular reorientation, the dipolar and quadrupolar splittings or the chemical shift anisotropy are respectively given by the following expressions:

$$\Delta\nu_q = \frac{3}{4}\ \frac{(3\cos\Theta - 1)}{I(2I - 1)}\ Q_o\ \overline{P}_2(\cos\beta')\left[\frac{1}{2}\ (3\cos^2\beta - 1) + \frac{\eta}{2}\ \sin^2\beta\cos^2\gamma\right] \qquad (4)$$

$$\Delta\nu_d = \frac{3}{8}\gamma_I^2 \frac{\hbar}{\pi r^3} (3\cos^2\Theta - 1)\, \overline{P}_2(\cos\beta')\, \frac{1}{2} (3\cos^2\beta - 1)$$

$$\Delta\sigma = \frac{3}{2}\sigma_{zz}\, \overline{P}_2(\cos\beta') \left[\frac{1}{2}(3\cos^2\beta - 1) + \frac{\eta}{2}\sin^2\beta\cos^2\gamma\right] \quad (4)$$

where Θ is the angle between Δ_L and the magnetic field B_o (figure 3), Q_o the quadrupolar coupling constant, γ_I a nuclear magnetogyric ratio, r the internuclear distance, σ_{zz} a principal component of the chemical shielding tensor and η the asymmetry parameter of a magnetic tensor.

Two basic models are generally considered for the motion of the Δ_M axis:

- Random libration within a cone of half aperture Γ [4] (figure 3).
- Orientation dependence on a potential of the form $U = -\lambda \cos^2\beta'$ [5].

Thus a molecular order parameter $P_2(\cos\beta') = 0.6$, typical of the systems under study corresponds to $\Gamma = 45°$ or $U = 3$ kcal mole^{-1}. The choice of the model of motion of the Δ_M axis is of importance in the calculation of the fourth rank order parameter: $P_4(\cos\beta') = 1/8 < 35\cos^4\beta' - 30\cos^2\beta' + 3 >$ which has to be taken into account in relaxation rate calculations.

4. NUCLEAR RELAXATION

The nuclear relaxation in liquid crystals has been subjected to many theoretical works (see for instance ref. 6-8). We shall give however some expressions relevant to the present work. For a quasi axial molecular reorientation, the spectral densities are of the form:

$$J_K^{(K')} = \left[A_K^{(K')}\cos^4\Theta + B_K^{(K')}\cos^2\Theta + C_K^{(K')}\right] \frac{\tau_K}{1+(\omega_{K'}\tau_K)^2} \quad (5)$$

with K or K' = 0, ± 1, ± 2. The A, B and C coefficients depend on $P_2(\cos\beta')$ and $P_4(\cos\beta')$ [6]. τ_K is a reorientation correlation time expressed as:

$$\tau_K = b[6D_{/\!/} + K^2(D_{/\!/} - D_\perp)]^{-1} \quad (6)$$

where b is a constant dependent of the model of motion and $D_{/\!/}$, $D_\perp$ the rotational diffusion coefficients. For the model of jump diffusion often adopted in liquid crystal studies, $\tau_0 = D_\perp^{-1}$.

At very high field, the chemical shift anisotropy is the main relaxation mechanism for ^{31}P, the longitudinal rate being:

$$(T_1^{-1})_{CS} = \omega^2 \sum_{-2}^{2} F_K'^2\, J_K^{(1)}(\omega) \quad (7)$$

with $F_K' = D_{KK'}(0,\beta,\gamma)F_K$

and: $F_0 = \sqrt{\frac{2}{3}}(\sigma_{zz} - \quad (\sigma_{xx} + \sigma_{yy}))$, $F_{\pm 1} = 0$, $F_{\pm 2} = \frac{1}{2}(\sigma_{xx} - \sigma_{yy})$

$D_{KK'}$ being the second rank Wigner matrix relating the principal axes of the σ chemical shielding tensor and of the rotational diffusion tensor.

The dipolar relaxation rate of ^{31}P induced by a nearby proton depends on the orientation and length of the internuclear vector r and is expressed as:

$$(T_1^{-1})_{dip} = \frac{1}{2}(\gamma_P \gamma_H \hbar)^2 \sum_{-2}^{2} F_K^2 \left[J_K^{(o)}(\omega_P - \omega_H) + 3J_K^{(1)}(\omega_P) + 6J_K^{(2)}(\omega_P + \omega_H) \right] \quad (8)$$

with $F_0^2 = \frac{1}{4} \langle (3 \cos^2 \beta' - 1) r^{-3} \rangle^2$, $F_{\pm 1}^2 = \frac{3}{2} \langle \sin \beta \cos \beta \, r^{-3} \rangle^2$ and $F_{\pm 2}^2 = \frac{3}{4} \langle \sin^2 \beta \, r^{-3} \rangle^2$, the averages being taken over all the molecular conformations.

The relaxation of nuclei of spin I $\geqslant$ 1 is governed by the quadrupolar coupling. For I = 1 (e.g. ^2H) one has for example:

$$(T_1^{-1})_q = C \sum_{-2}^{2} F_K'^2 \, [J_K^{(1)}(\omega) + 4 \, J_K^{(2)}(2\omega)] \quad (9)$$

with C = 1/4, $F_0 = \pi \sqrt{6} \, Q_o$, $F_{\pm 1} = 0$, $F_{\pm 2} = \pi \eta \, Q_o$. A similar expression holds for the dipolar relaxation of two equivalent 1/2 spins, taking C = 1, $F_0 = \gamma^2 \, h \sqrt{\frac{3}{2}} \, r^{-3}$, $F_{\pm 1} = F_{\pm 2} = 0$. For I > 1, the expressions of relaxation rates are dependent on the $\pm I \leftrightarrow \pm I-1$ transitions (see for instance [8]).

5. APPLICATIONS TO THE ORDERING AND DYNAMICAL BEHAVIOUR OF MOLECULES IN LYOTROPIC MESOPHASES

Interstitial water

The orientation and dynamical behaviour of water in the sodium dibutylphosphate/water system has been extensively investigated by ^1H, ^2H and ^{17}O NMR [1,9]. In this system, the domain of existence of the lamellar phase corresponds to a water/surfactant ratio between 4 and 6 while the maximum hydration number of phosphate is 6. This is a very favourable case where nearly all water molecules are bound to an anionic (phosphate) or cationic (Na^+) sites. We have considered three basic models of reorientation about an axis Δ_M joining the water molecule to its binding site [1,9]:

- I: Δ_M is in the HOH bisector plane making an angle φ with the molecule plane.
- II: Δ_M is aligned with the H-H direction.
- III: Δ_M coincides with one of the O-H bonds.

The z and x principal axes of the ^{17}O quadrupolar tensor are the normal to the molecular plane and the HOH bisector respectively, with

$Q_o = Q_{zz} = 6.67$ MHz and $\eta = (Q_{xx} - Q_{yy})/Q_{zz} = 0.93$. The ^2H quadrupolar tensor is nearly axially symmetric about the OH bond with $Q_o = 212$ kHz and $\eta = 0.11$. The value of the ratio of $^{17}O/^2H$ quadrupolar splittings calculated from equation 4 are given in Table I:

Table I

Ratios of $^{17}O/^2H$ quadrupolar splittings for different models of reorientation of water

				Model I							
φ	=	0	10	20	30	40	50	60	70	80	90°
$\left\|\Delta\nu(^{17}O)/\Delta\nu(^2H)\right\|$	=	1.13	0.15	13.4	10.4	7.1	6.2	5.9	5.8	5.7	5.6
			Model II					Model III			
$\left\|\Delta\nu(^{17}O)/\Delta\nu(^2H)\right\|$	=		6.5					6.0			

Figure 4. ^2H and ^{17}O NMR spectra in the dibutylphosphate/H_2O lamellar phase. The ^{17}O spectrum is obtained from a sample oriented in the magnetic field.

These ratios are to be compared with experimental values ranging from 6 to 6.5 according to the temperature and concentration. It is seen that model I with $50 < \varphi < 70°$ as well as models II and III are consistent with the observed ratio of quadrupolar splittings. On the other hand the reorientation about the HOH bisector seems ruled out.

Although several modes of water reorientation are likely to exist simultaneously, the predominant one is selected by T_1 or linewidths measurements on ^1H, ^2H and ^{17}O. For model I with $\varphi \simeq 60°$, only a reasonable agreement is achieved between the experimental and calculated relaxation rates using equ.9 (figure 5). This model corresponds to the reorientation of water about an axis directed along an oxygen lone pair orbital and is consistent with theoretical studies [10] and neutron diffraction experiments [11] on the hydration of alkaline cations in concentrated aqueous solutions.

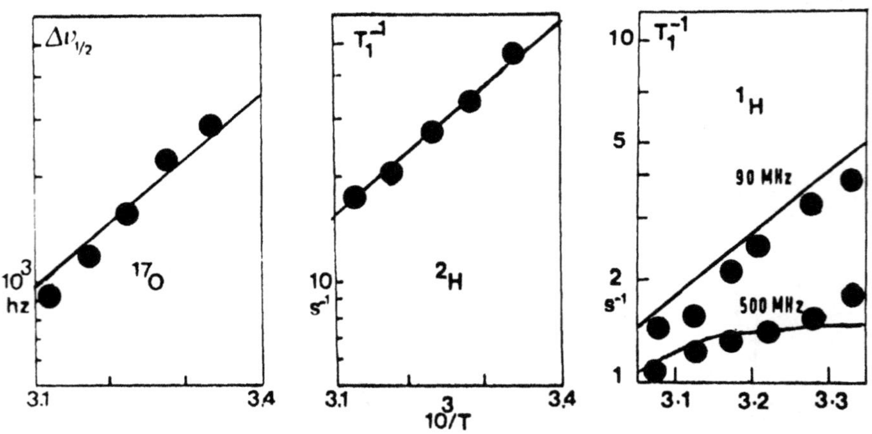

Figure 5. ^1H and ^2H longitudinal relaxation rates and ^{17}O linewidth of the $1/2 \leftrightarrow -1/2$ transition. The solid lines correspond to computed values.

Surfactant molecules

The study of the ordering and dynamical behaviour of surfactant molecules requires the statistical weight of all the molecular conformers and of the average orientation of the polar heads at the interface with water. These informations are given by the quadrupolar splittings of the alkyl chain deuterons [12]:

$$(\Delta\nu_i)_{^2H} = \frac{3}{8}(3\cos^2\Theta - 1)\, Q_o \left\{ \langle 3\ell^2_{z_iZ} - 1\rangle S_{ZZ} + (S_{XX} - S_{YY})\langle \ell^2_{z_iX} - \ell^2_{z_iY}\rangle \right\} \qquad (10)$$

where Q_o = 170 kHz z_i being the direction of a $C-^2H$ bond, and by the ^{31}P chemical shift anisotropy:

$$\Delta\sigma = \frac{3}{4} S_{ZZ}\,\sigma_{zz}\left[(3\ell^2_{zZ} - 1) + \eta(\ell^2_{xZ} - \ell^2_{yZ})\right] + \frac{3}{4}(S_{XX} - S_{YY})\,\sigma_{zz}\left[(\ell^2_{zX} - \ell^2_{zY}) + \frac{\eta}{3}(\ell^2_{xX} - \ell^2_{xY} + \ell^2_{yY} - \ell^2_{yX})\right] \qquad (11)$$

in equations 10 and 11, l_{vw} denote the direction cosines of the principal directions x,y,z of magnetic tensors in the X,Y,Z molecular frame. In the case of dibutylphosphate the principal directions of the ^{31}P chemical shielding tensor coincide with the molecular axes, the principal values being σ_{zz}=-17 ppm along the O_3PO_4 bisector, σ_{xx}=-77 ppm along the O_1O_2 direction and σ_{yy} = 110 ppm. The ^{31}P shielding tensor of the monobutyl-hydrogenophosphate (figure 6) is averaged by exchange of the phosphate proton among O_1, O_2 and O_3 with $\sigma_{/\!/}$ = - 38 ppm and $\sigma_{\perp}$ = + 19 ppm. The α angle between the cylindrical axis of this tensor and the Z molecular axis in the PO_1C_1 plane (figure 6) is an adjustable parameter.

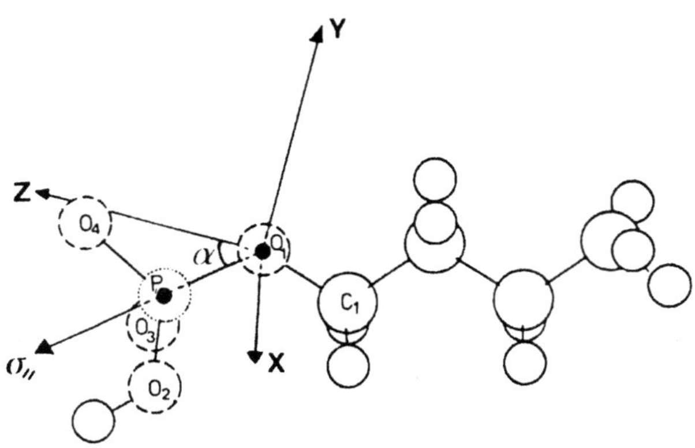

Figure 6. Molecular axes and principal directions of the ^{31}P chemical shielding tensor $\underset{\sim}{\sigma}$ in the monobutyl hydrogenophosphate (MBHP).

The averages appearing in equ. 10 are taken over all the molecular conformers occuring by trans (T) ⇄ gauche (G^+,G^-) isomerization of the chain. In the absence of sterical constraints, the G^+, G^- rotamers are equiprobable and the statistical weights of all the conformations are given by the vector:

$$W = \prod_{j=1}^{N-2} [(1-M)(U \otimes P_T) + M(U \otimes (1-P_T))] \qquad (12)$$

with $P_T = (1+2\exp(-\frac{\Delta E}{RT}))^{-1}$, P_T being the probabilities of a trans conformer about a bond j and ΔE the energy barrier between the T and G rotamers. For a chain of N atoms there are 3^{N-2} conformers with respect to a molecule fixed reference frame and U is a unit vector of dimension 3^{N-2}, M being a 3^{N-2}x(N-2) binary matrix giving the positions of the $G^{\pm}$ rotamers in the chain for all conformations. The effective values of P_T are somewhat different from those expected from the standard rotational barriers about the O-C and C-C bonds, of the order of 1 and 0.6 kcal mole^{-1}, respectively, because of intra and intermolecular constraints. The former are taken into account by surrounding the polar head by a hard sphere of 2 Å radius, preventing the folding back of the chain due to the occurrence of G^+G^- local forms. The intermolecular constraints are simulated by confining the molecule in a cylinder, a cylindrical sector or a cone, according to the curvature of the interface with water (see for instance ref. 13). These very simple models account for the disagreement between the rotamer populations observed or calculated for free molecules and those obtained by a least square adjustment of P_T by means of equations 10 and 12, as shown in table II on the example of the MBHP/H_2O system in the L_α and H_I phase.

Table II - Experimental and computed ^2H quadrupolar splittings and ^{31}P chemical shift anisotropy in monobutyl hydrogenophosphate/H_2O mesophases

		H_I phase , T = 300 K			
	O-C1		C1-C2		C2-C3
$P_T^{(a)}$	0.91		0.77		0.38
	H1	H2		H3	H4
$\lvert\Delta\nu_{^2H}\rvert$ obs.	15220	10520		7065	2965 Hz
$\lvert\Delta\nu_{^2H}\rvert$ calc.	15690	9946		7120	2591 Hz
		S_{XX} = 0.084,	S_{YY} = 0.257,	S_{ZZ} = -0.341	
$(\Delta\sigma)_{^{31}P}$		obs.: 6.2 ,	calc.: 6.2 ppm		
$P_T^{(b)}$		0.87	0.7	0.36	
$\lvert\Delta\nu_{^2H}\rvert$calc.	15730	9451		8070	1463 Hz

L_α phase , T = 340 K

	O-C1		C1-C2		C2-C3	
$P_T^{(a)}$	0.80		0.75		0.333	
	H1	H2		H3	H4	
$\lvert\Delta\nu_{^2H}\rvert$ obs.	27954	19816		12329	4842	Hz
$\lvert\Delta\nu_{^2H}\rvert$ calc.	28870	19100		10740	5589	Hz
	S_{XX} = -0.266,		S_{YY} = -0.297,		S_{ZZ} = 0.563	
$(\Delta\sigma)_{^{31}P}$	obs.: -18.6,		calc.: -15.1 ppm			
$P_T^{(c)}$		0.86	0.6	0.333		
$\lvert\Delta\nu_{^2H}\rvert$ calc.	30630	15200		10340	6130	Hz

(a) Least square adjustment of ^2H splittings, (b) confinement in a cylindrical sector of 47 $Å^2$ square basis and 7 Å radius, (c) confinement in a cylinder of 3.5 Å radius. All calculations are done with $\alpha = 35°$ (figure 6). For (b) and (c) the rotamer populations are initially taken as 0.8, 0.6, 0.6 like for the free molecule.

The conformation of the phosphate residues of the dibutylphosphate in the L_α phase differs strongly from that of the MBHP. The analysis of the deuterons quadrupolar splittings of the chains (Table III) shows that the total probability of the two equivalent molecular conformers $TG^+TG^-G^-TG^+T$ and $TG^-TG^+G^+TG^-T$ represents more than 80% of all the conformations. Because of the molecular shape (figure 7), these

Table III - Experimental and computed ^2H quadrupolar splittings and ^{31}P chemical shift anisotropy in the dibutylphosphate/H_2O lamellar phase (T = 300 K)

	O-C1		C1-C2		C2-C3	
P_T	1		0		0.86	
	H1	H2		H3	H4	
$\lvert\Delta\nu_{^2H}\rvert$ obs.	12139	20651		17626	7534	Hz
$\lvert\Delta\nu_{^2H}\rvert$ calc.	12380	20580		17230	8182	Hz
	S_{XX} = -0.372,		S_{YY} = -0.358,		S_{ZZ} = 0.73	
$(\Delta\sigma)_{^{31}P}$	obs.: -20 ,		calc.: -23.5 ppm			

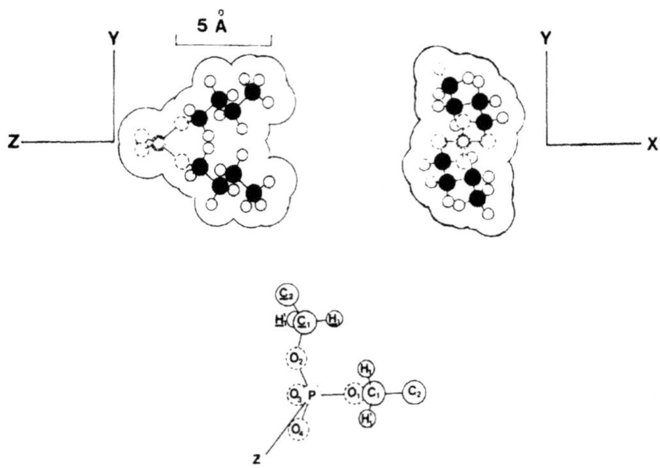

Figure 7. One of the equivalent main conformers of the dibutylphosphate molecule in the L_α phase. The effective molecular size is obtained from the Van der Waals radii of peripheral atoms.

conformers are nearly confined in a cylinder of 60 $\mathring{A}^2$ elliptic section which corresponds fairly well to the area per polar head obtained from X ray diffraction. Possibly because of the fast exchange between the two main conformers, the ordering tensor is nearly axial about the Z direction and the use of equations 7 and 8 in relaxation studies is justified.

The dynamical behaviour of DBP molecules in the L_α phase has been investigated by ^{31}P longitudinal relaxation at 36.5 and 202.6 MHz where the dipolar and chemical shift anisotropy mechanism predominate respectively. From inversion recovery experiments on perdeuterated samples at very high field, it appears that the CSA relaxation is strongly anisotropic (figure 8). To each point of the spectrum of a randomly oriented sample corresponds a value of the Θ angle between Δ_L and B_o:

$$\nu(\Theta) = \frac{1}{3}\,(3\cos^2\Theta - 1)(\sigma_{/\!/} - \sigma_\perp) \qquad (13)$$

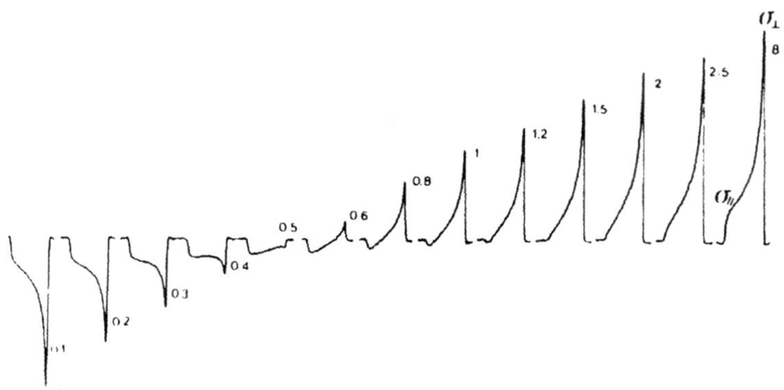

Figure 8. Inversion-recovery experiment on ^{31}P at 202.7 MHz on a dibutylphosphate/H_2O perdeuterated sample (T = 300 K).

The angular dependence of the $(T_1^{-1})_{CS}$ is given by equations 5 and 7 with:

$$A_0 = \frac{3}{2}(\overline{P}_2^2 - \overline{P}_4), \quad B_0 = \frac{1}{14}(\overline{P}_2 + 6\overline{P}_4 - 7\overline{P}_2^2),$$

$$C_0 = \frac{1}{70}(14 + 15\overline{P}_2 - 9\overline{P}_4) \tag{14}$$

$$A_2 = -\overline{P}_4/28, \quad B_2 = \frac{1}{70}(15\,\overline{P}_2 - \overline{P}_4),$$

$$C_2 = \frac{1}{140}(28 + 30\overline{P}_2 + 5\overline{P}_4)$$

where $\overline{P}_2$ and $\overline{P}_4$ denote the order parameters $P_2(\cos\beta')$ and $P_4(\cos\beta')$ defined above. The terms corresponding to K = ± 1 vanish because the Z axis of the $\underset{\sim}{\sigma}$ tensor coincide with the Δ_M molecular reorientation axis.

Z axis ($\beta = 0$) at an effective distance optimized to 3.1 Å. The adjustable parameters of relaxation rates are the correlation time τ_0 of reorientation of the Z (or Δ_M) axis (figure 9), and the anisotropy parameter $\rho = D_{/\!/}/D_{\perp}$ of the motion. Under most of our experimental conditions, ρ was found between 12 and 17. Such a large anisotropy is quite unexpected from the molecular dimensions of the most probable conformer: $d_z = 8$ Å, $d_y = 7.5$ Å, $d_x = 5$ Å. A similar observation made on the tempoamine spin probe dissolved in the same phase [14] may be significant of an anisotropic viscosity of the bilayers or possibly to collective modes of reorientation of the Δ_M axes (see for instance ref. 15). The latter assumption is supported by subsidiary ESR experiments on the VO^{2+} ion bound to 2-4 phosphate groups and oriented perpendicular to the bilayer. The ratio ot the total width of the VO^{2+} spectrum to its rigid limit yields indeed τ_0 values comparable to those given by ^{31}P longitudinal relaxation [14,16].

REFERENCES

[1] C. CHACHATY and J.P. QUAEGEBEUR, J. Phys. Chem., **87** (1983) 4341

[2] B. PERLY, J.P. QUAEGEBEUR, C. CHACHATY and B. GALLOT, Mol. Cryst. Liqu. Cryst., (in press)

[3] B.J. FORREST and L.W. REEVES, Chem. Rev. **81** (1984) 1

[4] O.H. GRIFFITH and P.C. JOST in 'Spin Labelling, Theory and Applications', Vol.I, chap. 12, edited by L.J. BERLINER, Academic Press, New York 1976

[5] H. SCHINDLER and J. SEELIG, J. Chem. Phys., **59** (1973) 1841

[6] J.H. FREED, J. Chem. Phys., **66** (1977) 4183

[7] M.F. BROWN, J. Chem. Phys., **77** (1982) 1576

[8] B. HALLE and H. WENNERSTROM,J.Chem.Phys., **75** (1981) 1928

[9] C. CHACHATY and J.P. QUAEGEBEUR, Molec. Phys., **55** (1976) 463

[10] K. HEINZINGER and P.C. VOGEL, Z. Natur., **A 31** (1976) 463

[11] J.E. ENDERBY, A. Rev. Phys. Chem., **34** (1983) 155

[12] N. BODEN, L.D. CLARK, R.J. BUSBY, J.W. EMSLEY, G.R. LUCKHURST and C.P. STOCKLEY, Molec. Phys., **42** (1981) 562

[13] J.N. ISRAELACHVILI, S. MARCELJA and G.N. HORN, Quart. Rev. Biophys. **13** (1980) 122

[14] C. CHACHATY, Can. J. Chem., **62** (1984) 1246

[15] J.S. HWANG, K.V.S. RAO and J.H. FREED, J. Phys. Chem., **80** (1976) 1490

[16] J.P. QUAEGEBEUR, B. PERLY and C. CHACHATY, 5th Intern. Symp. on Surfactants in Solution, Bordeaux (France), 9-13th July 1984 (to be published).

NMR STUDIES OF THE GEL PHASE OF LECITHINS AND CEPHALINS

Lutz Trahms
Physics Department
Freie Universität Berlin
Arnimallee 14
D-1000 Berlin 33
Fed. Rep. Germany

ABSTRACT. The gel state of fully hydrated lecithin and cephalin bilayers below the main transition temperature is subdivided into a number of distinct thermotropic phases. Some features of the crystalline structure of these states are discussed. This information is complemented by a large number of NMR studies of various authors. Measurements on different nuclei (^1H, ^2H, ^{13}C, and ^{31}P) reflect the effect of thermal fluctuations as a reduction of the anisotropy of the respective interaction tensor. In case of axial symmetric motions this averaging can be treated in a compact way in terms of the order parameter. With this concept, the interpretation of the NMR results provides a lot of additional knowledge about structure and dynamics at various molecular sites. Two different models of the mobility in the different gel states are compared with the experimental evidence.

1. INTRODUCTION

According to the fluid mosaic model of Singer and Nicolson /1/, biological membranes consist of membrane proteins embedded in a bilayer of lipid molecules. Depending on nature and function of the proteins, these macromolecules penetrate the bilayer in part or entirely. The biological function of this system is determined not only by the structure of the proteins, but also by the state of the lipid bilayer environment. As a matter of course the study of the lipid part of a membrane is essential for the understanding of the whole complex system.

In a first approach, a simplified model of a biological membrane can be investigated: a bilayer consisting solely of one kind of lipids. In consequent steps this model system can gradually be approximated to the natural biological membrane. This is done by mixing different sorts of lipids, incorporation of particular proteins, etc. It is well known, however, that upon addition of water even a pure lipid compound can spontaneously aggregate in bilayers. This phenomenon of self-organization is understood as a result of both, the amphiphilic character and the geometric structure of a single lipid molecule, /2/.

R. Daudel et al. (eds.), Structure and Dynamics of Molecular Systems, 203–224.

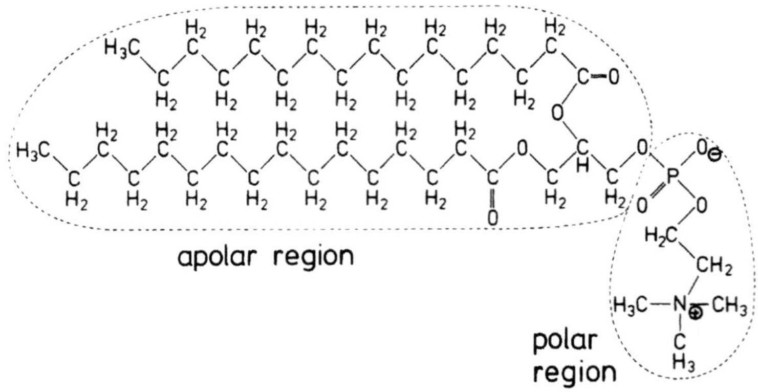

Figure 1. Chemical structure of a phospholipid molecule (DPPC).

Consider e.g. a phospholipid molecule such as 1,2-dipalmitoyl-sn-glycero-3-phosphocholine (DPPC) (fig. 1). Roughly, it can be divided into two parts, the hydrocarbon chain region and the phosphocholine headgroup. This latter moiety carries a permanent electric dipole and favours the contact with a polar solvent such as water. The two fatty acyl chains, in contrast, are nonpolar, and consequently hydrophobic. So, this amphiphilic nature of the lipid molecule requires a specific arrangement in an aqueous surrounding. If sufficient water is present, it is essentially the geometry of the molecule that determines the particular structure of this aggregation. The cross-sectional area of the headgroup of DPPC is only slightly larger than that of the hydrocarbon chains. In this case the favoured structure is the bilayer. Under other circumstances, various arrangements can be found, such as e.g. micelles, hexagonal phases, and their inverted counterparts /2,3/. The scope of this article will be confined to bilayer arrangements of phospholipids, i.e. in particular two classes, characterized by two different types of headgroups, phosphatidylcholine and phosphatidylethanolamine. For brevity their trivial names, lecithin and cephalin, will be used throughout this article. Tab. 1 provides a summary of nomenclature and chemical structure of these phospholipids.

Since the early experiments of Chapman and coworkers /4-6/, nuclear magnetic resonance (NMR) has played an important role in the elucidation of structure and dynamics of lipid bilayers. In the first period it was particularly the rather fluid like liquid crystalline state, which was subjected to various investigations. A number of review articles can be consulted as a guide through this voluminous field /7-17/. During the same time, NMR technique has improved in many respects. This progress comprises the growing skill of chemists in synthesizing compounds, labelled with particular isotopes, such as ^{2}H or ^{13}C, at certain molecular positions of interest, as well as many innovations in the design of NMR spectrometers. For instance the introduction of

TABLE I

Nomenclature and abbreviations of some cephalins and lecithins

headgroup \ acyl chains	$[CH_3-(CH_2)_{12}-COO-]_2$ dimyristoyl	$[CH_3-(CH_2)_{14}-COO-]_2$ dipalmitoyl
$- PO_4 - (CH_2)_2 - NH_3$ phosphatidyl-ethanolamine	DMPE	DPPE
$- PO_4 - (CH_2)_2 - N(CH_3)_3$ phosphatidyl-choline	DMPC	DPPC

superconducting magnets with static fields of 6 to 10 Tesla meant a drastic increase in resolution and sensitivity, compared to the earlier iron magnet devices of ~ 2 Tesla. However, this advance alone could improve the spectral resolution only for the observation of liquids, where intrinsic line broadening is averaged out by fast molecular motion. To eliminate line broadening interactions also in solids, special high power pulse methods were designed, such as the WaHuHa-sequence /18/ or the PENIS technique /19/. For a detailed discussion of these solid state NMR methods the reader is referred to the monographs of Mehring /20/ and Haeberlen /21/. A shorter summary with special emphasis on the application to lipid bilayers was given in a review article of Griffin /14/.

Prerequisite for the application of these techniques in a NMR experiment was the designing of efficient radiofrequency transmitters and probeheads, a development which is still in progress. Representing an intermediate state between fluid and solid, lipid bilayers are a comparatively less problematic object of solid state NMR investigations. During recent years the employment of this technique has allowed some interesting new insights into molecular structure and dynamics of these systems also in the rather solid like gel phases. It is the task of this article to discuss these results, including some investigations made in our laboratory.

2. THERMOTROPIC PHASES OF PHOSPHOLIPID BILAYERS

The thermotropic phase behaviour of bilayers of pure lipids is characterized by a first order phase transition. The temperature T_c of this so-called main transition depends on the chemical structure of both, fatty acyl chains and the headgroup. The phospholipids considered here consist of two saturated equal hydrocarbon chains of 12 to 14 methylene groups and a phosphatidylcholine or a phosphatidylethanolamine headgroup (Tab. I). Their main transition temperatures lie between 23°C and 63°C /22/.

Above T_c the bilayer is a two-dimensional fluid with lateral self-diffusion constants around 10^{-8} cm^2/s /23/. Also the intramolecular mobility of the hydrocarbon chains is increased considerably in this phase. This mobility occurs in form of rapid trans-gauche-isomerizations of the C-C bonds in the hydrocarbon chains. Raman and infrared studies yielded a large number of gauche conformers per chain /24/. ^{2}H-NMR studies have demonstrated, however, that angular fluctuations of the chain segments are concentrated at the end of the chains /25,26/. This finding implies that single gauche bonds, which lead to a bend of the chain ("jog"), occur only within the terminal third of the acyl chains. In the upper part, the trans-gauche-fluctuations are assumed to appear in pairs of opposite rotatory sense, separated by a single trans-bonds (g^-tg^+ = "kink") /27/. This configuration leaves the major part of the chain in an unaltered orientation, only the segments involved in the kink formation are disturbed with respect to the all-trans-configuration. In all segments of the chain the temporally averaged orientation is perpendicular to the bilayer plane.

In the nomenclature of Tardieu and coworkers /28/ the fluid phase above T_c is named Lα. The capital letter L stands for the lamellarity of this state, and the greek letter α shall describe both, the fluidity of the individual molecules and the intramolecular mobility of the fatty acyl chains. Dynamics and structure of the Lα-phase is similar in all pure phospholipids as well as in their mixtures, and it is certainly this phase, which is of major biological relevance. In contrast, the gel states below T_c of lecithins and cephalins exhibit some distinctions, raising a number of physically interesting problems. To avoid confusion it should be indicated, that the term "gel" will be used as a general name for all bilayer phases of these compounds below the main transition throughout this article.

A common feature of the gel state is the restricted inter- and intra-molecular mobility. This is evident from the existence of a two-dimensional lattice of the hydrocarbon chains, confirmed by many x-ray studies /29-34/. Note the pecularity of this arrangement that two lattice sites are occupied by one lipid molecule. The reduction of the intramolecular motion in the chain region is demonstrated by a drop in the average number of gauche conformers to $\lesssim 1$ per chain /35,36/.

In cephalins the chains remain perpendicular to the bilayer plane below T_c, and the chain lattice is hexagonal. In terms of the conventional nomenclature this phase is named Lβ, where β stands for the existence of a two-dimensional crystal and the all-trans-configuration of the chains. The corresponding phase in lecithins is termed Lβ'. The prime indicates a tilt angle of the acyl chains with respect to the bilayer normal. For DPPC this angle was observed directly to vary between 10° and 15° in the temperature range from 0°C to 42°C (= T_c) /34/. For steric reasons it is evident that the direction of the chain tilt must be uniform within certain domains. This preference of a particular axis in the bilayer plane corresponds to a break of symmetry in the two-dimensional lattice of the chains. X-ray studies have shown that this lattice is not hexagonal but orthorhombic, even though the deviation from hexagonality is only slight. This orthorhombic distortion can be visualized in a cross-section perpendicular to the long axis of the hydrocarbon chains (fig. 2a). In a perfect "pseudo"-hexagonal lattice[1] the ratio of the two unit vectors $a/b = \sqrt{3} = 1.73$. The values found in the Lß'-phase of DPPC range from 1.62 at 0°C to 1.69 at 34°C /33/.

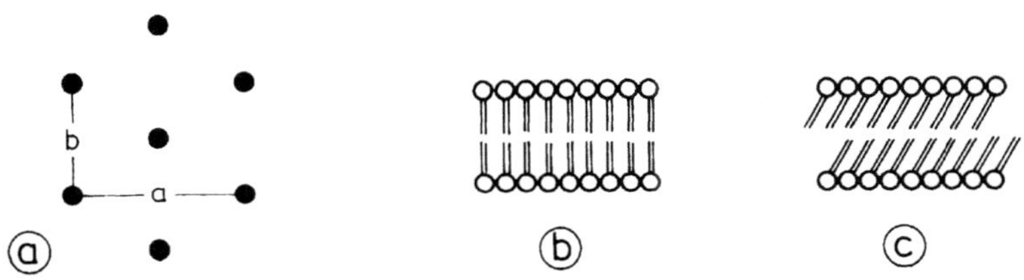

Figure 2. Bilayer organization in the gel phase. a. Cross-section perpendicular to the chain axes. b,c. Cross-section through the bilayer plane in case of the Lß- (b) and the Lß'- (c) phase.

The fact that two adjacent chains are linked in a lecithin molecule leaves several possibilities for the distribution of these pairs over the orthorhombic chain lattice. Three models are sketched in fig. 3a-c: 3a shows a random distribution, 3b indicates a herringbone structure, and 3c represents a linear arrangement. The latter structure implies that the distance between two chains of the same molecule is slightly larger than between neighboring chains of different molecules. In case of the herringbone structure this relation is reversed. Note also that in contrast to the first model, both, 3b and 3c, exhibit a symmetry axis in the bilayer plane. Even though no direct experimental evidence supporting or disproving any of these or other models has been reported to date, there is some plausibility in favour of the latter two: the existence of two more symmetry axes within the bilayer plane, i.e. the projection of (i) the chain tilt, and (ii) the symmetry axis of the orthorhombic lattice. In fact there is evidence, that these two symmetry axes are correlated /34/. A coincidence of the three symmetry axes would be the most satisfying model.

The Lß'-phase of lecithins does not transform directly into the Lα-phase. Some 8 K below the main transition temperature another endo-

[1] The adjunct "pseudo" is used to indicate that, for two reasons, this representation does not reflect the entire crystalline symmetry: 1. The cross-sectional plane is tilted with respect to the bilayer plane. Consequently, there is no translational invariance in this plane. 2. One molecule contains two chains. This molecular connection is omitted in the chosen representation.

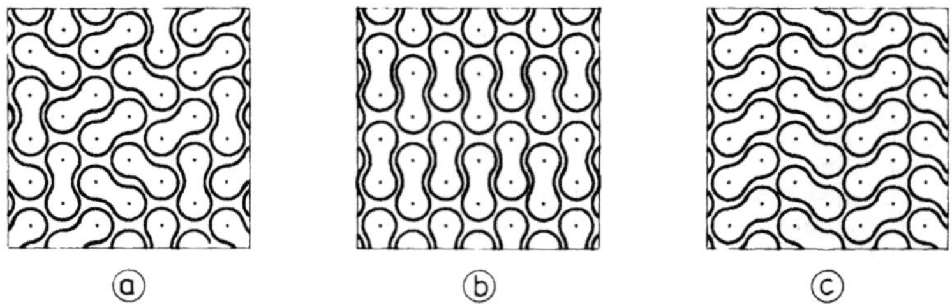

Figure 3. Distribution of lipid molecules over an orthorhombic chain lattice. a. Random distribution. b. Linear arrangement. c. Herringbone structure.

thermic phase transition was discovered by differential scanning calorimetry (DSC) /37/. It is the intermediate state between this so-called pretransition and the main transition, where the acyl chains arrange in a pseudo-hexagonal lattice /31-34/. Interestingly, this gain of symmetry in the microcrystalline structure is compensated by a loss of symmetry in the global structure of the bilayer. Electron microscopy revealed the existence of periodic undulations of the lamellae ("ripples"), that extend in one direction of the plane /38-42/. These findings could be confirmed by x-ray studies, which determined the wavelength and the amplitude of the ripples of ~150 Å and 25 Å, respectively /43/. In the basic work of Tardieu et al. /28/ this intermediate phase was referred to as Pß', where P stands for the periodicity of the undulation in the lamellar plane. It should be noted, however, that the idea of tilted hydrocarbon chains in the intermediate phase is not generally accepted. In an early x-ray investigation Rand et al. /29/ observed an increase in the bilayer spacing, when passing from Lß' to Pß', and concluded that the chains turn perpendicular to the bilayer plane above the pretransition. However, this finding can be explained alternatively by an increase in the amount of hydrated interlamellar water. More recently, Stamatoff et al. /43/ suggested a structure, in which the chains are tilted with respect to the normal of the locally

Figure 4. Different models of the intermediate state of lecithins. a. Pß. b. Pß'.

curved bilayer, but perpendicular to the average orientation of the lamellar plane (see fig. 4a). This is in contradiction to the direct observation of a tilt angle of 10° in Pß', measured in oriented samples /32-34/. It must not be taken as a decision in this controversy, if the notation Pß' for the intermediate phase is maintained for the rest of this article.

Another interesting aspect of the intermediate structure is the possible existence of bilayer regions along the ripples with increased molecular mobility. This idea has become new support by the finding of a pronounced anisotropy of the lateral self-diffusion rate within the lamellar plane /44/. With the method of "fast recovery after photobleaching" (FRAP) /45/, Schneider et al. /46/ could demonstrate with extremely low concentrations of the fluorescent label that the diffusion constant in the Pß' state differs between $D \approx 10^{-10}$ cm^2/s along the ripples and $D < 10^{-17}$ cm^2/s perpendicular to this direction.

Since its discovery, the "ripple" phase has challenged many workers to develop theories about the origin of this fascinating structure. As an adequate treatment of these studies is outside the scope of this article, the interested reader is referred to the original literature /47-55/. To an even larger extent experimentalists have applied a variety of methods to elucidate structural and dynamic details of the intermediate Pß' phase. Besides the already mentioned techniques of DSC /56-58/, electron microscopy, x-ray diffraction, and FRAP, other powerful methods have been applied such as neutron diffraction /59/, infrared and Raman spectroscopy /60-63/, and electron magnetic resonance /41,64,65/. A detailed discussion of these works must be omitted for the sake of an adequate treatment of the studies done by nuclear magnetic resonance (NMR).

Finally, the low temperature phase of DPPC is left to be discussed. Again, the existence of a fourth phase of this compound was first discovered by DSC /66/. A remarkable feature of this state is its temporal and thermotropic behaviour. This phase was found to grow slowly only within a narrow temperature range around 0°C /67/. The so-called subtransition into the Lß'-phase, in turn, takes place at 12°C with faster kinetics /68/. Infrared spectroscopy revealed a considerably restricted rotational mobility of the acyl chains below this transition /67/. To indicate this difference of the chain structure in an extension of the conventional nomenclature /28/, the notation Lσ' was introduced for the low temperature phase /68/. The knowledge about the Lσ' state is complemented by the finding of an increased packing density /69/. In crystallographic terms, the ratio a/b (fig. 2a) is substantially decreased. So the slow transition from Lß' to Lσ' can be visualized as a rearrangement from a slight distortion of a hexagonal acyl chain lattice (a/b = 1.61) towards an orthorhombic paraffin like lattice (a/b = 1.46) /70-72/. Again, details about interdependence of the three symmetry axes of the lattice (i.e. chain tilt, orthorhombic lattice, and molecular arrangement), are not known to date. Actually, not even the existence of a chain tilt has been proven directly, although the dense molecular arrangement is hard to imagine without it. So far, a Lσ'-phase is not known to occur in compounds different from

DPPC. However, in the related cephalins similar phases of reduced molecular mobility were found recently /73-75/. It is perhaps for this reason, that the interest in this low temperature state is increasing.

3. METHODICAL BACKGROUND

The Hamiltonian of a nuclear spin system can be summarized in the relation

$$H = H_0 + H_1 \quad . \tag{1}$$

$\mathbf{H}_0$ stands for the Zeeman interaction, which leads to a splitting of the nuclear states into two or more energy levels, if the sample is brought into a homogeneous static magnetic field. Radiofrequency induced transitions between these states are the observables of NMR investigations. The additional operator $\mathbf{H}_1$ shall represent the internal interactions of the nuclear magnetic moments with the molecular surrounding, which can cause shifts of the resonance frequency. In a usual laboratory magnetic field of B ≥ 2 Tesla these couplings can be considered as small perturbations of the Zeeman interaction. In particular, three different types of internal couplings will be of major importance in the context of this article: dipolar and quadrupolar interactions, and the chemical shift. Albeit distinct in magnitude and origin these interactions have in common, that generally they are not isotropic, but depend on the orientation of the molecule with respect to the external magnetic field $\vec{B}$. This can be expressed in terms of a symmetric 3-dimensional 2nd rank tensor with three principal components in its principal axis system (PAS). The angular dependence is given by the general relation

$$\mathbf{H}_1 = C^\lambda \cdot \frac{1}{2} \left(3\cos^2\theta - 1 - \eta^\lambda \sin^2\theta \cos 2\phi\right) \quad , \tag{2}$$

where θ and ϕ are polar and azimuthal angle of $\vec{B}$ in the PAS of the particular tensor, and C^λ and η^λ have particular meanings for the respective interactions.

3.1 Dipolar Interaction

This represents the perturbation of the Zeeman-levels due to the additional magnetic fields caused by the magnetic moments of the surrounding nuclei. For a pair of like nuclei

$$C^D = \frac{\gamma^2\hbar^2}{r_{12}^3} \left[I_1 I_2 - 3 I_{1z} I_{2z}\right] \tag{3}$$

$$\text{and} \qquad \eta^D = 0 \quad ,$$

where γ is the gyromagnetic ratio of the nuclei, and the I's represent the nuclear spin operators. Here, θ is taken with respect to the unique axis of this interaction, i.e. the direction of the internuclear vector $\vec{r}_{12}$. In contrast to the following two interactions, the dipolar coupling is a multi-spin interaction, where the contribution of all surrounding nuclei must be taken into account.

3.2 Quadrupolar Interaction

Nuclei with spin quantum numbers $I \geq 1$ carry an electric quadrupole moment. The presence of an electric field gradient (EFG) changes the energy levels of such a nucleus. In case of an $I = 1$ spin, which is oriented in an external magnetic field, the $I_z = \pm 1$ levels are shifted equally, and the $I_z = 0$ level is shifted twice as much into the opposite direction. The result is a splitting of the magnetic resonance frequency into two lines. Their shift with respect to the unperturbed Hamiltonian depends on the orientation of $\vec{B}$ in the PAS of the EFG tensor as described by eq. (2) with

$$c^Q = \frac{3}{4}\, eQ\, V_{33}$$
$$\eta^Q = \frac{V_{11} - V_{22}}{V_{11} + V_{22}} \quad . \tag{4}$$

Here eQ is the electric quadrupole moment, and $|V_{33}| \geq |V_{22}| \geq |V_{11}|$ are the principal values of the EFG tensor[1]. Roughly, quadrupolar interaction is at least an order of magnitude stronger than dipolar interaction.

3.3 Chemical Shift

Due to magnetic induction in the electronic surrounding, nuclei of a chemical compound experience a magnetic field, which is different from the external value. This shielding can be separated into an isotropic part σ_{iso}, which is the arithmetic mean of the three principal components σ_{11}, σ_{22}, and σ_{33}, and an angular dependent part. With the convention $|\sigma_{33}-\sigma_{iso}| \geq |\sigma_{11}-\sigma_{iso}| \geq |\sigma_{22}-\sigma_{iso}|$ this angular dependence is given by eq. (2) with

$$c^{CS} = \frac{2}{3}\, \Delta\sigma\, \gamma\hbar\, B_o = \frac{2}{3}[\sigma_{33} - \frac{1}{2}\, (\sigma_{11} + \sigma_{22})]\gamma\, \hbar\, B_o$$
$$\eta^{CS} = \frac{\sigma_{11} - \sigma_{22}}{\sigma_{33} - \sigma_{iso}} \quad , \tag{5}$$

---

[1] This definition of η^Q has the opposite sign of the conventional definition. This change was done for the sake of the general expression (2).

where $\Delta\sigma^{CS}$ and η^{CS} are called anisotropy and asymmetry of the chemical shift, respectively. The magnitude of the chemical shift is proportional to the applied static field, so it cannot be compared to the other two interactions without more ado. In the experiments considered in this article it will be less or equal to the dipolar interaction.

3.4 Motional Averaging

Eqs. (2-5) apply to immobile nuclei. To consider the influence of motions on the NMR spectra, one has to pay attention to two parameters: the temporal rate and the angular distribution of the molecular fluctuations. The first parameter must be compared to the magnitude of the observed interaction, i.e. the frequency shift. This implies that the result of motional averaging can be different on e.g. a quadrupolar splitting and a chemical shift anisotropy. If the correlation time of the motion is much longer than the timescale of the measured coupling, no effect on the rigid lattice spectrum will be observed. However, if the two timescales are of similar order of magnitude, the spectra change significantly. In this case the motions will reduce part of the anisotropy of the interactions and the particular effects will depend very sensitively on the ratio of the two timescales. In general, complicated computer calculations must be carried out to simulate these "slow motional" spectra. A detailed outline of these procedures was given in the monograph of Spieß /76/, and some particular applications on lipid bilayers and related systems were given in the recent literature /77,78/. The following treatment is confined to the other extreme, i.e. the fast motional limit with correlation times much shorter than the time domain of the interaction. In this case motional averaging will be complete, and its effect on the spectra will depend entirely on the other parameter, i.e. the angular distribution of the motion.

An isotropic reorientation, which occurs in liquids, leads to a vanishing of the anisotropic parts of the static interactions, and the result are narrow NMR lines. In mesophases of liquid crystals the motion is anisotropic. If confined to axial symmetric motions, the angular distribution of these fluctuations can be expressed in terms of the order parameter /79,80/

$$S = \frac{1}{2} \langle 3\cos^2\theta - 1 \rangle \qquad (6)$$

where θ is the angle between the instantaneous orientation of a molecule fixed axis and the symmetry axis of the fluctuation. The brackets indicate temporal averaging. This quantity has become a convenient measure of the average amplitude of angular fluctuations in liquid crystals and related systems. In absence of motion $S = 1$, and with complete isotropy of motion $S = 0$: so the value of S between 0 and 1 reflects somehow the "order" of the fluctuations. At this point it must be emphasized, that this is a very limited view of the problem. In mathematical terms the order parameter is the second Legendre coefficient of the angular distribution $f(\cos\theta)$. There is a multitude of functions

$f(\cos\theta)$ that result in any given value of S between -1/2 and 1. Consider e.g. a rotation about a fixed axis: Now $f(\cos\theta)$ is approximated by a delta function, and for $0° < \theta < 90°$ any value of S between -1/2 and 1 is possible. It is obvious that it does not make sense to interpret the order parameter as a measure of the "order" of this type of motion. To sum up, this crude interpretation of S is applicable only to a limited type of motions, i.e. solely to some kind of a "wobbling" about an average orientation.

The particular importance of the order parameter to the NMR experiment lies in its simple relation to the reduction of the anisotropy of the static interactions by motional averaging

$$\langle C^{\lambda}\rangle = C^{\lambda} \cdot S_{33}\ (1 - \frac{1}{3}\ \eta^{\lambda}\ \Delta S) \qquad (7)$$

with $\Delta S = (S_{22}-S_{11})/(S_{22}+S_{11})$. Here S_{ii} represent the three individual order parameters of the principal axes of the interaction tensor. Eq. (7) particularly applies for rotational diffusion about a fixed axis. It is a matter of straightforward calculations to derive the well known formula of Mehring et al. /81/ for the reduction of C^{λ}

$$\langle C^{\lambda}\rangle = C^{\lambda} \cdot \frac{1}{2}\ (3\cos^2\theta - 1 - \eta^{\lambda}\ \sin^2\theta\ \cos 2\phi) \qquad (8)$$

where θ and ϕ are polar and azimuthal angle of the rotation axis in the PAS of the interaction tensor. The latter two equations demonstrate, that the reduction in C^{λ} is a tool to monitor the second Legendre coefficient of the angular distribution of the motion. However, it does not give any information about the other coefficients of this expansion, or about the distribution function $f(\cos\theta)$ in general. In experimental terms this means, that e.g. the observation of a single narrow line can be due to a completely disordered motion, a rotation of the unique interaction axis about the "magic angle" θ_m (where $\cos^2\theta_m = 1/3$), or some other more complicated type of motion with a vanishing second Legendre coefficient. This should be kept in mind during the discussion of the NMR investigations of liquid crystals or lipid bilayers.

4. NMR STUDIES

4.1 ^{1}H-NMR

Experimentally, the abundant ^{1}H nuclei of lipids are not difficult to detect in NMR spectroscopy. This is due mainly to the high natural abundance of the ^{1}H-isotope and the high gyromagnetic ratio of its magnetic moment. As a matter of course, the earliest NMR investigations of lipid bilayers and related systems were carried out on these nuclei /4-6/. However, the experimental advantage of this method is more than compensated by the difficulty to derive quantitative information from generally broad, featureless proton spectra. Protons are spin-1/2 nuclei without a quadrupole moment. With their strong magnetic moments and their dense packing, particularly in the acyl chains of lipids, the dipolar interaction between these nuclei is the dominating source of broadening. Only at high fields of ≥ 4 Tesla the influence of the chemical shift becomes visible /82/. In an attempt to overcome the broadening effects, small vesicles were used, which tumble isotropically and fast enough to average out both, dipolar broadening and chemical shift anisotropy /83/. So the obtained spectra are fairly well resolved, but, unfortunately, these systems were found to be poor approximations of a model membrane. As a result of their high curvature, small vesicles exhibit an unusual phase behaviour, and represent an inadequate object for the study of the molecular structure in planar bilayers.

In large vesicles or multilamellar dispersions of lipids the dipolar interactions between many nuclei superimpose, so that an analysis of the motional state in terms of eq. (8) becomes hardly possible. Instead, it is appropriate to apply the method of moments /84/, to compare the experimental lineshape with suggested models of molecular motion. Within this framework it was pointed out by Wennerström /85/ and Bloom et al. /86/, that the anisotropy of the chain motion in the Lα-phase is clearly reflected in the shape of the ^{1}H-NMR line. For the gel phase of DPPC, the first NMR-observation of structural changes at the Lβ'-Pβ' pretransition was detected with a special ^{1}H-NMR pulse technique by a drop in the intermethylene part of the second moment /87,88/. The same method was applied by Volke et al. /89/, to determine the molecular motions as a function of the hydration state. Partially deuterated DPPC was used by MacKay /90/ to monitor the chains' second moment over a wide temperature range of the gel phase. Referring to model calculations of Jeffrey et al. /91/, a moment analysis was undertaken to compare several motional models. The low temperature range including the Lσ'-phase of DPPC was covered by a more recent ^{1}H-NMR study of our laboratory /68/. It turned out that the second moment of the acyl chains increases considerably in this state, indicating a substantial freeze of rotatory motion. This study was performed on fully protonated compounds with an analytical separation of the chain and headgroup signal, a method, which becomes possible by the different decay constants T_2 of the stiff chain and the relatively mobile headgroup protons. Interestingly, the relative contribution of the mobile protons was observed to drop from 16 % in Lβ' to 11 % in Lσ'. This was explained by a freeze of the headgroup rotation below the subtransition, which leaves the 4 methylene protons of the choline in an immobile configuration (see fig. 5). As for the chain mobility in the Lβ' and Pβ' state, the ^{1}H-NMR investigations agree in that there is fast rotation about the long axis /87-92/. It is not undisputed, however, if this motion is performed by the whole molecule or by the chains alone. This important point of discussion will be picked up at the end of this article.

Recently, this type of investigation was extended in our laboratory to the phase behaviour of cephalins /93/. Between ~ 0°C and the main transition temperature the second moments of these compounds were found to be similar to those of lecithins. Towards lower temperatures there is a broad transition to a state with a high second moment. In contrast to the subtransition of DPPC, this transition exhibits no hysteresis and no temporal effects. There is, however, a temporal transition of DMPE to a dehydrated state (referred to as "subgel" by Wilkinson and Nagle /75/), observed at room temperature. Though also characterized by a low molecular mobility, this state is markedly different from the low temperature state in terms of longitudinal relaxation time T_1, and a number of structural x-ray parameters. It will be the task of further investigations, i.e. in particular by NMR on other nuclei, to elucidate more about dynamics and structure of these bilayer phases.

4.2 ^{2}H-NMR

As mentioned above, this method has played an important role in the determination of the chain mobility in the Lα-phase /25,26/. For the gel state, the ^{2}H-NMR experiment is complicated by the increased spectral broadening, which is difficult to cover with the transmitter power usually available. Using the quadrupolar echo technique, Davis recorded ^{2}H-NMR spectra of DPPC with perdeuterated hydrocarbon chains over a wide temperature range between -7°C and 50°C /94/. Since then a number of studies have been performed in the gel phase of lecithin, cephalin, and related compounds, selectively deuterated at particular positions in the center of the chain /95-102/. At low temperatures (~ 0°C) these spectra begin to exhibit the full quadrupolar splitting of ~ 120 kHz, indicating a cease of motion in the chain region. The Lσ'-phase has not yet been investigated by this method, but one may expect that a similar freeze will be displayed in the corresponding ^{2}H-spectra. At higher temperatures of the gel phase, thermal motion leads to an averaging of the spectra, approaching splittings near 60 kHz. The interpretation of these features was done, in accordance to the ^{1}H-NMR work, in terms of rotational motion about the long molecular axis. The static EFG tensor has its unique axis aligned parallel to the C-D bond, and a negligible asymmetry ($\eta^Q = 0$). According to eq. (8) a fast rotation of the chains about their long axis ($\theta = 90°$), would lead to a reduction of the quadrupole splitting by a factor 2. This motion is assumed to occur in discrete jumps in the slow motional limit, where computer simulations must be used to fit the experimental data. With consideration of additional trans-gauche-isomerization, the spectra of DPPE can roughly be approximated with axial hopping rates between 2.1×10^5 Hz (0°C) and 7.5×10^5 Hz (55°C). Interestingly, no distinct spectral changes were observed at the pretransition Lß' - Pß' of DPPC. This is different for observations at the terminal CD_3-group /103,104/. At the pretransition Meier et al. /104/ observed a change in the rotational correlation time of about an order of magnitude. Additionally, a second chain population of Lα-like mobility was found in the Pß' phase.

The higher mobility of the headgroup below T_c is evident from the small quadrupolar splittings, found for specifically ^{2}H-labelled phospho-cholines and -ethanolamines /105-107/. Interestingly, also the pretransition of DPPC is clearly reflected by a distinct step of the $(CD_3)_3$-splitting /108/. Only few data concerning the C-D_2 resonances in the glycole moiety have been reported to date. In the Lα-phase, Seelig et al. have observed different splittings for the two C3-D_2 deuterons /105,109-111/, indicating that the axis of rotational diffusion in this state (i.e. the bilayer normal) is not within the symmetry plane of this segment. In the gel state a pronounced broadening to some 80kHz has been found /16/.

4.3 ^{13}C-NMR

As an isotope of intermediate abundance (~ 1.1 %), this spin-1/2 nucleus is suited for investigations in both, isotopically enriched and natural compounds. In lipids the problem of dipolar coupling with the abundant protons can be overcome by the irradiation of a decoupling rf-field. The resulting spectra basically display the chemical shift of the different ^{13}C positions, which allows a separate observation of the acyl chains, the carbonyl positions and the headgroup with the glycerol moiety /112/. The measurement of the coupling to the proton spin reservoir, in turn, can provide a wealth of dynamic information. With this method it was possible to detect the increased headgroup mobility in the Pß'-phase, as well as the change of rotational correlation times in the acyl chain region /113,114/, in agreement with the findings mentioned in the preceding section /104/.

Some new insight into the structure of the gel phases of both, lecithins and cephalins, was given by ^{13}C-NMR measurements on compounds, that were isotopically enriched at the carbonyl position of the 2-chain /98-100,115/. In dry lecithin the corresponding chemical shift tensor is asymmetric (η^{CS} = 0.2) with an anisotropy of $\Delta\sigma$ = -137 ppm /98/. In the Lß'-phase an axial symmetric tensor is observed with $\langle\Delta\sigma\rangle$ = -112 ppm. In terms of eq. (8) this can be explained by the presence of axial diffusion about an axis inclined ~ 20° with respect to the 3-axis of the tensor. In the Lα-phase the anisotropy is reduced to ~ 0 ppm, corresponding to a change of the inclination angle to 55° ± 2°. Most interestingly, this narrow line appears superimposed on the Lß' spectrum already in coincidence with the onset of the pretransition ~ 7 K below its midpoint. With increasing temperature the peak grows in intensity until it has reached 100 % of the carbonyl signal above the main transition /98,115,116/. So these studies again give clear evidence of a coexistence of two lipid populations in the Pß'-phase of lecithin. In cephalins, the corresponding spectra in Lß and Lα are similar. However, no superposition of the two spectra is observed at any temperature, in accordance with the absence of an intermediate phase in these systems /98,99/. An ultimate discussion of these results will have to be postponed, until the orientation of the chemical shift tensor of the carbonyl position in the molecular frame is available.

4.4 ^{31}P-NMR

^{31}P is a spin-1/2 nucleus with 100 % natural abundance, occurring at one specific position in the phospholipid molecule. Like in ^{13}C-NMR, the dipolar interaction with the protons can be decoupled, leaving essentially the chemical shift anisotropy as a parameter of investigation. These properties make ^{31}P-NMR a very attractive method for the study of model membranes. An important result of a number of investigations, performed in the second half of the 1970's /117-124/, was the finding that the headgroup is aligned approximately parallel to the bilayer plane in a state of fast rotation about an axis close to the bilayer normal. Further support of this view was given by a measurement of the ^{1}H-^{31}P-coupling /125,127/, and by neutron diffraction studies /59/. The rotational motion is evident from the axial symmetry of the ^{31}P spectra observed throughout all phases of cephalins (Lβ, Lα) and lecithin, (Lβ', Pβ', Lα) without any discontinuous changes at the pretransition and main transition. Only upon reduction of the water content or at temperatures below -10°C, the observed tensor changes to an asymmetric shape /120,128/.

Recent measurements of our laboratory have shown, however, that the static, asymmetric tensor is clearly displayed in the Lσ'-phase of DPPC /129,130/. This is a direct corroboration of the view that the subtransition Lσ' - Lβ' coincides with an onset of a headgroup rotation. In time dependent experiments at 0°C the slow (~ 48 h) reversal of this transition Lβ' - Lσ' is demonstrated. In addition, the increased accuracy of these measurements allowed a detection of distinct steps in the plot of $\Delta\sigma$ and the intrinsic spectral line width vs. temperature at both upper phase transitions. But contrary to the ^{2}H and ^{13}C findings, mentioned in the previous sections, no superpositions of two populations were visible in the ^{31}P-spectra at any temperature.

In a final consideration, an attempt was made to determine the axis of rotation, which causes the reduced axial symmetric spectrum. Owing to the single crystal measurements of Herzfeld et al. /121/ and

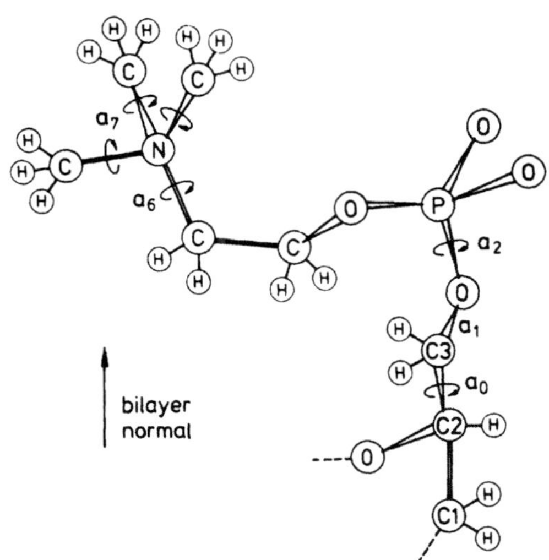

Figure 5. Motional state of the headgroup of lecithin. A rotation about the bonds a_0 and a_1 can be excluded by the NMR findings. Rotational diffusion about a_2 is present above the subtransition. The terminal amine-group remains mobile also at lower temperatures.

crystallographic studies /131-133/, several molecular bonds can be tested as candidates for the rotation axis of the choline headgroup, inserting the respective coordinates θ and ϕ in eq. (8). It turns out, that particularly the P-O bond (a_2 in fig. 4) is consistent with the observed reduction of the chemical shift anisotropy. Also a rotation about the C2-C3 bond could account for the observed data within the error margins, but this model can be discarded for two reasons: First, the change in the number of mobile protons at the subtransition amounts 4 per molecule. A rotation about the C2-C3 bond would involve the two C3-H_2 protons in addition to the four methylene protons of the choline group (see § 4.1). Second, the observation of different quadrupolar splittings for the two C3-D_2 deuterons in the Lα-phase is not consistent with a rotation about the C2-C3 axis (see § 4.3). In summary, the only consistent interpretation of the NMR data is a high rotational mobility of the choline headgroup about the P-O-bond to the glycerole moiety /130/. Below the subtransition this motion is frozen on the NMR timescale.

5. CONCLUSION

In the course of the interpretation of the preceding NMR findings two different views of the molecular mobility in the gel state of cephalins and lecithins have emerged. Some authors relate the rotational mobility, found in many positions of the molecule, with a rotational diffusion of the entire molecule. This idea is supported by the observation of axial symmetric chemical shift tensors at the ^{13}C-carbonyl and ^{31}P-phosphate positions, and by the magnitude of both, quadrupole splittings and ^{1}H-second moments of the acyl chains. On the other hand, this model has some implications concerning the crystalline structure, which appear to contradict the experimental evidence. A rotational diffusion of a molecule occupying two adjacent sites in a two-dimensional lattice must occur in steps of 60° rotations, where one of the two chains jumps into a vacant lattice site. In this type of defect diffusion rotational and lateral motion are linked in a particular manner. Evidently this model is consistent only with the random molecular arrangement illustrated in fig. 3a. As discussed in section 2 this structure is barely consistent with the crystallographic knowledge of the Lβ'-phase of lecithins, because it implies that the orthorhombic unit cell of the chain lattice does not correspond to a regular motif of the molecular packing. In addition, a consideration of the diffusion mechanism argues against this model. With tilted chains, the molecule must deform considerably in the glycerole moiety at every diffusion step. Indeed, if the slower diffusion constant measured in the Pβ' phase of $D < 10^{-17}$ cm^2/s /46/ is taken as valid also in Lβ', one easily finds that this motion is far too slow to affect NMR spectra.

In an attempt to link the lattice symmetry with the symmetry of motion, a different model was suggested recently by our group /68,114/. It considers the position of the molecules as fixed in an arrangement similar to fig. 3b or 3c. The rotational motion of the chains in the Lβ' phase is visualized as sketched in fig. 6a. This 90°-jump between

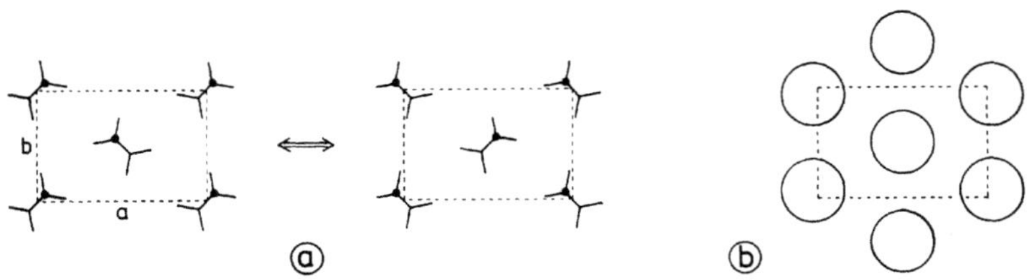

Figure 6. Model of the chain rotation in the orthorhombic lattice. a. 90°-flip-flop (Lβ'). b. Free rotation (Pβ').

two orientations is symmetric with respect to the b-axis of the orthorhombic unit cell, and accounts equally good for the ^{1}H- and ^{2}H-NMR observations. At the pretransition to Pβ' this type of mobility is assumed to change into a quasi-free rotation about the long chain axis. The interchain part of the ^{1}H second moment would be reduced by this kind of motion, in accordance to the experimental evidence. The ^{2}H quadrupole splitting, in turn, is insensitive to this mobility change /15,114/, again in consistency with the experiment. As pointed out in the preceding section, the axial symmetry of the ^{31}P chemical shift tensor is well understood as a result of a rotational motion of the headgroup about the P-O-bond. The only discrepancy of this model lies in the axial symmetry of the ^{13}C-carbonyl chemical shift tensor. Clearly, this position does not undergo a rotational motion, which could explain this feature immediately. However, it is evident that the suggested 90°-flip-flop of the chains requires some flexibility at the carbonyl position, and this could be sufficient to account for the observed spectrum. A quantitative discussion of this point is not possible without the knowledge of the orientation of the ^{13}C-tensor in the molecule. So an ultimate decision between the two models must be postponed until more experimental data are available. In particular, a ^{13}C-NMR experiment on compounds, labelled in the glycerole moiety could be very useful in this respect.

For cephalins, the situation is even less known. Clearly, the crystallographic arguments from above do not apply, since the chains are not tilted and their lattice is hexagonal. So the mobility in the gel phase of these compounds could be markedly different from the case of lecithins. However, as judged from the NMR results, this does not seem to be evident. Again, a decision between the two models is not possible on the basis of the data available today.

Even though a number of questions could not be answered to date, the employment of solid state NMR has provided a wealth of structural and dynamical information about the gel state of phospholipid bilayers. In particular, the complementary knowledge of the crystallographic parameters has turned out to be essential for a comprehensive understanding.

ACKNOWLEDGEMENT

Stimulating discussions with W.D. Klabe are greatfully acknowledged. This work was supported by the Sfb 161 of the Deutsche Forschungsgemeinschaft.

REFERENCES

1. S.J. Singer, and G.L. Nicolson. 1972. Science **175**, 720.
2. C. Tanford. 1973. The Hydrophobic Effect (Wiley, New York).
3. K. Fontell. 1981. Mol. Cryst. Liq. Cryst. **63**, 59.
4. N.J. Salsbury, and D. Chapman. 1968. Biochim. Biophys. Acta **163**, 314.
5. Z. Veksli, N.J. Salsbury, and D. Chapman. 1969. Biochim. Biophys. Acta **183**, 434.
6. J.T.Daycock, A. Darke, and D. Chapman. 1971. Chem. Phys. Lipids **6**, 205.
7. A.H. Mantsch, H. Saito, and I.C.P. Smith. 1977. Progr. NMR Spectr. **11**, 211.
8. J. Seelig. 1977. Quart. Rev. Biophys. **10**, 353.
9. J. Seelig. 1978. Biochim. Biophys. Acta **515**, 105.
10. J. Seelig, and A. Seelig. 1980. Quart. Rev. Biophys. **13**, 19.
11. J. Charvolin, and A. Tardieu. 1978. Solid State Phys. Suppl. **14**, 209.
12. J.L. Browning. 1981. Liposomes: From Physical Structure to Therapeutic Applications. pp. 189-242 (Elsevier, North Holland Biomedical Press).
13. R.E. Jacobs, and E. Oldfield. 1981. Progr. NMR Spectr. **14**, 113.
14. R.G. Griffin. 1981. Meth. Enzymol. **72**, 108.
15. D.M. Rice, A. Blume, J. Herzfeld, R.J. Wittebort, T.H. Huang, S.K.D. DasGupta, and R.G. Griffin. 1981. Biomolecular Stereodynamics, Vol. II, p. 255 (Adenine Press, New York).
16. J.H. Davis. 1983. Biochim. Biophys. Acta **737**, 117.
17. R. Kimmich, G. Schnur, and A. Scheuermann. 1983. Chem. Phys. Lipids **32**, 271.
18. J.S. Waugh, L.M. Huber, and U. Haeberlen. 1968. Phys. Rev. Lett. **20**, 180.
19. A. Pines, M.G. Gibby, J.S. Waugh. 1973. J. Chem. Phys. **59**, 569.
20. M. Mehring. 1976. NMR Basic Principles and Progress **11** (Springer Verlag, Berlin, Heidelberg, New York).
21. U. Haeberlein. 1976. Advances in Magnetic Resonance, Suppl. 1 (Academic Press, New York, San Francisco).
22. A.G. Lee. 1977. Biochim. Biophys. Acta **472**, 237.
23. E.S. Wu, K. Jacobson, and D. Papahadjopoulos. 1977. Biochemistry **16**, 3935.
24. D.F.W. Wallach, S.P. Verma, and J. Fookson. 1979. Biochim. Biophys. Acta **559**, 153.
25. J. Seelig, and J.L. Browning. 1978. FEBS Lett. **92**, 41.
26. E. Oldfield, M. Meadows, D. Rice, and R. Jacobs. 1978. Biochemistry **17**, 2727.

27. J.F. Nagle. 1973. J. Chem. Phys. **58**, 252.
28. A. Tardieu, V. Luzzati, and F.C. Reman. 1973. J. Mol. Biol. **75**, 711.
29. R.P. Rand, D. Chapman, and K. Larsson. 1975. Biophys. J. **15**, 1117.
30. M.J. Janiak, D.M. Small, and G.G. Shipley. 1976. Biochemistry **15**, 4575.
31. M.J. Janiak, D.M. Small, and G.G. Shipley. 1979. J. Biol. Chem. **254**, 6068.
32. M. Hentschel, R. Hosemann, and W. Helfrich. 1980. Z. Naturforsch. **35a**, 643.
33. M. Hentschel. 1981. Thesis. Freie Univ. Berlin.
34. M. Hentschel. 1983. Mol. Cryst. Liq. Cryst. **94**, 291.
35. D.G. Cameron, H.L. Casal, H.H. Mantsch, Y. Boulanger, and I.C.P. Smith. 1981. Biophys. J. **35**, 1.
36. D.A. Pink, T.J. Green, and D. Chapman. 1980. Biochemistry **19**, 349.
37. H.J. Hinz, and J.M. Sturtevant. 1972. J. Biol. Chem. **247**, 6071.
38. P. Pinto da Silva. 1971. J. Microsc. **12**, 185.
39. A.J. Verkleij, P.H.J. Ververgaert, L.L.M. v. Deenen, and P.F. Elbers. 1972. Biochim. Biophys. Acta **288**, 326.
40. P.H.J. Ververgaert, P.F. Elbers, A.J. Luitingh, and H.J. v.d. Berg. 1971. Cytobiology **6**, 86.
41. E.H. Luna, and H.M. McConnell. 1977. Biochim. Biophys. Acta **466**, 381.
42. C. Gebhardt, H. Gruler, and E. Sackmann. 1977. Z. Naturforsch. **32c**, 581.
43. J. Stamatoff, B. Feuer, H.J. Guggenheim, G. Tellez, and T. Yamane. 1982. Biophys. J. **38**, 217.
44. H.G. Kapitza, D.A. Rüppel, H.J. Galla, and E. Sackmann. 1984. Biophys. J. **45**, 577.
45. B.A. Smith, and H.M. McConnell. 1978. Proc. Natl. Acad. Sci. USA **75**, 2759.
46. M.B. Schneider, W.K. Chan, and W.W. Webb. 1983. Biophys. J. **43**, 157.
47. W. Helfrich. 1974. Z. Naturforsch. **29c**, 692.
48. K. Larsson. 1977. Chem. Phys. Lipids **20**, 225.
49. S. Doniach. 1979. J. Chem. Phys. **70**, 4587.
50. W.K. Chan, and W.W. Webb. 1981. Phys. Rev. Lett. **46**, 39.
51. H.L. Scott. 1981. Biochim. Biophys. Acta **643**, 161.
52. P.A. Pearce, and H.L. Scott. 1982. J. Chem. Phys. **77**, 951.
53. H.L. Scott. 1984. J. Chem. Phys. **80**, 2197.
54. M.S. Falkovitz, M. Seul, H.L. Frisch, and H.M. McConnell. 1982. Proc. Natl. Acad. Sci. USA **79**, 3918.
55. V.A. Parsegian. 1983. Biophys. J. **44**, 413.
56. A. Blume. 1983. Biochemistry **22**, 5436.
57. C. Grabielle-Madelmont, and R. Perron. 1983. J. Coll. Interf. Sci. **95**, 471.
58. C. Grabielle-Madelmont, and R. Perron. 1983. J. Coll. Interf. Sci. **95**, 483.
59. G. Büldt, H.U. Gally. A. Seelig, J. Seelig, and G. Zaccai. 1978. Nature **271**, 182.

60. B.P. Gaber, and W.L. Peticolas. 1977. Biochim. Biophys. Acta **465**, 260.
61. N. Yellin, and I.W. Levin. 1977. Biochim. Biophys. Acta **489**, 177.
62. D.G. Cameron, H.L. Casal, E.F. Gudgin, and H.H. Mantsch. 1980. Biochim. Biophys. Acta **596**, 463.
63. D.G. Cameron, H.L. Casal, and H.H. Mantsch. 1980. Biochemistry **19**, 3665.
64. D. Marsh. 1980. Biochemistry **19**, 1632.
65. P. Meier, A. Blume, E. Ohmes, F.A. Neugebauer, and G. Kothe. 1982. Biochemistry **21**, 526.
66. S.C. Chen, J.M. Sturtevant, and B.J. Gaffney. 1980. Proc. Natl. Acad. Sci. USA **77**, 5060.
67. D.G. Cameron, and H.H. Mantsch. 1982. Biophys. J. **38**, 175.
68. L. Trahms, W.D. Klabe, and E. Boroske. 1983. Biophys. J. **42**, 285.
69. J.F. Nagle, and D.A. Wilkinson. 1982. Biochemistry **21**, 3817.
70. H.H. Füldner. 1981. Biochemistry **20**, 5707.
71. M.J. Ruocco, and G.G. Shipley. 1982. Biochim. Biophys. Acta **684**, 59.
72. M.J. Ruocco, and G.G. Shipley. 1982. Biochim. Biophys. Acta **691**, 309.
73. H. Chang, and R.M. Epand. 1983. Biochim. Biophys. Acta **728**, 319.
74. H.H. Mantsch, S.C. Hsi, K.W. Butler, and D.G. Cameron. 1983. Biochim. Biophys. Acta **728**, 325.
75. A. Wilkinson, and J.F. Nagle. 1984. Biochemistry **23**, 1538.
76. H.W. Spiess. 1978. NMR Basic Principles and Progress **15**, 55 (Springer Verlag, Berlin, Heidelberg, New York).
77. R. Ebelhäuser, T. Fahmy, and H.W. Spiess. 1984. Macromol. Chem. **5**, 333.
78. R. Ebelhäuser, H.W. Spiess. 1984. Macromol. Chem. **5**, 403.
79. A. Saupe. 1964. Z. Naturforsch. **19a**, 161.
80. P.G. deGennes. 1974. The Physics of Liquid Crystals, pp. 23-34 (Clarendon Press, Oxford).
81. M. Mehring, R.G. Griffin, and J.S. Waugh. 1971. J. Chem. Phys. **55**, 746.
82. P. Lichtenberg, N.O. Petersen, J. Giradet, M. Kainoshe, P.A. Kroon, C.H.A. Seiter, G.W. Feigenson, and S.I. Chan. 1975. Biochim. Biophys. Acta **382**, 10.
83. M. Bloom, E.E. Burnell, A.L. MacKay, C.P. Nichols, M.I. Valic, and G. Weeks. 1978. Biochemistry **17**, 5750.
84. A. Abragam. 1961. The Principles of Nuclear Magnetism, pp. 88-132, (Clareandon Press, Oxford).
85. H. Wennerström. 1973. Chem. Phys. Lett. **18**, 41.
86. M. Bloom, E.E. Burnell, S.B.W. Roeder, and M.I. Valic. 1977. J. Chem. Phys. **66**, 3012.
87. L. Trahms, and E. Boroske. 1978. Ann. Phys. **3**, 289.
88. L. Trahms, and E. Boroske. 1979. Biochim. Biophys. Acta **552**, 189.
89. F. Volke, K. Arnold, and K. Gawrisch. 1982. Chem. Phys. Lipids **31**, 179.
90. A.L. MacKay. 1981. Biophys. J. **35**, 301.
91. K.R. Jeffrey, T.C. Wong, E.E. Burnell, M.J. Thompson, T.P. Higgs, and N.R. Chapman. 1979. J. Magn. Res. **36**, 151.

92. A. Peters, and R. Kimmich. 1978. Biophys. Struct. Mech. **4**, 67.
93. M.P. Hentschel, S. Braun, R. Dietrich, and L. Trahms. 1984. submitted to Mol. Cryst. Liq. Cryst.
94. J.H. Davis. 1979. Biophys. J. **27**, 339.
95. S.Y. Kang, H.S. Gutowsky, J.C. Hsung, R. Jacobs, T.E. King, D. Rice, and E. Oldfield. 1979. Biochemistry **18**, 3257.
96. T.H. Huang, R.P. Skarjune, R.J. Wittebort, R.G. Griffin, and E. Oldfield. 1980. J. Am. Chem. Soc. **102**, 7377.
97. A. Bienvenue, M. Bloom, J.H. Davis, and P.F. Devaux. 1981. J. Biol. Chem. **257**, 3032.
98. R.J. Wittebort, A. Blume, T.H. Huang, S.K. DasGupta, and R.G. Griffin. 1982. Biochemistry **21**, 3487.
99. A. Blume, and R.G. Griffin. 1982. Biochemistry **24**, 6230.
100. A. Blume, R.J. Wittebort, S.K. DasGupta, and R.G. Griffin. 1982. Biochemistry **24**, 6243.
101. M.F. Brown. 1982. J. Chem. Phys. **77**, 1576.
102. D.J. Siminovitch, M.J. Ruocco, E.T. Olejniczak, S.K. DasGupta, and R.G. Griffin. 1984. Proc. 22nd Congress Ampere, Zürich. p. 463.
103. P.W. Westermann, M.J. Vaz, L.M. Strenk, and J.W. Doane. 1982. Proc. Natl. Acad. Sci. USA **79**, 2890.
104. P. Meier, E. Ohmes, G. Kothe, A. Blume, J. Weidner, and H.J. Eibl. 1983. J. Phys. Chem. **87**, 4904.
105. H.U. Gally, W. Niederberger, and J. Seelig. 1975. Biochemistry **14**, 3647.
106. J. Seelig, and H.U. Gally. 1976. Biochemistry **15**, 5199.
107. M.F. Brown, and J. Seelig. 1978. Biochemistry **17**, 381.
108. H.H. Füldner. 1980. Thesis. Univ. Ulm.
109. H.U. Gally, G. Pluschke, P. Overath, and J. Seelig. 1980. Biochemistry **19**, 1638.
110. H.U. Gally, G. Pluschke, P. Overath, and J. Seelig. 1981. Biochemistry **20**, 1826.
111. F. Borle, and J. Seelig. 1983. Biochemistry **22**, 5536.
112. J. Urbina, and J.S. Waugh. 1974. Proc. Natl. Acad. Sci. USA **71**, 5062.
113. B.A. Cornell, M. Keniry, R.G. Hiller, and R. Smith. 1980. FEBS Lett. **115**, 134.
114. E. Boroske, and L. Trahms. 1983. Biophys. J. **42**, 275.
115. R.J. Wittebort, C.F. Schmidt, and R.G. Griffin. 1981. Biochemistry **20**, 4223.
116. B.A. Cornell. 1981. Chem. Phys. Lipids **28**, 69.
117. S.J. Kohler, and M.P. Klein. 1976. Biochemistry **15**, 967.
118. S.J. Kohler, and M.P. Klein. 1977. Biochemistry **16**, 519.
119. J. Seelig, H.U. Gally, and R. Wohlgemuth. 1976. Biochim. Biophys. Acta **467**, 109.
120. R.G. Griffin. 1976. J. Am. Chem. Soc. **98**, 851.
121. J. Herzfeld, R.G. Griffin, and R.A. Haberkorn. 1978. Biochemistry **17**, 2711.
122. R.G. Griffin, L. Powers, and P.S. Pershan. 1978. Biochemistry **17**, 2718.
123. R.F. Campbell, E. Meirovitch, and J.H. Freed. 1979. J. Phys. Chem. **83**, 525.

124. R. Skarjune, and E. Oldfield. 1979. Biochemistry **18**, 5903.
125. P.L. Yeagle, W.C. Hutton, C.H. Huang, and R.B. Martin. 1975. Proc. Natl. Acad. Sci. USA **72**, 3477.
126. P.L. Yeagle, W.C. Hutton, C.H. Huang, and R.B. Martin. 1976. Biochemistry **15**, 2121.
127. P.L. Yeagle, W.C. Hutton, C.H. Huang, and R.B. Martin. 1977. Biochemistry **16**, 4344.
128. K. Arnold, E. Loebel, F. Volke, and K. Gawrisch. 1981. Stud. Biophys. **82**, 207.
129. L. Trahms, and W.D. Klabe. 1984. submitted to Mol. Cryst. Liq. Cryst.
130. L. Trahms, and W.D. Klabe. 1984. Proc. 22nd Congress Ampere, Zürich, p. 465
131. M. Sundaralingam. 1972. Ann. N.Y. Acad. Sci. **195**, 324.
132. P.B. Hitchcock, R. Mason, K.M. Thomas, and G.G. Shipley. 1974. Proc. Natl. Acad. Sci. USA **71**, 3036.
133. R.H. Pearson, and I. Pascher. 1979. Nature **281**, 499.

INTERPRETATION OF PROTON N.M.R. SPIN-LATTICE RELAXATION TIME MINIMA IN HETEROGENEOUS SYSTEMS BY THE EFFECTS OF BOUNDED TWO-DIMENSIONAL DIFFUSION. COMPARISON WITH NEUTRON SCATTERING MEASUREMENTS

J. Tabony
CEA-IRDI-DESICP
Département de Physico-Chimie
Service Chimie Moléculaire
Centre d'Etudes Nucléaires de Saclay
91191 GIF SUR YVETTE CEDEX
France

ABSTRACT. The n.m.r. relaxation times of liquids in heterogeneous systems, where a molecule is either close to an interface, or has diffusive motions which are in some way geometrically bounded, often show features such as T_1 minima, resembling those found in bulk solids. These measurements have nearly always been interpreted using the theory of relaxation for bulk materials. This frequently leads to the conclusion that, as a result of proximity to the interface, there is a reduction of the molecular mobility of several orders of magnitude compared to the bulk liquid. This result often conflicts with those of other measurements such as quasielastic neutron scattering. Another interpretation which takes into account the restricted nature of the diffusion and which gives the correct molecular mobilities is given. These results place a doubt on the correctness of the conclusions of the majority of proton n.m.r. relaxation time measurements in heterogeneous systems.

I. INTRODUCTION

The measurement of n.m.r. relaxation times is one of the most commonly used approaches for studying molecular motions in liquids and solids. Nuclear spin lattice relaxation is most efficient when the frequency of some molecular motion coupled to a magnetic interaction, is comparable to the Larmor frequency (10-500 MHz). Changing the temperature, changes the motional frequency spectrum and the variation of the spin lattice relaxation time (T_1) with temperature can show a minimum. In bulk materials T_1 minima are frequently observed in solid phases but are rarely seen in liquids.

A particular field of application is heterogeneous systems where one is interested in the motion of a molecular liquid in the vicinity of either a solid or a liquid interface. Various examples are ; molecules physisorbed on solid surfaces, molecules diffusing in

R. Daudel et al. (eds.), Structure and Dynamics of Molecular Systems, 225–243.

is relatively few, and this sometimes makes an exact comparison with n.m.r. relaxation time measurements difficult. One example of where systematic neutron and n.m.r. measurements have been made is that of molecules physisorbed on solid surfaces and much of this paper will be concerned with this case. Other systems where similar contradictions appear, but where either the n.m.r. or the neutron studies are incomplete, or where the measurements have not been made on exactly the same samples, will also be considered. These cases, which are illustrative rather than exhaustive are, water adsorbed in clays ; the hexagonal lyotropic liquid crystal mesophase ; water adsorbed in a biological fibre.

In this paper we wish to draw attention to the contradictions which arise between n.m.r. T_1 relaxation and neutron quasielastic scattering measurements when the former are interpreted using bulk theories, and that to point out these contradictions disappear when a relaxation theory which takes into account bounded two dimensional motion is used. This result places a doubt on the correctness of the conclusions of the majority of proton n.m.r. relaxation time measurements in heterogeneous systems. In particular, the conclusions concerning 'bound water' at biological interfaces[5] should be reexamined.

Section II briefly discusses some of the differences which arise in the application of quasielastic neutron scattering and n.m.r. relaxation time measurements to heterogeneous systems. Section III discusses the contradictions between the conclusions of these methods when used on molecules physisorbed at homogeneous surfaces. Section IV points out some similar contradictions for some other systems. Finally, V presents the evidence in favour of an interpretation of the T_1 minimum in terms of a bounded two dimensional motion, and qualitatively discusses the effects of the latter. A complete theoretical description and a quantitative comparison with experimental results is given by Korb in the next paper in this book.

II. EXPERIMENTAL METHODS

The incoherent neutron energy spectrum contains information about the motion of the scatterer over distances and frequencies comparable respectively to the wavelength and energy of the neutron beam [14,15]. For the study of heterogeneous materials, one of the advantages of neutron scattering over n.m.r. is that it can probe fast (10^{10} - 10^{13} sec^{-1}) diffusive motions directly[16]. Another advantage, is that the distance over which motions are observed is well defined and variable. This dimension, which is in general between 2 A° and 100 A°, is related to the scattering vector Q. Where $Q = 4\pi/\lambda . \sin 2\theta$, and 2θ is the scattering angle and λ the neutron wavelength. Thus, molecular diffusion can be studied over distances which may be either smaller or larger than the characteristic size of the heterogiety. For sizes less than that of the heterogeniety, the diffusion will resemble that of the bulk liquid ; whereas for distances greater than the size of the heterogeniety, the effects of the restriction will become apparent.

In practice, the sample is placed in the incident neutron beam and the scattered neutrons analysed in energy at different angles. Because of molecular motions, the neutrons both gain and lose energy resulting in a 'quasielastic' broadening of the incident beam. Fickian diffusion causes a Lorentzian broadening which increases linearly with Q^2. As energy spectra are normally measured simultaneously for at least nine different scattering angles, plots of the observed broadening against Q^2 can be used to deduce a translational diffusion constant. As already mentioned, the values obtained this way are not very sensitive to the effects of either reduced dimensionality or of other spatial restrictions. This is not the case for n.m.r. spin lattice relaxation times. Not only is the observation distance ill defined but also fast motions are not probed directly. What is measured is the intensity of the motional spectral density at the Larmor frequency (10^8 sec^{-1}). Provided the form of the spectral densities is known, then the rate of molecular motions can be calculated. In bulk liquids the spectral densities are Lorentzians. However in heterogeneous materials this is not always so. Complications arise because of the long time scale of the T_1 measurement. For a liquid having a translational diffusion coefficient of 10^{-9} m^2 sec^{-1}, then in 10^{-8} sec, a molecule diffuses through a root mean square distance of several hundred Angstroms. Since in general this distance is larger than the size of the heterogeniety, the relaxation phenomena are time averaged over the topological constraints of the diffusive motions in question. One of the consequences of this is that the molecules do not sample all the space equally. In practice 'isotropic motional narrowing' no longer occurs and the spectral densities often show an increased intensity at the Larmor frequency compared to the bulk case. This behaviour is easily mistaken for a slow motion and a correct analysis must take into account the bounded nature of the diffusion. The reencounter probability for two molecules on a surface has a rather large finite value of exactly 1. This not only means that the molecules never completely lose their correlations but also that reencounters will continue almost indefinitely, and it is this process, which under suitable circumstances, can give rise to an additional T_1 minimum, which is absent in bulk materials.

Another method for measuring translational diffusion constants is the n.m.r. pulsed field gradient experiment[27,28]. Unlike T_1 measurements, the molecular displacement is measured without recourse to a microscopic theory. The most important difference with neutron scattering is that both the timescale (a few milliseconds) and hence the observation distances (several microns) are long.

III. GASES PHYSISORBED ON HOMOGENEOUS SURFACES

The adsorption of small molecules upon homogeneous surfaces, as characterised by the adsorption isotherm[29,34,39], gives rise to various quasi two dimensional (2D) phases. In different regions of the phase diagram, 2D solids, liquids and gases are known to exist and have been studied by methods such as : - X ray diffraction[31], neutron scattering[32], n.m.r. and adsorption isotherm [29,30,39] measurements. Because it has a relatively large surface area and because it provides a clean homogeneous surface which is readily prepared and stable, the most commonly used substrate has been graphite. Two types of graphite are currently used ; exfoliated graphite and graphitised carbon blacks[33,34]. Graphitised carbon blacks are made by graphitising at high temperatures carbon black powders. The resulting quasi spherical particles have, depending on the carbon black and graphitisation temperature used, radii varying from 100 A° to 1000 A°[33]. The surface of these particles is locally flat, and consists of polyhhedra of the exposed graphite basal plane of dimensions 80 A° x 80 A°[35]. For systems showing B.E.T. type II isotherms[29,34,39], such as methane on graphite or neopentane on graphite, the molecules wet the surface, and well defined quasi two dimensional phases are formed. Systems showing B.E.T. type III[29,34] isotherms, such as ammonia on graphite or water on graphite, do not wet the surface but tend to form islands of multilayers before completion of the first monolayer.

There are several n.m.r. papers concerning molecules adsorbed on graphitised carbon blacks. Relaxation times show the features previously mentioned : - short T_1's ; a T_1 minimum with changing temperature ; T_2 less than T_1. To date three systems have been studied, methane on graphite[9], neopentane on graphite[6,7,12], and ammonia on graphite[4,6,8,12]. Neutron scattering measurements have also been made on the same systems[10,11,30,32,36,37,38,40].

a) Methane on graphite.

In different regions of temperature and coverage, two dimensional solids , liquids, and gases, have been identified [11,30,31,37-39]. For a coverage of 0.4 monolayers, neutron diffraction studies of the 2D solid shows that a continual melting to a 2D liquid occurs between 40 °K and 60 °K[37,38]. This should be compared with the bulk melting point of methane (91 K). This behaviour is common in adsorbed molecules and results from the 2D nature of the phases formed. At 60 °K and above, both quasielastic neutron scattering[11] and the n.m.r. pulsed field gradient method[12] have been used to measure the translational diffusion coefficient (D) of the adsorbed liquid phase. In both case values faster than $10^{-9} m^2 sec^{-1}$ were obtained. These are shown in Tables 1 and 2.

Riehl and Koch have reported proton n.m.r. relaxation times for methane on graphite [9]. For a statistical coverage of 0.4 monolayers, they observed a T_1 minimum at 50 °K which minimum disappeared with dilution in CD_4 (figure 1). The latter observation shows that the relaxation minimum arises from the intermolecular dipole-dipole

Table 1 Values of the diffusion coefficient (10^{-9} m^2 sec^{-1}) for methane on graphite determined by neutron scattering[10].

Coverage	D
0.30	2.2 ± 0.4
0.4	2.5 ± 0.3
0.55	2.6 ± 0.3
0.63	2.7 ± 0.3

Coverage	D
0.45	13 ± 1.3
0.7	7 ± 0.7
0.9	2.2 ± 0.3

Table 2 Values of the diffusion coefficient for methane on graphite determined by the n.m.r. pulsed field gradient method[11].

Coverage	D
0.4	7.5 ± 1.0
0.65	3.8 ± 1.0
0.9	3.5 ± 0.9
1.2	5.0 ± 1.2
1.4	5.2 ± 1.3

Table 3 Values for the diffusion coefficient (10^{-9} m^2 sec^{-1}) for 1.25 monolayer of neopentane adsorbed upon graphite as determined by the n.m.r. pulsed field gradient method.

T (K)	D
253	9.3 ± 2.3
239	5.2 ± 0.9
228	3 ± 0.8

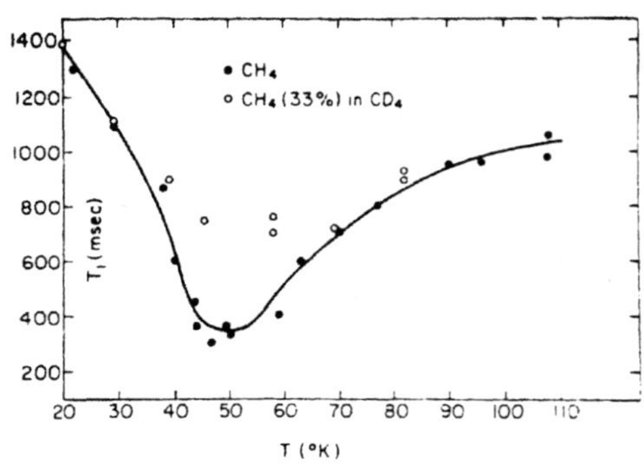

Figure 1 Spin lattice relaxation times T_1 versus temperature (^oK) for methane (•) and CH_4 diluted in CD_4 (o). The concentration of CH_4 was 33% in the diluted sample. The statistical coverage was 0.4 monolayer. Reproduced from reference 8 by permission of J. Chem. Phys.

interaction, hence, the motion associated with H is translational and not rotational. At 50°K, neutron diffraction shows the presence of two phases, a liquid and a solid[37,38]. Evidence that the n.m.r. T_1 minimum comes from the adsorbed liquid and not the solid, is that the value of T_2 at 50° K corresponds to a linewidth of only 100 Hz. This is far too low for a solid, but is of the order expected for a 2D liquid[21]. Using the bulk theory of relaxation, Riehl and Koch interpreted the T_1 minimum, as arising from slow molecular translation having correlation times (τ) of 3×10^{-9} sec and 1×10^{-9} sec at 50°K and 60°K respectively. In contradiction with this, quasielastic neutron scattering measurements[11] at 60°K gave a diffusion constant (D) of 2.5×10^{-9} $m^2 sec^{-1}$. Assuming two dimensional Fickian diffusion and a jump length of 3 A° gives a translational correlation time of 10^{-11} sec. This is two orders of magnitude faster than that deduced by n.m.r.. It may be argued that to compare measurements of D and one is forced to introduce an arbitrary jump distance. However in this case, to equate the two measurements, the jump distance would need to be 30 A°. Not only is this physically unreasonable, but also the quasielastic neutron measurements shown no evidence for such long jump lengths. Additional support for the neutron measurements comes from the concurrence with the values measured by the n.m.r. pulsed field gradient method[12]. In this n.m.r. experiment the diffusion coefficient is measured directly, without recourse to a relaxation model. The agreement between these n.m.r. results and those from neutron scattering, suggests that the disagreement with the conclusions of the n.m.r. relaxation times measurements arises from the microscopic theory used to interpret the latter.

b) Neopentane on graphite.

For a statistical coverage of 1.25 monolayers, both n.m.r. and neutron diffraction show a continual melting between 150 °K and 220 °K[5,6,39]. Proton relaxation time measurements on the liquid phase show features similar to that of methane on graphite : - short T_1's, T_2 less than T_1, and a T_1 minimum with temperature. Also like methane, measurements with mixtures of protonated and deuterated neopentane, showed the minimum to arise from intermolecular dipole-dipole interactions.[6,7] This is shown in Figures 2 and 3. Applying the bulk theories for relaxation, yields a translational correlation time (4×10^{-9}sec) at the minimum (228 °K) typical of that found in solids. In contrast to this, measurements of the diffusion constant using the n.m.r. pulsed field gradient method, gave a value of 3×10^{-9} m^2 sec^{-1} at 228 °K. Taking a jump distance of 3.5 A° and assuming Fick's law gives a translational correlation time of 10^{-11}sec, i.e. two orders of magnitude faster than that deduced by applying the bulk theories of relaxation.

c) Ammonia on graphite.

Ammonia on graphite has a B.E.T. type III adsorption isotherm

and forms islands of multilayers before completion of the first monolayer. The proton n.m.r. relaxation time measurements[8], over the temperature range where thermodynamic[41] and neutron diffraction[36] experiments show the existence of a liquid phase, are shown in figure 4.They are extremely short, being comparable to the values found in the bulk solid, and show signs of passing through a minimum.

Measurements of the translational diffusion constant by quasielastic neutron scattering[10] and the n.m.r. pulsed field gradient method[12], agree and give values greater than $10^{-9}m^2sec^{-1}$ (tables 4, 5). There is once again, a contradiction between the conclusions of the n.m.r. relaxation times interpreted using bulk theories and the results of other experiments.

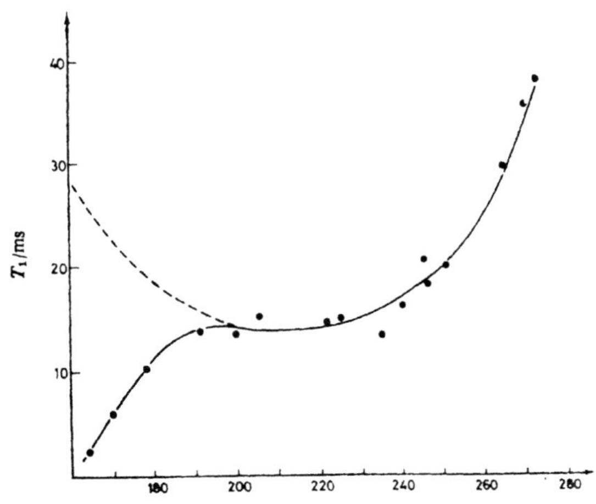

Figure 4 Variation of the spin lattice relaxation time with temperature for ammonia on graphite. The statistical coverage is 1.1 monolayers and the spectrometer frequency is 60 MHz.

Coverage	D			
	T = 163 K	T = 185 K	T = 205 K	T = 245 K
0.5			13 ± 2	
1.0	3.0 ± 1	6.0 ± 1	11.5 ± 2	21 ± 4
2.0			10 ± 3	

Table 4 Values for the diffusion coefficients ($10^{-9}m^2s^{-1}$) for ammonia adsorbed upon graphite as determined by quasi-elastic neutron scattering[9].

Table 5 Values for the diffusion coefficient ($10^{-9}m^2s^{-1}$) for ammonia adsorbed upon graphite as determined by the n.m.r. pulsed field gradient method[11].

Coverage	D		
	$T = 180$ K	$T = 200$ K	$T = 230$ K
0.5		24 ± 6	
1.0	6.0 ± 1.5	11.7 ± 3	60 ± 15
2.0		8.4 ± 2	

IV. OTHER SYSTEMS

a) Water in clays.

Although the experimental results are rather less complete than for molecules physisorbed on graphite, similar contradictions between n.m.r. and quasielastic neutron scattering exist for molecules in lamellar systems. One example is the diffusion of water in clays. Clays are alumina silicate sheets which depending on the relative humidity, contain water layers of between one and several tens of molecules thick.

Quasielastic neutron scattering experiments have been made, at temperatures close to ambient, on clays containing different cations. In all cases the water molecules undergo fast translational diffusion, with rates comparable to those in bulk electrolytic solutions. For a one layer hydrate of lithium montmorillonite, Cebula et al[42] determined the water diffusion constant to be $3 \times 10^{-10}m^2sec^{-1}$. Proton n.m.r. spin lattice relaxation time measurements[43] on a one layer hydrate of a similar clay, lithium hectorite, show a double T_1 minimum (figure 5). Since measurements with different deuterated mixtures were not made, it is not known whether the minimum arises from an inter on an intra molecular effect. The minimum was attributed to a slow rotation of the lithium hydration shell. However, the neutron scattering measurements show no evidence for such a slow motion and the interpretation of the two methods are in contradiction.

b) Lyotropic liquid crystals.

Restricted diffusion also occurs in liquid crystal mesophases. For the lyotropic lamellar phase, where there are alternating water and soap bilayers, the situation resembles that in clays. A slightly different situation occurs in the neighbouring hexagonal phase. Here, the structure consists of cylindrical aggregates of soap molecules dispersed in water. The polar heads of the surfactant molecules are constrained to the outer surface of the cylindrical aggregate with the result that they undergo a curved quasi 2D motion.

Temperature dependent proton T_1 measurements at 60 MHz[44] show a minimum at 283 K. An interpretation in terms of a bulk theory would lead to the conclusion that the soap molecules are relatively immobile. This is not borne out by direct measurement of the diffusion constant. Both neutron scattering[45] and the n.m.r. pulsed field gradient method[44] gave values of D of 2×10^{-10} m^2 sec^{-1} at 283 K. Also neutron scattering measurements[46] in related micellar and microemulsion phases show a fast translational motion of the surfactant polar heads around the surface of the aggregates.

c) Water in biological materials.

A more complicated, but nevertheless important class of heterogeneous media, is that of water adsorbed in materials of biological significance. Like the adsorbed molecules already mentioned, a part of the water in these systems does not freeze until temperatures well below the bulk freezing point. At the same time, this non-frozen liquid water is often stated to be partially immobilised or 'bound' to the substrate. Since some of the evidence for this comes from n.m.r. relaxation time measurements[5], it is important that they should have been correctly interpreted. The situation resembles the other cases presented, in that the n.m.r. results are frequently analysed using bulk theories for relaxation and this leads to conclusions which are in contradiction with quasielastic neutron scattering measurements.

For example, n.m.r. relaxation time measurements of water adsorbed on collagen fibres[47], show a T_1 minimum at 238 K. Applying a bulk theory for relaxation would give correlation times of 3×10^{-9} sec at 238 K and 5×10^{-10} sec at 293 K. Quasielastic neutron scattering spectra[48,49] at 293 K have energy broadenings characteristic of a diffusion constant of 5×10^{-10}m^2 sec^{-1}, corresponding to correlation times of 5×10^{-11} sec. Once again the interpretation of n.m.r. relaxation times yields molecular motions an order of magnitude slower than those deduced from neutron scattering.

For the cases presented it is clear that the n.m.r. relaxation times are not satisfactorily interpreted in terms of a bulk theory. As the motions of these adsorbed molecules all have bounded two dimensional properties it is appropriate to consider how this feature might effect their n.m.r. relaxation times.

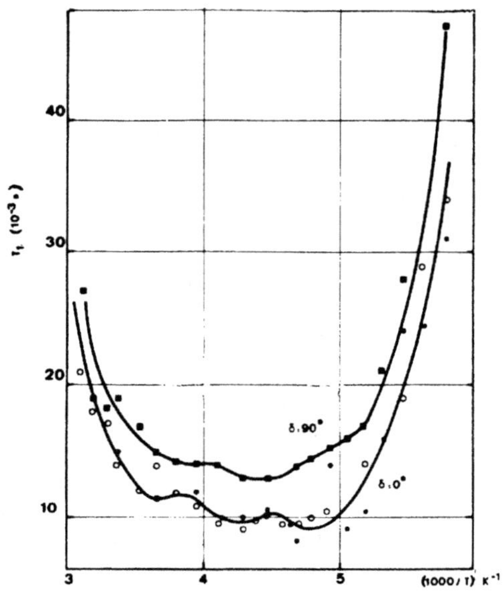

Figure 5 Variation of T_1 with temperature for one layer of water in lithium hectorite. The two curves correspond to different orientations with respect to the magnetic field [43]. Reproduced from reference 43 by permission of J. de Phys.

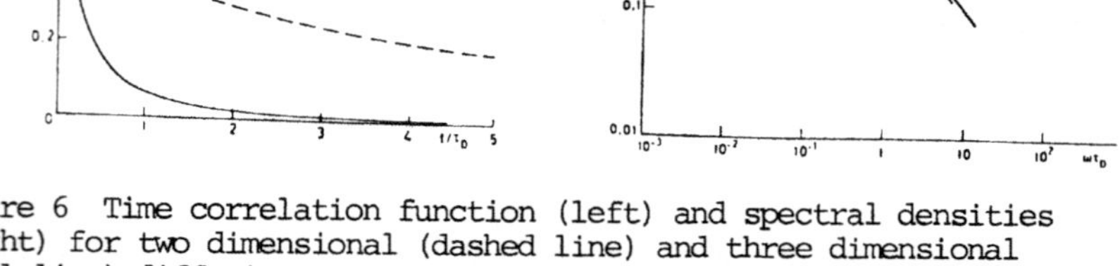

Figure 6 Time correlation function (left) and spectral densities (right) for two dimensional (dashed line) and three dimensional (full line) diffusion. Reproduced from reference 21 by permission of J. Chem. Phys.

V. EFFECTS OF BOUNDED TWO-DIMENSIONAL DIFFUSION ON N.M.R. RELAXATION TIMES

Kokin and Ismest'ev[19] first pointed out that the time correlation function in two dimensions is significantly different from that in three dimensions, and that this should effect relaxation times in adsorbed molecules. This work was followed up by Beckert[20] and more recently by Avogadro and Villa[21] and by Fries[22,23]. At short times or high frequencies the 2D and 3D correlation functions almost coincide, whilst at long times and low frequencies, the 2D correlation function diverges logarithmically. This is shown in Figure 6. The extremely long time tail of the 2D correlation function resulting from the incomplete averaging of the intermolecular dipole-dipole interaction has several consequences which for a given rate of diffusion are qualitatively different from those in bulk liquids. These are

a) a reduction of the intermolecular relaxation time by about an order of magnitude.

b) the spin-spin relaxation time, T_2, is always much less than the spin lattice relaxation time, T_1.

c) The intermolecular spin lattice relaxation rate has a logarithmic dependence upon spectrometer frequency.

This explains many of the effects which are not satisfactorily accounted for by relaxation theories of bulk liquids. Two dimensional motion will effect principally the intermolecular contribution to the relaxation time ($T_{1\ inter}$) and it is for this reason that the experimental separation of this contribution is of importance. For both methane on graphite and neopentane on graphite, isotopic dilution experiments, show $T_{1\ inter}$ to be the principal contribution to the total relaxation time. Further evidence that a two dimensional relaxation theory is the correct approach is provided by the logarithmic dependence on frequency of $T_{1\ inter}$ for neopentane on graphite (figure 7 and Table 6) and the semi quantitative agreement of the experimental relaxation times with those calculated using a two dimensional theory[22,23].However, one property of two dimensional relaxation that appers to have been overlooked is the way it can give rise to a T_1 minimum. This is most easily understood by first considering 'bound' two dimensional motion. Here the molecules cannot escape from the island on which they diffuse, and a 'to and fro' motion is established which is absent in bulk materials. The frequency (f) of this motion, which depends upon the rate of diffusion and the size of the island, can be approximated as $f = 2D/\pi l^2$ where D is the rate of diffusion and l the size of the island). For a restriction of 50 A° and a rate of diffusion of 1 x $10^{-9}m^2sec^{-1}$, then f equals 25 MHz. Changing the temperature will change D, and hence the variation of T_1 with temperature can show a minimum. Of course, for such an effect to be observed requires that the molecules retain a significant proportion of their correlations until they have crossed the island. For a molecule having a jump distance of 5 A° diffusing on an island of 50 A°, the crossing time is about ten times the molecular correlation time. Figure 6 shows that for a 2D liquid, correlations do persist until such long times, and hence such a mechanism can give rise to a T_1 minimum.

For molecules adsorbed on graphitised carbon blacks, bounded diffusion occurs because the adsorbate is restricted to the polyhedral raft of the graphite surface. Increasing this raft size will increase the distance of bounded diffusion and hence shift the T_1 minimum to higher temperatures or lower frequencies. Figure 2 shows the T_1 temperature dependence for 1.25 monolayers of neopentane adsorbed upon graphitised carbon blacks of different sizes. The lower set of curves show the T_1 minimum obtained using graphon, a particle having raft sizes of 80 A°[35]. Measurements at several frequencies were necessary, since with only one frequency the temperature window was insufficient to clearly identify the minimum. The upper curve shows the T_1 temperature dependence obtained using Stirling FT, a particle having a somewhat larger raft size (140 A°)[33]. In this case, although the minimum is not completely defined it is clearly shifted towards higher temperatures. This result strongly suggests that the T_1 minimum arises from fast molecular diffusion on a two dimensional island. Such an explanation requires no slow molecular motions and hence raises the contradictions presented in the preceeding sections. In the next paper in this book, Korb presents a theoretical description of two dimensional bounded relaxation[25,26]. Using the rates of diffusion measured either by neutron scattering or by the n.m.r. pulse field gradient method, and the known size of the restriction, he has calculated the variation of T_1 with temperature for both methane on graphite and for neopentane on graphite. The calculated T_1 minima are in good agreement with the experimental results and this substantiates the interpretation of these T_1 minima in terms of a bounded 2D motion.

Although Korb gives a full treatment of bounded 2D motion in his paper, a few qualitative remarks are in order. Firstly, because of the different shapes of the time correlation functions, bounded two and three dimensional motions will differ considerably. The exponential decay of the three dimensional function means that for distances greater than 10 A°, the influence of bounded diffusion should be minimal. On the contrary, the extremely slow decay of the two dimensional time correlation function means that effects will be observed for restrictions as large as several hundred Angstroms.

Bounded diffusion has been presented as establishing a "to and fro" motion, absent in bulk materials, which can coincide with the Larmor frequency. In fact, harmonics of this frequency also exist in the same way as striking a chord establishes different harmonics. The first harmonic is given by the size of the restriction,whilst the highest harmonic is given by the size of the molecule. The relative contributions of these harmonics depends upon the size ratio of the molecule to the restriction. For small restriction the first harmonic will dominate over the others. It is this contribution which is often closest to the Larmor frequency and dominates the observed effect. Hence, as might be expected the spin lattice relaxation times T_1 become shorter and the minimum deeper as the size of the restriction decreases.

A good illustration of this, is ammonia on graphite. Unlike methane or neopentane, ammonia does not wet the graphite surface but forms small multilayer clusters on the surface. The observed values of T_1 (Figure 4) are now an order of magnitude shorter than for the cases where wetting occurs. This difference in behaviour has been observed on a number of occasions and can now be attributed to the small size of the clusters formed in the non wetting case.

As the size of restriction becomes larger, the minimum become weaker until it eventually disappears. As in many cases, the size of the restriction is of the order of 100 A°, this explains the shallowness of so many of the T_1 minima reported in the literature. Any polydispersity in the size of the restriction will only serve to accentuate this effect.

Under some circumstances, it is possible to have two or three harmonic contributions which dominate with approximately equal contributions. This then gives rise to double or even triple minima. Some illustrations of this are given by Korb in the next paper in this book. They show a strong qualitative resemblance to the water in clay results shown in Figure 5. Another qualitative concurrence of theory with experiment is that the shape of the minimum is often unsymmetrical.

Increasing the size of the restriction displaces T_1 minima to higher temperatures or lower frequencies. For two dimensional diffusion, although the minimum becomes weaker as the size of the restriction increases, it never completely vanishes. This may be understood in the following way. As the size of the surface increases, the number of harmonic modes increases and although their intensity decreases, it decreases to a finite value and the harmonics never completely disappear. As already mentioned, for molecules diffusing on a hypothetically infinite 2D surface, the reencouter probability has a relatively large finite value, so that encounters will continue almost indefinitely. For temperatures where the frequency of these encounters is the Larmor frequency, then a T_1 minimum will occur. The effect of reducing the size of the plane to finite values, is to increase the frequency and effectiveness of this process to an observable level. It does not occur in infinite 3D systems because here the reencounter probability is negligibly small. In fact an infinite 2D system is only a particular case of a bounded 3D motion.

Finally, some investigators[1-6] have chosen to describe the T_1 minima observed in heterogeneous systems by a distribution of correlation times. In the strict meaning of the word this is correct, because if each separate harmonic is considered as contributing its own correlation time, then a distribution of correlation times does exist. However the majority of authors were thinking in terms of a distribution of molecular correlation times, and their interpretation still led to the conclusion that there was a slowing down of

Table 6 Values of the intermolecular relaxation rate $(T_{1\ inter})^{-1}\ sec^{-1}$, and the total relaxation rate $(T_{1\ total})^{-1} sec^{-1}$, for 1.25 monolayers of neopentane on graphite at different temperatures and resonance frequencies[24].

Temperature/K	221		238		253	
Spectrometer frequency/MHz	T_1 total^{-1}/ s^{-1}	T_1 inter^{-1}/ s^{-1}	T_1 total^{-1}/ s^{-1}	T_1 inter^{-1}/ s^{-1}	T_1 total^{-1}/ s^{-1}	T_1 inter^{-1}/ s^{-1}
10	2·40	2·40	1·90	1·80	1·20	0·90
25	1·85	1·65	1·60	1·30	1·30	0·90
60	0·95	0·10	1·15	0·45	1·35	0·75

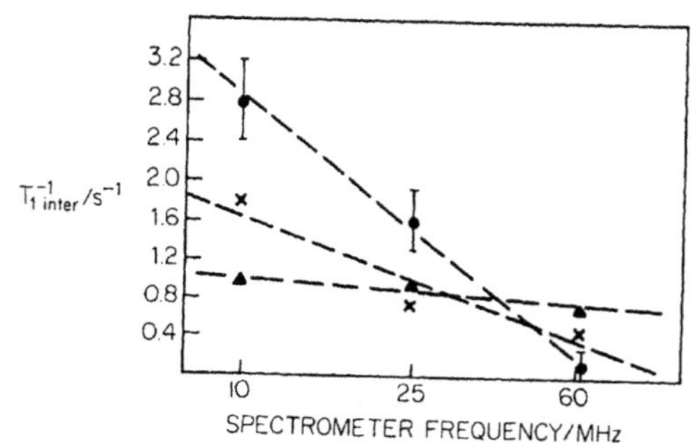

Figure 7 Variation of the intermolecular relaxation rate $T_{1\ inter}^{-1}$, with the logarithm of the spectrometer frequency for 1.25 monolayers of neopentane on graphite at different temperatures. 221 K(o), 240 K(x), 253 K(o).

molecular motion because of proximity to the surface. It is this conclusion which is at variance with other measurements and with which we do not agree.

VI. CONCLUSIONS

Many heterogeneous phases, show n.m.r relaxation times which are characteristic of solids and have normally been interpreted as such. This is nearly always in contradiction with neutron scattering measurements where they exist. For molecules adsorbed on graphite surfaces there is a mass of thermodynamic, structural, and dynamic measurements which show the mobile liquid nature of the phases formed in the temperature range where n.m.r. T_1 minima are observed. Interpretation of the minima using the theories for relaxation in bulk materials leads to mobilities which are to slow by two orders of magnitude. For the case of neopentane on graphite and methane on graphite, where rather more extensive T_1 measurements have been made, the relaxation behaviour is in agreement with that from a finite two dimensional liquid having the correct rate of diffusion. This interpretation is substantiated by the displacement of the minimum towards higher temperatures as the size of the restriction is increased.

Since in many heterogeneous systems diffusion is spatially bounded, these results show that one cannot apply the bulk theories of relaxation indescriminately. Failure to consider the effects of restricted diffusion may result in erroneous conclusions. It seems highly likely that the majority of proton n.m.r. relaxation times in heterogeneous systems have been misinterpreted. In particular the conclusions regarding 'bound' water in biological systems should be reexamined.

Apparently proton n.m.r. relaxation time measurements in heterogeneous systems yield almost no information about the dynamics of interfacial molecules. However, provided the rate of diffusion is known, they may provide information about the topology of the diffusive motions over distances which can extend to several hundred Anstroms. It is possible that measurements of this kind might eventually be used to obtain structural information where more direct methods fail. One such possible field of application is heterogeneous nucleation.

ACKNOWLEDGMENTS

I would like to acknowledge discussions with R.K. THOMAS, J.W. WHITE, P. FRIES and J.P. KORB.

REFERENCES

1. K.J. PACKER, Progress in NMR spectroscopy, Vol.3 Pergamon (London) 1969.
2. H.A. RESING, Advances in Molecular Relaxation Processes, 1, 109, (1967) and 3, 199 (1972).
3. J.J. FRIPIAT ; Catalysis Review, 5, 269 (1971).
4. H. PFEIFER, NMR basic principles and progress, Vol. 7, Springer-Verlang ; Physics Reports, 7, 293, (1976).
5. R. COOKE, I.D. KUNTZ, Ann. Rev. Biophy. Bioeng. 3, 95, (1974).
6. J. TABONY, Progress in N.M.R. spectroscopy, Vol. 14, (1980).
7. J. TABONY, J.W. WHITE, Mol. Phys. 40, 1209 (1980).
8. J. TABONY, G. BOMCHIL, N. HARRIS, M. LESLIE, J.W. WHITE, P. GAMLEN, R.K. THOMAS, T.D. TREWERN, J. Chem. Soc., Far. Trans. I, 75, 1570, (1979).
9. J.W. RIEHL, K. KOCH, J. Chem. Phys., 57, 2199 (1972).
10. P. GAMLEN, R.K. THOMAS, T.D. TREWERN, G. BOMCHIL, N. HARRIS, M. LESLIE, J. TABONY, J.W. WHITE, J. Chem. Soc., Far. Trans. I, 75, 1553, (1979).
11. J.P. COULOMB, M. BIENFAIT, P. THOREL, Phys. Rev. Lett, 42, 733, (1979).
12. J. TABONY, T. COSGROVE, Chem. Phys. Lett. 67, 103, (1979).
13. J. TABONY, S. ROSER, Mol. Phys. 44, 329, (1981).
14. T. SPRINGER, In 'Springer tracts in modern physics', n° 64, (1972), Springer Verlag.
15. F. VOLINO, In 'Microscopic structure and dynamics of liquids', (Ed. J. Dupuy, A.J. Dianoux) NATO Series B33, Plenum (1978).
16. R.K. THOMAS, 'Inelastic and quasielastic neutron scattering spectroscopy', Mol. Spectroscopy (Chem. Soc. London) 6, 232, (1979).
17. F. VOLINO, A.J. DIANOUX, Mol. Phys.,41, 271, (1980).
18. P. HALL, D.K. ROSS, Mol. Phys. 42, 673, (1981).
19. A.A. KOKIN, A.A. IZMEST'EV, Russ. J. Phys. Chem., 39, 309, (1965).
20. D. BECKERT, Ann. Physik, 20, 220, (1967).
Ann. Physik, 23, 98, (1969).
21. A. AVOGADRO, M. VILLA, J. Chem. Phys. 66, 2359, (1977).
22. P.H. FRIES, J. Phys. Lett. 42, L257 (1981); 42, L513 (1981).
23. P.H. FRIES, Mol. Phys. 48, 503 (1983).
24. P. BRULET, H.M. Mc CONNELL, Proc. Nat. Acad. Sci. U.S.A., 72, 1451, (1975).
25. J.P. KORB, D.C. TORNEY, H.M. Mc CONNELL, J. Chem. Phys.

78, 5782 (1983).
26. J.P. KORB, M. WINTERHALTER, H.M. Mc CONNELL, J. Chem. Phys. 80, 1059 (1984).
27. E.O. STEJSKAL, J.E. TANNER, J. Chem. Phys. 42,288, (1965).
28. J.E. TANNER, J. Chem. Phys. 52, 2523, (1970).
29. S. ROSS, J.P. OLIVIER, 'On physical adsorption', Interscience. (1964).
30. 'Phases bidimensionnelles', J. Physique (Paris) Colloq. C4 (1977).
31. P.M. HORN, R.J. BIRGENAU, P. HEIMEY, E.M. HAMMONDS, Phys. Rev. Lett. 41, 961 (1978).
32. J.W. WHITE, R.K. THOMAS, T. TREWERN, I. MARLOW, G. BOMCHIL, Surface Science, 76, 13, (1978)
33. N.N. AVGUL, A.V. KISELEV, 'Chemistry and Physics of Carbon', Vol. 6, Marcel Dekker, New York, (1970).
34. J.G. DASH, 'Films on solid surfaces', Academic Press, (1975).
35. P.H. GAMLEN, J.W. WHITE, J. Chem. Soc. Faraday II, 72, 446, (1976).
36. P.H. GAMLEN, R.K. THOMAS, T.D. TREWERN, G. BOMCHIL, N.M. HARRIS, M. LESLIE, J. TABONY, J. Chem. Soc. Faraday I, 75, 1542, (1979).
37. I. MARLOW, R.K. THOMAS, T.D. TREWERN, J.W. WHITE, J. Physique (Paris), C4, 38, 14, (1977).
38. G. BOMCHIL, A. HULLER, T. RAYMENT, S. ROSER, M. SMALLEY, R.K. THOMAS, J.W. WHITE, Phi. T. Roy. B 290, 537, (1980).
39. A. THOMY, X. DUVAL, J. Chim. Phys. 66, 1966 (1969) and 67, 1101 (1970).
40. T. RAYMENT, R.K. THOMAS , (private communication).
41. G. BOMCHIL, N. HARRIS, J. TABONY, J.W. WHITE, P. GAMLEN, R.K. THOMAS, T. TREWERN, J. Chem. Soc. Faraday I, 75, (1979)
42. D.J. CEBULA, R.K. THOMAS, J.W. WHITE, Clay and Clay Minerals, 29, 241, (1981).
43. J.J. FRIPIAT, J. Physique, C4, 38, 44, (1977).
44. J. CHARVOLIN, P. RIGNY, J. Chem. Phys., 58, 3999 (1973).
45. A.J. DIANOUX, Chemical information, ILL Internal report, (1978), Institut Laue Langevin, Grenoble, France
46. J. TABONY, Chem. Phys. Lett. (in press).
47. B. FUNG, J. WITSCHEL, L. Mc AMIS, Biopolymers, 13, 1767, (1974).
48. J.W. WHITE, In 'Water at Interfaces'. I.L.L. report 81T055S, Institut Laue Langevin, Grenoble, France, (1981).
49. A. MILLER, G.T. JENKIN, J.W. WHITE, Experimental report no. 0804 004, I.L.L. Annual Reports (1976), Institut Laue Langevin, Grenoble, France.

THEORY OF SPIN RELAXATION BY TRANSLATIONAL DIFFUSION IN TWO-DIMENSIONAL SYSTEMS

J.-P. Korb
Centre de Mécanique Ondulatoire Appliquée, 23, rue du Maroc
75019 Paris, France

ABSTRACT. A statistical treatment is used to calculate the correlation functions and the spectral densities for the secular and nonsecular parts of the dipolar relaxation processes when molecular diffusion is bounded to finite two-dimensional planar surfaces. This leads to convergent formulas for spin-spin T_2^{-1} and spin-lattice T_1^{-1} relaxation rates, no matter how large these surfaces are. This contradicts the logarithmic divergences found for these rates in previous theories for unbounded diffusion. The theoretical diffusion rates T_1^{-1} and T_2^{-1} for finite planar two-dimensional systems are found to have a number of unique properties that should facilitate the interpretation of magnetic resonance spectra of molecules physically adsorbed on solid surfaces. For exemple the important size effect in the diffusion (or temperature) and frequency dependences of T_1^{-1}, has proven useful in the interpretation of the T_1 minima observed in physisorbed heterogeneous systems. The theory presented here appears directly applicable to the experiments described by J. Tabony in the preceding paper.

1. INTRODUCTION

It is well known that a number of physical and chemical properties of two-dimensional (2D) systems are expected to be remarkably different from those of three-dimensional (3D) systems. Such properties include chemical reaction rates (1), radical recombination (2), heterogeneous catalysis (3), transport phenomenae (4), the nature and order of phase transitions (5), lateral diffusion (6) and various correlation functions (7-17) especially those involved in the calculation of magnetic resonance line shapes.

It is also well known that magnetic resonance can be used to obtain dynamical information on two-dimensional systems. Systems of interest include atomic (18) or molecular (19) fluids on solid surfaces, grafted polymers on silicagel (20), graphite intercalation compounds (21), electrolytes layers of hydrated chalcogenides (22), clay-water compounds (23), ionic and superionic conductors (24), smectic liquid crystals (25), micelles (26) and model (27) and natural (28) biological membranes.

The purpose of the present paper is to provide a theoretical inter-

R. Daudel et al. (eds.), Structure and Dynamics of Molecular Systems, 245–267.

pretation of spin-relaxation rates for two-dimensional systems. In evaluating such theoretical work the question is often raised as to whether or not a particular physical system is truly two-dimensional. Certain problems as those described in the preceding paper by J. Tabony (29) are either strictly 2D, or sufficiently close to 2D that 2D theory must be used. This problem is troublesome since, as is well known, the average correlation time for secular dipolar interactions is infinite for an infinite 2D system, even though the system is in every respect a 2D fluid where there is rapid molecular motion (8, 29). Consequently it might appear that one cannot use the well-known theory of Kubo (30) and Abragam (31) to calculate relaxation rates for such systems. Several attempts have been made to eliminate such divergences in the theory (15-17, 32). For instance, if there is some rapid spin-lattice relaxation process not explicitly dependent on 2D diffusion, then this provides a natural cutoff of the otherwise long-time decay of pairwise dipolar correlations. This is sometimes an ad hoc procedure which has to be checked a posteriori (32). An other procedure, which is applied in the present work, consists in a restriction of the diffusive motions in the case of finite surfaces (15-17). This enhances the time decay of the secular dipolar correlation function and leads to a finite value of the corresponding spectral density even at zero frequency. Thus the theory of Kubo (30) and Abragam (31) can be applied to a two-dimensional planar spin system providing certain conditions involving the area of the 2D surface, the spin correlation, and the translational diffusion coefficient (15).

In the present paper we develop the theory of the relaxation behavior of an ensemble of spins undergoing dipolar interactions, which diffuse on a finite but arbitrarily large planar 2D surface. We provide equations for both spin-lattice T_1^{-1} and spin-spin T_2^{-1} relaxation rates. These theoretical relaxation rates are found to have a number of unique properties that should facilitate the interpretation of experimental data. A particularly significant result of these calculations is the large absolute values of the relaxation rates T_1^{-1} and T_2^{-1} compared to 3D systems for a given diffusion coefficient. Another characteristic is that T_1^{-1} is smaller than T_2^{-1} for fast diffusion, whereas these rates are equal in 3D systems. Under certain conditions the relaxation rate T_1^{-1} is strongly dependent on the size of the 2D surface on which atoms or molecules diffuse. Moreover the shape of the surface (planar or spherical), which is of particular importance in the description of the 2D dynamics, can profoundly alter the frequency and temperature dependences of the spin relaxation rates. Some of these theoretical characteristics of 2D motion have been observed many times experimentally. For instance the preceding paper by J. Tabony use these characteristics to give an interpretation of the spin relaxation of molecules physisorbed on solid surfaces. The angular anisotropies of the relaxation rates for planar systems have also proven useful for the evidence of spin diffusion in 2D systems (9,33 ,34).

In the final part of the paper we test our theoretical expressions of spin relaxation rates against nuclear relaxation experiments of molecular fluids physisorbed on graphite beads (19).

2. THEORY OF SPIN RELAXATION IN 2D SPIN SYSTEMS

We consider a large number M of molecules, each having N_o nuclear spins I of the same species, which diffuse on a finite 2D surface of area A in presence of a strong constant magnetic field. We study the case of a relaxation process due to the modulation of dipolar interactions by the translational diffusion of spins. In the rapid modulation Kubo approximation the spin-lattice T_1^{-1} and spin-spin relaxation rates are expressed in terms of secular $J^{(o)}(o)$ and non-secular $J^{(\lambda)}(\lambda\omega_o)$ ($\lambda = 1,2$) spectral densities according to the well-known (30,31) equations

$$T_1^{-1} = \frac{3\hbar^{-2}}{2} I(I+1)\left[J^{(1)}(\omega_o) + J^{(2)}(2\omega_o)\right], \qquad (1)$$

$$T_2^{-1} = \frac{3\hbar^{-2}}{8} I(I+1)\left[J^{(o)}(o) + 10\, J^{(1)}(\omega_o) + J^{(2)}(2\omega_o)\right]. \qquad (2)$$

An expression for the spin-lattice relaxation rate in the rotating frame $T_{1\rho}^{-1}$ can be obtained by replacing $J^{(o)}(o)$ in Eq. (2) by $J^{(o)}(2\omega_1)$, ω_1 being the angular frequency of the rotating field inducing the transitions (35). We have shown previously (15) that the Kubo theory (30) which leads to Eqs. (1) and (2) is applicable to finite 2D spin systems, providing that the translational diffusion is sufficiently rapid for given values of the spin concentration N/A ($N \equiv MN_o$) and the area A.

In Eqs. (1) and (2) the spectral densities are proportional to the exponential Fourier transforms of the corresponding dipolar correlation functions $G^{(\lambda)}(\tau)$

$$J^{(\lambda)}(\omega_o) = \int_{-\infty}^{+\infty} G^{(\lambda)}(\tau) \exp(-i\omega_o\tau)\, d\tau, \qquad (3)$$

where

$$G^{(\lambda)}(\tau) = \frac{N}{N_o^2} \sum_{i=1}^{N_o}\sum_{j=1}^{N_o} \left\{ \langle E_{ij}^{(\lambda)}(o)\, E_{ij}^{(-\lambda)}(\tau)\rangle - \langle E_{ij}^{(\lambda)}(o)\rangle^2 \right\}. \qquad (4)$$

These latter functions describe the persistence of the fluctuations of the λ component of a second rank tensor $E_{ij}^{(\lambda)}$ characterizing the dipolar interactions between two arbitrary spins i and j belonging to two different test molecules and given in the laboratory frame by

$$E_{ij}^{(\lambda)} = \gamma^2\hbar^2 C_\lambda Y_2^\lambda(\Theta_{ij}, \phi_{ij})/r_{ij}^3, \qquad (5)$$

where the z axis is parallel to the applied field and where r_{ij}, Θ_{ij}, ϕ_{ij} are the spherical coordinates of the interspin vector $\tilde{r}_{ij}$ at time o or

τ, $C_o = -\sqrt{16\pi/5}$, $C_1 = -\sqrt{8\pi/5}$, $C_2 = \sqrt{32\pi/15}$, and Y_2^λ are the normalized spherical harmonics. The first average in Eq. (4), noted < > corresponds to the ensemble averages over all the positions of spins i and j at times o and τ. The second average in Eq. (4) corresponds to the ensemble average over the stationary distributions of spins achieved at equilibrium, which gives a non zero contribution for a finite 2D system. This latter average has to be substracted in order that all the correlations disappear when $\tau \to \infty$. There are some great simplificaions in the dipolar correlation functions presented in Eq. (4). Mainly when one considers spherical or quasi-spherical molecules in liquid phases. Basically this allows the stochastic independence between the translational and rotational molecular motions (36). The rotation of a spherical molecule having practically no effect on the local structure of the liquid considered, one can also express the correlation function given in Eq. (4) as a product of a centered-spin correlation function $G_t^{(\lambda)}(\tau)$ for the translational motion and an excentred-spin correlation function $G_r^{(\lambda)}(\tau)$ for the rotational motions (37-39). Since this latter function is not modified by the spatial restriction of the diffusion, a simple application of the expressions previously derived (37,39) will be considered. On the contrary, $G_t^{(\lambda)}(\tau)$ is very dependent on the spatial restriction through the ensemble averages in Eq. (4) usually defined in terms of integral averages over the normalized conditional probability $P(\underset{\sim}{r}_1, \underset{\sim}{r}_2 | \underset{\sim}{r}_1(o), \underset{\sim}{r}_2(o))$, by

$$G_t^{(\lambda)}(\tau) = \frac{N}{A^2}\{\int d^2\underset{\sim}{r}_1(o) \int d^2\underset{\sim}{r}_2(o)\, g\,(\underset{\sim}{r}_2(o) - \underset{\sim}{r}_1(o))\, E_{12}^{(\lambda)}(o)\, \times$$

$$\times \int d^2\underset{\sim}{r}_1 \int d^2\underset{\sim}{r}_2\, P(\underset{\sim}{r}_1, \underset{\sim}{r}_2 | \underset{\sim}{r}_1(o), \underset{\sim}{r}_2(o)) E_{12}^{(-\lambda)}(\tau) - \bar{E}^{(\lambda)^2}\}. \quad (6)$$

Here $g\,(\underset{\sim}{r}_2(o) - \underset{\sim}{r}_1(o))$ is the initial radial distribution function (rdf) which describes the very non uniform radial molecular distribution or local order observed at short intermolecular distances (40,41). The conditional probability P describes the dynamics of the spins-beared molecules in a Markovian process. Its expression depends on the geometry, dimensionality, and on the initial and boundary conditions assumed for the diffusion. Finally $\bar{E}^{(\lambda)}$ is the integral average of the dipolar interactions over the limiting P value achieved when $\tau \to \infty$ (i.e. the rdf g). In the following subsections we shall propose analytical expression for the rdf and the calculations of dipolar correlation functions (Eq. (6)), spectral densities and spin relaxation rates for planar systems.

2.1. Radial molecular distribution function

Experimentally one obtains this function by Fourier transform of the measured structure factors of X-ray (41) scattering intensities.

Theoretically several attempts have been proposed, for bulk liquids, to calculate this rdf on the basis of hard-sphere solutions to the Percus Yevick equation (42) given by Throop and Bearman (43) and corrected near to the distance of minimal approach δ between two molecules (for dense fluids) by the computer dynamics studies of Verlet ans Weiss (44). For 2D systems there are only some Monte Carlo computations of hard-disk (45) or Lennard Jones (46) rdf at various reduce densities $\rho_r \equiv \rho/\rho_o$ compared to the hexagonal close-packed density of hard disks (ρ_o = 0.9069). Since we consider two dimensional fluids where the local order is close to the hexagonal close-packing such as encountered in physisorbed molecular fluids (19, 29), we use a simple theoretical expression for the rdf as a sum of an exponential near to the core (intermolecular distance $r \sim \delta$) (44) and of a superposition of pondered Gaussians centered on the sucessive shells of neighbours (14). The relative positions, widths and intensities of these Gaussians depend for a given δ, on the relative density ρ_r as shown in Fig. 1. Our expression follows relatively close to the values obtained from the Monte Carlo g(r) (45), especially in the spatial range $1 \leq r/\delta \leq 2$ where the effects of g (r) on the correlation function $G(\tau)$ are the most important (see the insert of Fig. 1).

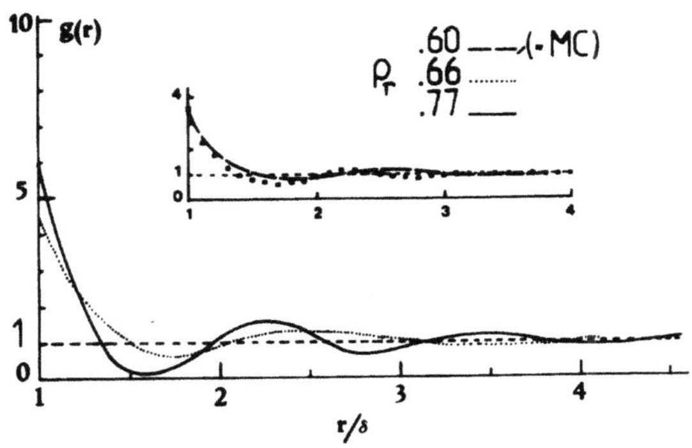

Figure 1. Plots of the radial molecular distribution function g (r) vs the number of intermolecular distances r in unit of distance of minimal approach δ for three different reduce densities $\rho_r \equiv \rho/\rho_o$ (compared to the close-packed density for hard disks ρ_o = 0.907). The dotted line represents the case of uniform density. In insert we compare our theoretical expression of g(r) for ρ_r = 0.6 against the values obtained from the Monte Carlo (from Ref. 45).

2.2. Dipolar correlation functions, spectral densities and spin relaxation rates for planar spin systems

The dynamics of the spins is described by the conditional probability P. For planar fluids with a local order close to the hexagonal closed packing, one can make the assumption of Oppenheim and Bloom (47) that the molecular translational motions take place via very small individual jumps in straight line with a constant acceleration (i.e. with intermolecular forces $F(r) \propto \nabla \ln g(r)$ independent of time). This allows to approximate P in Eq. (6) by gP_B where P_B is the solution of a Brownian (force free) diffusion equation. Now the explicit form of P_B depends on the geometry and boundary conditions used. Let consider as a finite planar surface, a square of side a ($0 \leq x,y \leq a$). We take a unit initial instantaneous material line (parallel to the normal $\underset{\sim}{n}$ to the surface) passing through the point $x_i(o), y_i(o)$ ($i = 1$ or 2) at time $\tau = o$, and no diffusive flow out of the square limits. In that case, analytical solution of the diffusion equation (48) leads to

$$P_B = \prod_{i=1,2} P_{Bi}, \tag{7a}$$

$$P_{Bi} = a^{-2}\left[1 + 2\sum_{\ell=1}^{\infty} \exp(-\ell^2\pi^2 D\tau/a^2)\cos\left(\frac{\ell\pi x_i(\tau)}{a}\right)\cos\left(\frac{\ell\pi x_i(o)}{a}\right)\right] \times \tag{7b}$$

$$\times \left[1 + 2\sum_{\ell'=1}^{\infty} \exp(-\ell'^2\pi^2 D\tau/a^2)\cos\left(\frac{\ell'\pi y_i(\tau)}{a}\right)\cos\left(\frac{\ell'\pi y_i(o)}{a}\right)\right]. \tag{7c}$$

This form of the Green's function P_B assumes that volume exclusion is not significant in the correlation of the interparticle distance. The different terms in the series present in Eqs. (7a) and (7b) can be regarded as coming from the superposition of the normal modes of diffuse relaxation in a square. To calculate the dipolar correlation functions, we substitute P_B given in Eqs. (7a)-(7c)) into Eq. (6) and we express the dipolar interactions (Eq.(5)) in terms of the spherical harmonics in the reference frame having its $\underset{\sim}{z}'$axis along the normal $\underset{\sim}{n}$ to the surface (introducing the angle $\beta \equiv (\underset{\sim}{z},\underset{\sim}{z}')$). After some algebraic manipulations (see the Appendixes of Refs. 15 and 16), one finds

$$G_t^{(\lambda)}(\tau) = 2\pi^2 \frac{N}{A}\frac{\gamma^4\hbar^4}{a^2\delta^2} \times$$

$$\times \left\{u_\lambda(\beta)\left[\sum_{\ell=1}^{\infty} b_\ell^2 \exp(-\tau/\tau_\ell) + \frac{1}{2}\sum_{\ell=1}^{\infty}\sum_{\ell'=1}^{\infty} b_{\ell\ell'}^2 \exp(-\tau/\tau_{\ell\ell'})\right] + \right.$$

$$\left. + v_\lambda(\beta)\sum_{\ell=1}^{\infty} a_\ell^2 \exp(-\tau/\tau_\ell)\right\}, \tag{8}$$

with the angular factors

$$u_o(\beta) = (1-3\cos^2\beta)^2 \;,\; u_1(\beta) = \sin^2\beta\cos^2\beta,\; u_2(\beta) = \sin^4\beta,$$

$$v_o(\beta) = 9\,u_2(\beta), \quad v_1(\beta) = u_1(\beta) \;,\; v_2(\beta) = (1+\cos^2\beta)^2, \quad (9)$$

responsible to the important angular anisotropies in the spin relaxation in 2D systems. $G_t^{(\lambda)}(\tau)$ is expressed as sums of single- (ℓ) and double -($\ell\ell'$) mode contributions. Each of these contributions is a product of a time-decaying exponential and an intensity factor. The decay constants of these modes are given in terms of the area and the diffusion coefficient by

$$\tau_\ell = a^2/(2\pi^2 D\ell^2) \;,\; \tau_{\ell\ell'} = a^2/[2\pi^2 D(\ell^2 + \ell'^2)]. \quad (10)$$

The intensity factors b_ℓ^2 , $b_{\ell\ell'}^2$, c_ℓ^2 are given by the following integrals

$$b_\ell = \delta \int_\delta^{\sim n\delta} \frac{g(r)}{r^2} J_o(\pi\ell \frac{r}{a})\, dr, \qquad (n \sim 10) \quad (11a)$$

$$b_{\ell\ell'} = \delta \int_\delta^{\sim n\delta} \frac{g(r)}{r^2} J_o(\pi\sqrt{\ell^2+\ell'^2}\, \frac{r}{a})\, dr, \quad (11b)$$

$$c_\ell = \delta \int_\delta^{\sim n\delta} \frac{g(r)}{r^2} J_2(\pi\ell \frac{r}{a})\, dr, \quad (11c)$$

where J_o and J_2 are Bessel functions of integer orders 0 and 2, respectively. Numerical calculations of these integrals show that the decreases of b_ℓ^2, $b_{\ell\ell'}^2$, and c_ℓ^2 with ℓ (or $\sqrt{\ell^2 + \ell'^2}$) are size-dependent (Fig.2). These decreases are sufficient to give a convergent result (from Eq. (8)) for the total power $G_t^{(\lambda)}(o)$ available in the relaxation process. Substituting the dipolar correlation functions (Eq.(8)) into Eq. (3) gives the following spectral densities

$$J^{(\lambda)}(\omega_o) = 2\pi^2 \frac{N}{A} \frac{\gamma^4\hbar^4}{a^2\delta^2} \times$$

$$\times \{u_\lambda(\beta) [\sum_{\ell=1}^{\infty} b_\ell^2 L_\ell(\omega_o) + \frac{1}{2}\sum_{\ell=1}^{\infty}\sum_{\ell'=1}^{\infty} b_{\ell\ell'}^2 L_{\ell\ell'}(\omega_o)] +$$

$$+ v_\lambda(\beta) \sum_{\ell=1}^{\infty} c_\ell^2 L_\ell(\omega_o) \}, \quad (12)$$

where L_ℓ and $L_{\ell\ell'}$ are the Lorentzian spectral densities associated to each single-or double-mode contributions

$$L_\ell(\omega_o) = 2\tau_\ell/(1+\omega_o^2\tau_\ell^2) \;,\; L_{\ell\ell'}(\omega_o) = 2\tau_{\ell\ell'}/(1+\omega_o^2\tau_{\ell\ell'}^2). (13)$$

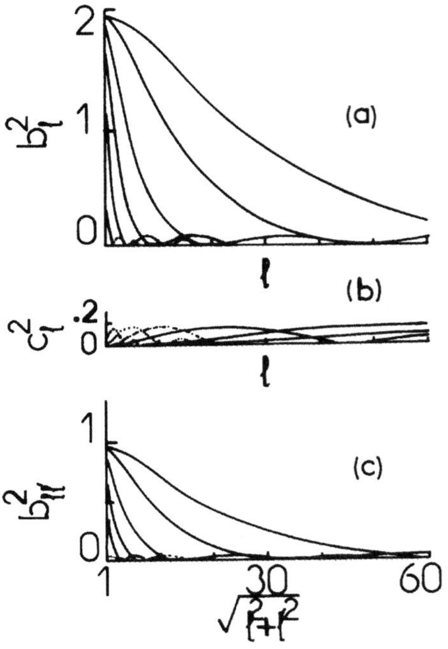

Figure 2. Plots of the intensity factors of the single-diffusive modes b_ℓ^2 (a), c_ℓ^2 (b) and double-diffusive modes $b_{\ell\ell'}^2$ (c) vs the index ℓ (or $\sqrt{\ell^2+\ell'^2}$, here $\ell = \ell'$) of the modes for δ = 6.5Å and different values of a (20, 50, 100, 200, 500 and 1000 Å from left to right) ($g(r)\neq 1$, ρ_r=0.77).

It is well known in spin relaxation theory for bulk systems that, for a given Larmor frequency ω_o, the spectral density is largest when one has a single correlation time τ_c such as $\tau_c = 1/\omega_o$. thus for a given total power $G(o) = 1/2\pi\int_{-\infty}^{+\infty} L(\omega) d\omega$ available in the relaxation spectrum the transition probability of inducing relaxation is also maximum when $\tau_c \sim 1/\omega_o$ (31). In the present work, according to Eqs. (8) and (12), we do not have a single correlation time but rather a number of correlation times corresponding to the different modes of diffusion on a finite planar surface (Eq.(10)). The single-modes of relaxation $b_\ell^2 L_\ell(\omega_o)$ come from the products of non uniform cosine distributions for both particles along a given direction (say x) $\cos \pi\ell \frac{x_1}{a}(\tau) \cos \pi\ell \frac{x_2}{a}(\tau)$ at time τ by a uniform distribution along the other direction. The double-modes of relaxation $b_{\ell\ell'}^2 L_{\ell\ell'}(\omega_o)$ come from the products of non uniform cosine distributions for both particles and directions $\cos \pi \ell x_1(\tau)/a \cos \pi \ell x_2(\tau)/a \cos \pi \ell' y_1(\tau)/a \cos \pi \ell' y_2(\tau)/a$. Due to the symmetry of the square, the exchange of ℓ and ℓ' in these products will not change the spectral densities in Eq. (12) ($b^2_{\ell\ell'} L_{\ell\ell'}(\omega_o) = b^2_{\ell'\ell} L_{\ell'\ell}(\omega_o)$). One then notes a formal analogy with these results and those of the energetic solutions of the Schrödinger equation for a particle in a square well potential. The double-mode of relaxation are then composed of a superposition of "doubly degenerate states of relaxation" $\ell\ell'(\ell \neq \ell')$ and of "non degenerate states of relaxation" $\ell = \ell'$. Naturally for a rectangular planar system, this double degeneracy disappears ($b^2_{\ell\ell'} \neq b^2_{\ell'\ell}$). Now let vary the size of the square system through the size parameter a/δ (fixed δ a varies) for given frequency ω_o and diffusion coefficient D. For very small sizes (a of the order of δ) , the intensity factors b^2_ℓ, $b^2_{\ell\ell'}$ decrease rapidly (Fig.2) and $\omega_o\tau_\ell$ (or $\omega_o\tau_{\ell\ell'}$) stays <<1. This restricts the series giving the spectral density (Eq.(12)) on a small number of low-frequency discrete modes of relaxation whose frequencies correspond to the backwards and forwards molecular motions on the finite area of the system. When the size of the system increases, many more modes are involved in the relaxation process (Fig. 2). One sees in Figs.3 and 4 that this produces a longer persistence of the dipolar correlations at long times (Fig. 3) and consequently a modification in the frequency dependence of the spectral density (Fig. 4). Introducing in Eq. (12) the usual translational correlation time $\tau_t = \delta^2/2D$, one sees in Fig. (4) that when a increases, $J^{(\lambda)}(\omega_o)/\tau_t$ increases more and more rapidly for $\omega_o\tau_t$<<1 and less and less rapidly for $\omega_o\tau_t$>>1 (in order to conserve the total power $1/2\pi \int_{-\infty}^{+\infty} J(\omega)$. That effect is almost equivalent for λ = o and for λ = 1 or 2. This reveals the difficulty to average out the $1/r^3$ dipolar contribution in a 2 D system. For almost infinite size (a>>δ) , one can replace in Eq. (8) and (12) the sums by integrals over a continuous distribution of available modes. Assuming some quasi-constant values for the intensity factors b^2_ℓ, $b^2_{\ell\ell'}$ and c^2_ℓ from ℓ = 1 to the mode $\ell_{max} \sim a/\pi\delta$ associated to the highest possible frequency $1/\tau_t$ necessary for a single molecular step of length δ, one has

$$\lim_{\tau \gg \tau_t} G_N^{(\lambda)}(\tau) \propto 2\tau_t/\tau, \tag{14}$$

$$J^{(\lambda)}(\omega_o)/\tau_t \to \frac{\pi}{4}\frac{N}{A}\frac{\gamma^4\hbar^4}{\delta^4}\bar{b}_d^2\, u_\lambda(\beta)\ln(1 + 1/\omega_o^2\tau_t^2). \tag{15}$$

These <u>limit-expressions</u> are only due to the double-mode terms assuming a constant intensity factor $\bar{b}^2_d$ ($\sim$1)(Fig.2) for $b^2_{\ell\ell'}$. They correspond to the results of previous theories of relaxation for 2D infinite systems (13,14). In reality the intensity factors are non constant over that range of ℓ,ℓ' values. Our numerical expressions of spectral densities (Eq.(12)) lead to finite results, even at zero frequency, no matter how large the areas of the systems are. In consequence Eqs.(14) and (15) must be considered only as analytical limit-expressions.

In order to obtain the expressions for the spin-relaxation rates in finite planar systems, one substitutes Eq. (12) into Eqs. (1) and (2):

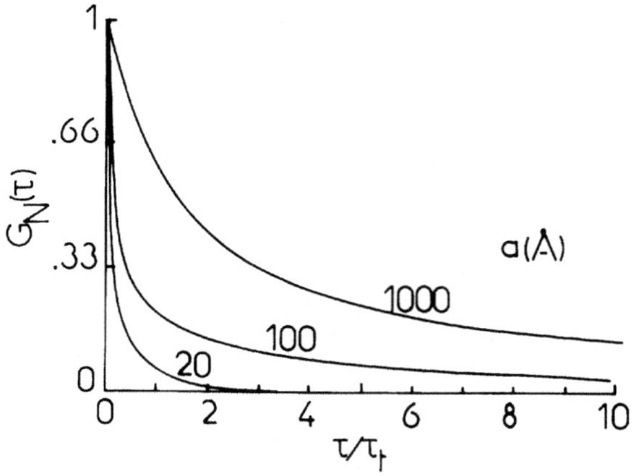

Figure 3. Semilogarithmic plots of the time decays of the reduced dipolar correlation function, $G_N^{(o)}(\tau) = G_t^{(o)}(\tau)/G_t^{(o)}(o)$, for $\beta = 90°$, $\delta = 6.5Å$, but varying the size of the system through the parameter a in Angstroms.

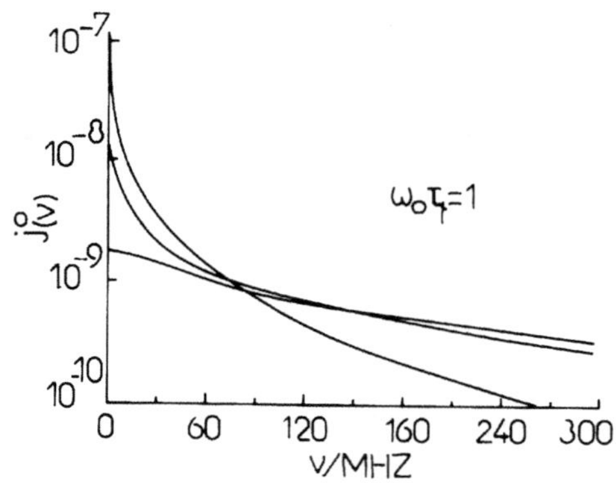

Figure 4. Semi logarithmic plots of the variations of the reduced spectral density with ν_o for $\beta = 90°$, $\delta = 6.5A$ but varying the size of the system through the parameter a in Angstroms.

$$T_1^{-1} = [3 \frac{N}{A} \frac{\gamma^4 \hbar^2}{\delta^2} I(I+1)] (\pi/a)^2 \times$$

$$\times \sum_{\lambda=1,2} \{u_\lambda(\beta)[\sum_{\ell=1}^{\infty} b_\ell^2 L_\ell(\lambda\omega_o) + \frac{1}{2} \sum_{\ell=1}^{\infty} \sum_{\ell'=1}^{\infty} b_{\ell\ell'}^2 L_{\ell\ell'}(\lambda\omega_o)] +$$

$$+ v_\lambda(\beta) \sum_{\ell=1}^{\infty} c_\ell^2 L_\ell(\lambda\omega_o)\} , \qquad (16)$$

$$T_2^{-1} = [3 \frac{N}{A} \frac{\gamma^4 \hbar^2}{\delta^2} I(I+1)] (\pi/a)^2 \times$$

$$\times \sum_{\lambda=0,1,2} k_\lambda \{u_\lambda(\beta)[\sum_{\ell=1}^{\infty} b_\ell^2 L_\ell(\lambda\omega_o) + \frac{1}{2} \sum_{\ell=1}^{\infty} \sum_{\ell'=1}^{\infty} b_{\ell\ell'}^2 L_{\ell\ell'}(\lambda\omega_o)] +$$

$$+ v_\lambda(\beta) \sum_{\ell=1}^{\infty} c_\ell^2 L_\ell(\lambda\omega_o)\} , \qquad (17)$$

where $k_o = 1/4$, $k_1 = 5/2$, $k_2 = 1/4$.

2.3. Angular anisotropies of the spin relaxation rates

The anisotropies of these spin relaxation rates (due to the angular factors $u_\lambda(\beta)$ and $v_\lambda(\beta)$ given in Eq. (9)) are illustrated in Figs. 5 a-d. In Figs. 5a and 5b we have plotted the angular variations of $T_1^{-1}(\beta)$ and $T_2^{-1}(\beta)$ calculated from Eqs. (16) and (17) for fixed values of the size parameter $\delta/_a = 0.05$, taking g (r) = 1, D = 10^{-8} cm^2/s but varying ω_o through the dimensionless coefficient $\omega_o\tau_t$.

Even in the fast motion-low field case ($\omega_o\tau_t \ll 1$) T_2 is less than T_1 for most orientations in contrast to their equality in the familiar 3D case. In Figs. 5c and 5d we have shown the angular variations of these rates for fixed values of $\omega_o\tau_t = \pi/10$ but varying $\delta/_a$. For a fixed orientation, say $\beta = 0°$, and when $\delta/_a$ decreases, one sees that $T_1^{-1}(0°) \propto 4 \sum_{\ell=1}^{\infty} c_\ell^2 L_\ell (2\omega_o)$ tends to small but finite values whereas T_2^{-1} (0°) tends to a much higher finite values.

2.4. Diffusion dependences of the spin relaxation rates

We have displayed in Fig.6(a)-(d) the diffusion dependences of T_1^{-1} calculated from Eqs. (16) and (17) for δ = 6.5Å, ν_o = 10MHz, β=0° and 90°, when varying the size of the system through the parameter a (in Å) in

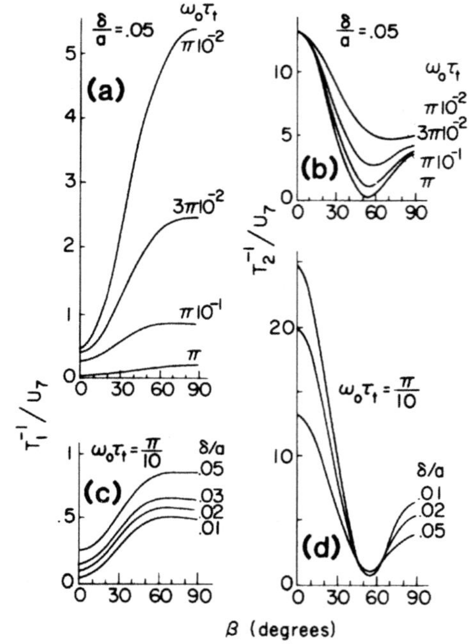

Figure 5. Angular variations of the spin-lattice T_1^{-1}/U_7 and spin-spin T_2^{-1}/U_7 relaxation rates with β (planar case). Up is a parameter that depends on the spin density, values of I and δ and on a scale factor related to the value of D in cm^2/s : Up $(s^{-1}) = [3(N/A)\gamma^4\hbar^2/\delta^2 I(I + 1)$ (unit of D)$]10^p$ with p = 7 in this figure. In all the diagrams (a)-(d) $D = 10^{-8}$ cm^2/s and g (r) = 1 (uniform radial molecular distribution).

the two cases of uniform radial molecular distribution (g(r) = 1, Figs. (6a),(6c))and non uniform radial molecular distribution (Figs. (6b) , (6d)). A simple comparison of Figs. (6a),(6b) and Figs. (6c), (6d) shows that g(r) enhances significantly the absolute values of T_1^{-1}. That increase being more important for β = 90° than for β = 0° because of the non contribution of the double-diffusion modes of relaxation in Eq. (16) for this latter orientation. This explains also why the diffusion dependences are symmetric for β = 0° and highly non symmetric for β = 90°. One of the main results of this study consists in the remarkably large size effects seen on these figures. Numerical calculation of the single- and double-modes contributions to Eq. (16) shows that the single-modes are effective only for very small sizes of the system ($a \geq \delta$) and disappear

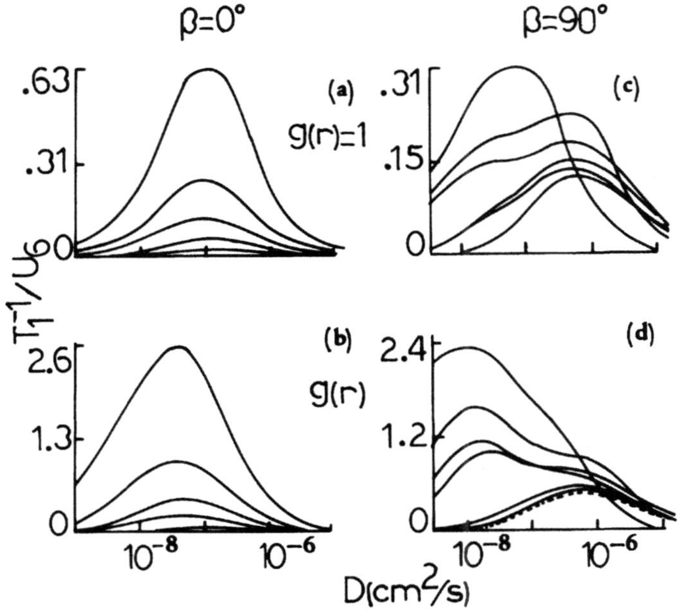

Figure 6. Semilogarithmic plots of the variations of the spin-lattice relaxation rate T_1^{-1}/U_6 with the diffusion coefficient D, for δ=6.5Å, ν_o=10MHz, β=0° and 90°, but varying the size of the system in the two cases of g(r)=1 ((a) and (c)) and non uniform g(r)((b) and (d)) (given in Fig. 1 with ρ_r=0.77) ; (a = 20, 50, 100, 200, 500 and 1000 Å downwards).

completly when $a \gg \delta$ in opposition to the double modes which give a larger contribution for any sizes and tend to limit-values for $a \ggg \delta$. We have already described in section 2.2. the important size effects on the dipolar correlation functions (Fig.3) and spectral densities (Fig. 4). Here we analyse these effects on the diffusion dependences of the spin-relaxation rates. For a very small size of the system (a of the order of δ) only a few number of modes are available (Fig. 2), the dipolar correlation functions decay rapidly almost in an exponential way (Fig.3) leading to a quasi Lorentzian behaviour of the spectral density up to $\omega_o \tau_t \sim 1$ (Fig.4). This explains the well pronounced maximum of T_1^{-1} observed in Figs. 6(a) - (d) at low values of D. When the size increases ($50\ \text{Å} \leq a \leq 200\text{Å}$), many more harmonics are involved. This explains the non-exponential decay of the dipolar correlation functions observed at longer times than for $a \sim 20\text{Å}$ (Fig.3). The total power spectrum, which is constant, is then shared among more modes. In consequence, the previous T_1^{-1} maximum decreases and the variations with D spread out over a wide range of D values. One notes the appearance of an other swallow maximum at higher values of D. It corresponds to the maximum of the superposition of the double-mode contributions. For almost infinite size ($a > 500\text{Å}$), one can replace in Eqs. (16) and (17) the sums by integrals over a quasi continuous distribution of available modes and obtains the <u>limit-expressions</u> (17)

$$T_1^{-1} \to [\ ]\ \frac{\pi}{8}(\bar{b}_d/\delta)^2\ \tau_t\{\sin^2\beta\ \cos^2\beta\ \ln(1 + 1/\omega_o^2\tau_t^2) + \sin^4\beta\ \ln(1 + 1/4\omega_o^2\tau_t^2)\}, \tag{18}$$

$$T_2^{-1} \to [\ ]\ \frac{\pi}{16}(\bar{b}_d/\delta)^2\ \tau_t\{4(1-3\cos^2\beta)^2\ln(\frac{a}{\pi\delta}) + 10\sin^2\beta\cos^2\beta\ \ln(1 + 1/\omega_o^2\tau_t^2) + \sin^4\beta\ \ln(1 + 1/4\omega_o^2\tau_t^2)\}. \tag{19}$$

Here the notations [ ] stands for the factors in brackets at the beginning of Eqs. (16) and (17). We have already noted in section 2.2 (Eqs. (14) and (15)) that these expressions must be considered only as analytical limit-expressions (represented in dotted line in Fig. (6d)). In reality one must use the equations (16) and (17) which give convergent results no matter how large the areas of the system are.

Finally, in Fig. 7 (a)-(c) we have displayed the variations of T_2^{-1} with D, using Eq. (17), for different values of β when varying the size of the system. These figures show a significant size effect in magnitude but not in shape (with exception of the case where β equals the magic angle (Fig. 7(b))in the motional narrowing in two dimensions. A direct comparison of Figs.6(b),6(d)and 7(a),7(c) shows that the values of T_2 stays much shorter than T_1 even in the fast motion-low field case ($\omega_o\tau_t^2 \ll 1$). This is mainly due to the zero-frequency part of Eq. (17)

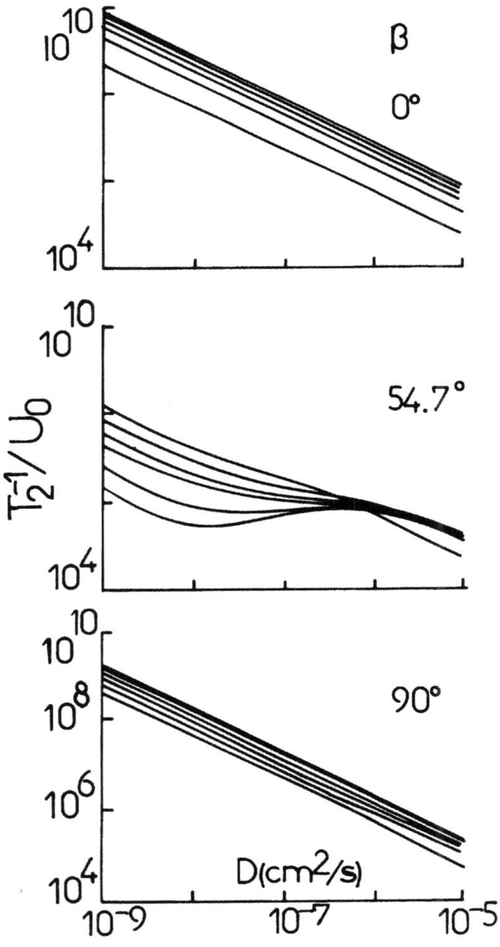

Figure 7. Logarithmic plots of the spin-spin relaxation rates T_2^{-1}/U_o with the diffusion coefficient D for δ = 6.5Å, ν_0 = 10MH$_z$, β=0°, 54.73° and 90° but varying the size of the system through the parameter a in Angstroms (case of non uniform) g (r)) ; a = 20, 50, 100, 200, 500 and 1000 Å, upwards (β = 0° and 90°), downwards (β = 54.73°).

which present a quasi logarithmic size dependence for infinite system (Eq.(19)).

3. COMPARISON WITH EXPERIMENT

In this section we test our theoretical expression of intermolecular spin-lattice relaxation rate given in Eq. (16) against proton relaxation experiments of molecular fluids physisorbed on graphite beads (19,29). Experimental details concerning these 2D systems have been given in the preceeding paper by J. Tabony. Here we just remind that graphitised carbon blacks are quasi spherical particles of radii from 100Å to 1000Å (R∿280Å for the graphon), whose the surface is locally flat and consists of polyedra of the exposed graphite basal plane of dimensions 85Åx85Å. the adsorption of molecules such as methane (49-51) and Neopentane (19a) upon these homogeneous surfaces gives rise to various 2D fluid phases for given coverages Θ and temperature ranges.

The use of this theoretical expression (Eqs.(16), restricted to the intermolecular dipolar contribution to the relaxation process, requires that an isotopic dilution method should be used in order to separate unambiguously the intra-and intermolecular contributions of this process. For instance, measurements of this rate with various mixtures of protonated and deuterated neopentane have shown that the minima of T_1 arised almost completely from the intermolecular dipole-dipole interactions (19a) For the same system, measurements of the diffusion coefficient D at different temperatures by the NMR pulsed field gradient method are fitted by an Arrhenius law of the form D $(cm^2/s) = 2.9 \exp(-2620/T(^{\circ}K))$ (19a) This corresponds, in the temperature range (200K-240K) where the minima of T_1 stands, to translational motions compatible with those of a liquid. The use of a complementary technique such as incoherent quasi-elastic neutron scattering can give precise information on these very fast translational motions. For instance Coulomb and Bienfait have applied this technique to study the microdynamics of physisorbed methane on graphite (Θ=0.4) and found that $D(cm^2/s)=3.55\ 10^{-3} \exp(-306.6/T(K))$(50).

Due to the particular structure of the beads, a powder average of Eq. (16) over all possible values of β is needed. This gives the theoretical diffusion dependences displayed in Fig. 8 for ν_0=10MHz, δ=6.5Å and for increasing values of a. The large size effects shown in these diagrams have already been explained in the previous sections. They will facilitate the interpretation of experimental data in 2D heterogeneous system providing that one knows the temperature dependences of the diffusion coefficient.

The following figures show the good accuracy between our theoretical expressions and the observations concerning the temperature (Fig. 9) and the frequency (Fig. 10) dependences of the spin-lattice relaxation times for neopentane. We assume a diffusion process with a single activation energy and two different bounded motions. The first one corresponds to the translational motions restricted inside the area of a facette (a∿85Å). The second one corresponds to the interfacette motions on the total area of the bead (with $a=2\sqrt{\pi}R$∿1000Å). The contribution of the intrafacette motions, displayed in dots in Figs (9a) and (9b), gives the

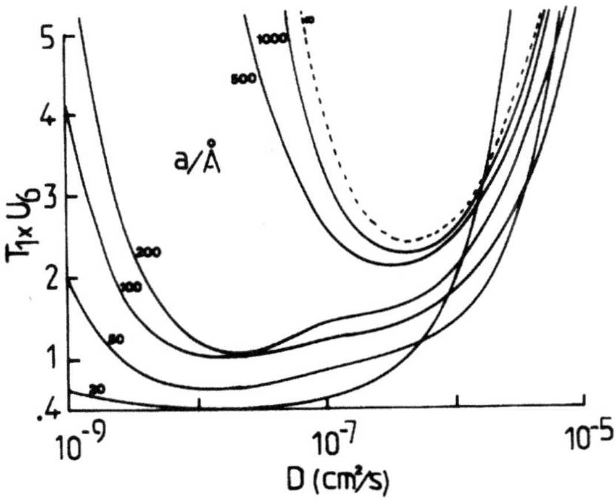

Figure 8. Semilogarithmic plots of the variations of the spin-lattice relaxation rate $T_1 x U_6$ (powder case) with the diffusion coefficient D for δ=6.5A, ν_o = 10MHz and varying the size of the system through the parameter a in Angstroms.

correct values of T_1 for the temperatures below 200K but a minimum of T_1 standing in a too low temperature range. On the contrary, the inter-facette motions give much higher values of T_1 and a minimum of T_1 standing at higher temperatures (Fig.8). Thus the superposition of both motions, displayed in dotted lines in Figs (9a) and (9b), shows a better agreement with the experimental points. The deviations observed at high temperatures decrease when one adds to these two latter bounded motions the observed temperatures dependences of the intramolecular contribution (52). This gives the continuous lines in Figs. (9a) and (9b) which show a reasonable agreement with the experimental points. In Fig.10 we have compared our theoretical variations of T_1^{-1} with the frequency ν_o (Eq.16) against the observations (29). Just as for the temperature variations, one notes a significative improvement when one considers both intra-facette and interfacette bounded motions in the relaxation process. For other physisorbed molecular fluids like methan a similar agreement is obtained (Fig.11).

In all these calculations we have considered the excentricity using expressions previously derived (37,39). At the values of $\omega_o \tau_t$ considered here ($\sim 10^{-2}$) Fries has shown that these excentricity effects increase the spectral densities J approximatively by a factor 3 (39).

The good accuracy between our theoretical expression and the observations concerning the frequency and temperature dependences of the spin-lattice relaxation time T_1 proves that the minima of T_1 and the other relaxation features can be explained by a restricted diffusion

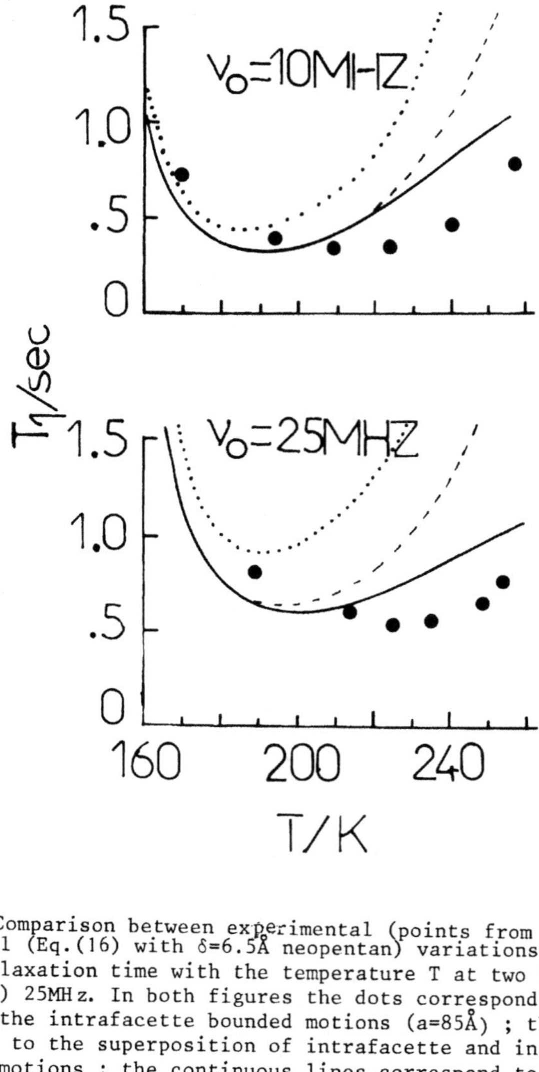

Figure 9. Comparison between experimental (points from Ref 29) and theoretical (Eq.(16) with δ=6.5Å neopentan) variations of the spin-lattice relaxation time with the temperature T at two frequencies (a) 10MHz ; (b) 25MHz. In both figures the dots correspond to the contribution of the intrafacette bounded motions (a=85Å) ; the dotted lines correspond to the superposition of intrafacette and interfacette (a=1000Å) motions ; the continuous lines correspond to the superposition of intrafacette, interfacette and intramolecular contributions (from Ref. 52).

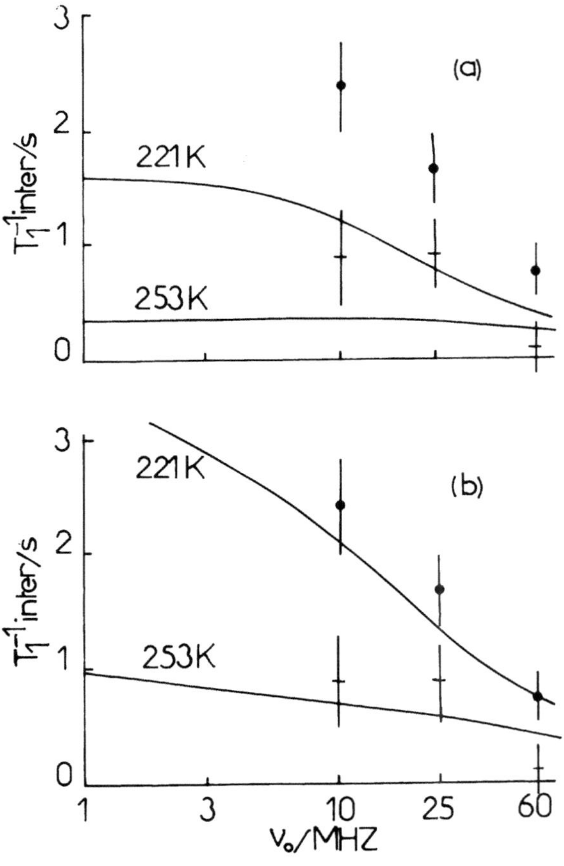

Figure 10. Comparison between experimental (points from Ref. 29) and theoretical (Eq.(16) with δ=6.5Å neopentan) variation of the spin-lattice relaxation rate with the frequency ν_o (semilogarithmic scales) at two temperatures. (a) contribution of intrafacette bounded motions (a=85Å) ; (b) contribution of intrafacette and interfacette (a=1000Å) bounded motions.

For physisorbed molecular fluids on homogeneous surfaces, this latter theoretical characteristic of 2D motion has proven useful to show that the minima of T_1 and the other relaxation features, described here and in the preceeding paper, can be explained by a restricted diffusion motion in a 2D liquid instead of a slow motion due to the proximity of the solid interface. Further detailed comparisons between theory and experiment will be given elsewhere (54).

The use of these theoretical formulas, restricted to the intermolecular dipolar contribution to the relaxation process, requires that an experimental method should be used in order to separate unambiguously the intra-and intermolecular contribution of this process. For instance, this is the case of the isotopic dilution method (29). Measurements of spin-relaxation rates T_1^{-1} and T_2^{-1} should be made over a range of temperatures and at different frequencies to enhance the information on the restricted 2D motion. Other measurements can be useful in the two limiting cases of very slow or very rapid molecular motions. In the former case, measurement of the temperature dependence of the spin-lattice relaxation rate in the rotating frame $T_{1\rho}^{-1}$ has proven useful to study slow molecular motions in the gel phases of dipalmitoyl lecithin (55,56). In the latter case, the use of a complementary technique such as incoherent quasi elastic neutron scattering can give information on very fast translational motions (57).

We have been limited in this paper on the dipolar relaxation processes when molecular diffusion is bounded to finite planar surfaces. Of course a straightforward calculation can be made to deal with an other geometry. We have previously considered the case of diffusion at the surface of a spherical system (16). More generally, our method can be adapted to other relaxation processes involving restricted motional averages (e.g. anisotropic molecular reorientations in lipid bilayers (58) providing that a restricted diffusion model is implied).

ACKNOWLEDGEMENTS

I wish to thank Pr.H.M. McCONNELL for helpful discussions concerning the model and subsequent treatment and M. WINTERHALTER for advice in computational matters.

REFERENCES

1. D.C. Torney and H.M. McConnell, Proc. R. Soc. London A 387, 147 (1983).
2. J.M. Deutch, J. Chem. Phys. 56, 6076 (1972).
3. R. Gomer, *Interaction on Metal Surfaces* (Springer, Berlin, 1975).
4. M.B. Salomon, *Physics of Superionic Conductors* (Springer, Berlin, 1979).
5. S. Ostlund and B.I. Halperin, Phys. Rev. B23, 335 (1981).
6. P.G. Saffman and M. Delbruck, Proc. Natl. Acad. Sci. U.S.A. 72, 3113 (1979).
7. A.A. Kokin and A.A. Izmest'ev, Theor. Exp. Chem.(USSR) 1, 242 (1965).

8. P. Brulet and H.M. McConnell, Proc. Natl. Acad. Sci. U.S.A. 72, 1451 (1975).
9. A. Avogadro and M. Villa, J. Chem. Phys. 66, 2359 (1977).
10. D. Beckert, Ann. Physik 20, 220 (1967) ; *ibid.* 23, 98 (1969).
11. B. Cowan, J. Phys. C13, 4575 (1980).
12. C.A. Sholl, J. Phys. C14, 447 (1981).
13. P.M. Richards and M.B. Salamon, Phys. Rev. B9, 32 (1974).
14. P.H. Fries, J. Phys. Lett. 42, L257, L513 (1981) ; Mol. Phys. 48, 503 (1983).
15. J.-P. Korb, D.C. Torney, and H.M. McConnell, J. Chem. Phys. 78, 5782 (1983).
16. J.-P. Korb, M. Winterhalter, and H.M. McConnell, J. Chem. Phys. 80, 1059 (1984).
17. J.-P. Korb, J. Chem. Phys. (15 Jan. 1985).
18. W. Hardy, M. Morrow, R. Jochensen, B.W. Sttat, P.R. Kubik, R.M. Marsolais, J.A. Berlinsky, and A. Landesman, Phys. Rev. Lett. 45 453 (1980).
19. (a) J. Tabony, Progr. NMR Spectrosc. 14, 1 (1980) ; (b) J.W. Ri hl and K. Koch, J. Chem. Phys. 57, 2199 (1972).
20. L. Facchini, T. Tajouri, A.P. Legrand, H. Balard and P. Tougne, Proceedings of the XXIInd Congress Ampere, (Zurich, 1984, pd 5).
21. J. Connard, H. Estrade-Szwarckopf, P. Lauginie and G. Hermann, *Physics of intercalation Compounds* (Spinger series Solid State Science 38, 264, 1981).
22. U. Röder, W. Müller-Warmuth, H.W. Spiess, and R. Schöllhorn, J. Chem. Phys. 77, 4627 (1982).
23. J.J. Fripiat, J. Phys. C4, 44 (1977).
24. P.M. Richards, *Physics of Superionic Conductors, Topics in Current Physics* (Springer-Verlag, Berlin, 15, 141, 1979) ; J.L. Bjorkstam and M. Villa, Magnetic Res. Rev. 6, 1 (1980) ; P.P. Man, H. Theveneau, P. Papon, and J.L. Fourquet, J. Chem. Phys. 80, 1272 (1984).
25. M. Vilfan and S. Zumer, Phys. Rev. A 21, 672 (1980).
26. M. Ueno, H. Kishimoto and Y. Kyogoku Chemistry Lett. (Chem. Soc. of Japan), 599 (1977).
27. R. Kimmick, G. Schnur and A. Scheuermann, Chemistry and Physics of Lipids 32, 271 (1983).
28. P. Devaux, C.J. Scandella, and H.M. McConnell, J. Magn. Reson. 9, 474 (1973).
29. see the preceding paper by J. Tabony.
30. R. Kubo and K. Tomita, J. Phys. Soc. (Jpn.) 9, 888 (1954).
31. A. Abragam, *The principles of Nuclear Magnetism* (Oxford University, London, 1961), Chap. VIII, Sec. II Ec.
32. T.T.P. Cheung and Z.G. Soos, J. Chem. Phys. 69, 3845 (1978).
33. O. Takizawa, R. Srinivasan, and E. de Boer, Mol. Phys. 44, 677 (1981).
34. R. Murugesan and E. de Boer, Chem. Phys. Lett. 95, 301 (1983).
35. D.C. Look and I. J. Lowe, J. Chem. Phys. 44, 2995 (1966).
36. R.A. Assink, J. Dezwaan, and J. Jonas, J. Chem. Phys. 56, 4975 (1971).
37. P.S. Hubbard, Rev. of Mod. Phys. 33, 249 (1961) ; Phys. Rev. 131, 275 (1963).

38. L.P. Hwang and J. Freed, J. Chem. Phys. 63, 4017 (1975).
39. P.H. Fries, Ph. D. Thesis, University of Grenoble (1982).
40. S.A. Rice and P. Gray, *The Statistical Mechanics of Simple Liquids*, (Interscience, New York, 1965).
41. A.H. Norten, M.D. Danford, and H.A. Levy, Discus. Faraday Soc. 43, 97 (1967).
42. J.K. Percus and G.J. Yevick, Phys. Rev. 110, 1 (1958).
43. G.J. Throop and R.J. Bearman, J. Chem. Phys. 42, 2408 (1965).
44. L. Verlet and J.J. Weiss, Phys. Rev. A5, 939 (1972).
45. D.G. Chae, F.H. Ree, and T. Ree, J. Chem. Phys. 50, 1581 (1969).
46. F. Tsien and J.-P. Valleau, Mol. Phys. 27, 177 (1974).
47. I. Oppenheim and M. Bloom, Can. J. Phys. 39, 845 (1961).
48. H.S. Carslaw and J.C. Jaeger, *Conduction of heat in Solids* (Oxford University, London, 1959), Chap. XIV.
49. J.-P. Coulomb and M. Bienfait, Phys. Rev. Lett. 42, 733 (1979).
50. J.-P. Coulomb, M. Bienfait and P. Thorel, J. Phys. 42, 293 (1981).
51. J. H. Quateman and M. Bretz, Phys. Rev. B23, 1159 (1984).
52. From ref. 19 (a) one has : T_1^{-1} intramolecular (T) = 0.0113T-2.47 for ν_0 = 10MHz and T_1^{-1} intramolecular (T) = 0.0067T-1.26 for ν_0 = 25MHz.
53. From ref. 19 (b) one has T_1^{-1} intramolecular (T) = 0.0022T + 0.716.
54. J. Tabony and J.-P. Korb, J. Chem. Phys. (to be published).
55. E. Boroske and L. Trahms, Biophys. J. 42, 275 (1983).
56. See the paper of L. Trahms in this book.
57. P. Gamlen, R.K. Thomas, T.D. Trewern, G. Bomchil, U.M. Harris, M. Leslie, J. Tabony, and J.W. White, J. Chem. Soc. Faraday Trans. I 75, 1533 (1979).
58. M.F. Brown, J. Chem. Phys. 77, 1576 (1982).

INDEX OF SUBJECTS

Acyl chain 171, 187, 204, 218
Activation
-- energy 43, 47, 126-128, 130, 133
thermal -- 125, 127-130, 135
Adsorption 229, 243
Aggregate 88, 129,130, 187
Alumina silicate 234
Ammonia 231
Angle-exchange operation 27
Anisotropy 201, 244, 255
Annihilation 72, 80, 81, 113
Anthracene 101, 102, 104, 111, 112
Attractor 7, 8, 12

Basin 7-12, 47, 49, 64, 65
Bilayer 171-173, 203-205, 219, 226, 234, 255
Bond 12, 16, 18, 41
Borane 15, 17
Boundary conditions 78, 80
Broadening
homogeneous -- 137-141, 145, 149
inhomogeneous -- 116, 149, 159
line -- 205
Bulk 225, 226

^{13}C 187, 203, 204, 216, 219
Carbon black 229, 260
Catalysis 243
Catchment region 42, 45-51, 64-66
Cephalin 203-205, 218, 219
Chalcogenide 243
Chemical
-- form 9, 10, 20
-- process 41, 57, 60
-- reaction 20, 41, 42, 47, 57-59, 82, 243
-- reactivity 10, 23
-- shielding 188
-- shift 187, 188, 196, 197, 211, 218, 219
-- structure 47, 62, 65
Clay 227, 234, 243
Cluster 111-113, 138, 140
-- ing 137
-- state 108
percolation -- 88
Cog-wheel effect 26, 34
Collagen 235
Collective state 119
Collision
-- frequency 171, 172
-- rate 178, 180, 183
Concentration
-- dependence 113, 119, 135, 137, 143
effective -- 137-142, 144, 146
sample -- 137-142, 145, 146
Condensed media 84, 116, 137
Conformation (al) 187, 189
-- analysis 41, 47, 49
-- change 41-43, 47, 51
-- distribution 23, 39
-- process 41, 47
Conformer 41, 43, 45, 49-52, 189, 206
Connectivity matrix 19, 20
Coordinate
internal -- 3, 4, 7, 9
local -- 7
Coordination number 140-142
Core 12, 167, 168
Correlation
-- function 243-246, 262
-- time 239, 244
Coupling
dipolar -- 134, 135, 187, 210, 244
exciton-phonon -- 120, 121, 130
multipolar -- 131
spin-orbit -- 119, 121, 123, 124,

133, 135
superexchange -- 131, 135
Coverage 229
Critical point 12, 42, 45-47, 51, 52, 64-68
Crystal 88
mixed -- 101, 102, 111, 116, 119, 121-124, 127-130, 135
molecular -- 101, 111, 119 137, 138
Curvature 3, 9
Cycloaddition 19

Deformable molecule 23, 24, 37, 39
Delocalization 106, 108, 116 129, 133, 135
Dialkylphosphate 187
Diffusion 71-73, 75, 77, 82, 84, 88
-- coefficient 77, 187, 191, 206, 226, 228, 233, 234, 244
-- dependence 255
-- equation 81
-- mechanism 218
-- rate 184
bounded -- 225, 226, 237, 238
jump -- 191
lateral -- 206, 243
molecular -- 243, 264, 265
restricted -- 234, 260, 265
rotational -- 187, 191, 217, 218
spectral -- 149, 152
spin -- 172, 244
translational -- 226, 228, 233, 234, 243, 244
two-dimensional -- 225,237, 244
Diffusive motion 71, 225
Dimensionality 226
Dipolar
-- correlation 244
-- coupling 134, 135, 187, 210, 244
-- interaction 104, 105, 108-111, 137-141
---ophile 19
-- relaxation 243, 264, 265
-- reservoir 152, 168
Dipole 16, 19, 23, 38, 39
Director 189
Disorder 72, 73, 84, 87
diagonal -- 101-116
off-diagonal -- 101-106, 108, 114
structural -- 101, 102
substitutional -- 101
Disordered 71, 88, 101, 114 116, 145
Dissociation 47, 49
Distribution 71, 137
homogeneous -- 73, 145
inhomogeneous -- 127
pair -- 137-139, 146
radial --137, 139-141, 146
reduced -- 137, 139-141, 146
trap -- 123
Double-switch operation 27
Doublet 15
DPPC 204

ELDOR 138, 144, 145, 182
Electrolyte layer 243
Electron (ic)
-- density 10-12
-- equation 1, 2
-- motion 1, 2, 13
-- state 42, 43, 46, 57
-- structure 10, 12, 14-16, 20
ENDOR 168
Energy functional 41-47, 51, 52, 60
EPR 23, 144
Equivalence class 43-47, 57-66
ESR 141, 149, 171, 172, 184
-- line 150, 153, 155
-- sensitivity 184
Ethylene 13
Excentricity effect 260
Exchange
-- frequency 171, 172
-- interaction 105, 106
-- mechanism 184
-- model 172, 173
-- rate 178
Exciton 119-124, 127, 130, 137 138
Excitonic eigenstate 101, 102, 111, 114

Fatty acid 171, 184
Fibre 227, 235
Flexible molecule 189

Fluctuation 149, 152, 203
Fluid 243, 244, 265
Fluorescence 104, 125
delayed -- 119-121, 124-130, 135, 138
Force
-- constant 3, 9
-- field 3, 7, 8
Fractal 71, 72, 76, 79-83, 87-95, 99
Fracton 92
Free rotation states 33
Fusion 119, 121, 124, 137, 138

Gel 203, 205, 218, 219, 226, 265
Glass 138, 140
Gradient 3, 7-11
Graph 1, 10, 12-20, 42
Graphite 226, 229, 231, 243, 244, 260
Group 57-64

^{1,2}H 187, 192, 193, 203, 204, 213, 215, 219, 225
Hessian matrix 45
Heterogeneous system 225, 226, 243
Hexagonal phase 204
Hole burning 167, 168
Homogeneous
-- broadening 137, 139-141, 145, 149
-- distribution 73, 145
-- line 138, 140-143, 150
Homology 51
Homotopy 52, 57-66
Hopping model 131
Hydratation shell 234
Hydroboration 17, 19
Hydrocarbon chain 187, 204
Hydrogen bridge 16
Hyperfine 149, 152, 159

Infrared (IR) 23, 206
Inhomogeneous
-- broadening 116, 149, 159
-- distribution 127
-- line 104, 113, 131, 150, 153, 155
Interaction
exponential -- 104-108
long-range -- 71, 75, 101, 103, 108, 114, 119
microscopic -- 73
short-range -- 101-106, 114-116, 120
Intercalation compound 243
Interconversion process 41, 42, 49
Interface 225-227, 265
Intermolecular interaction 187
Internal 23, 24, 41, 42
Intramolecular mobility 206
Inverse participation ratio 101-103, 114
Ionic conductor 243
Isodynamic 27, 28
Isomerization 206

Lamellar 187, 226
Lattice
1D -- 88
2D -- 88, 122, 137-146
3D -- 88, 137-146
cubic -- 80, 81, 96, 97
fractal -- 84, 91, 98, 99
regular -- 76, 80, 82, 87-95, 99
square -- 80, 81, 96, 97
Lecithin 203-205, 218, 219, 265
Level set 43, 48, 61-66
Lipid 203
-- bilayer 205, 265
-- chain 178
labeled -- 171
Lipid-protein contact 171
Liquid crystal 187, 204, 226, 227, 234-243
Lithium hectorite 234
Localization 101-106, 109-112, 114, 116
Localized orbitals (LO) 12-15
Loge 12, 13
Lone pair 11, 12, 16
Lyotropic 187, 192, 227, 234

Macromolecule 203
Membrane 171, 178, 184
biological -- 203, 243
model -- 171, 243
Mesophase 187, 192, 212, 226, 227
Methane 14, 229, 260
Metric 44-46, 52

Micelle 204, 226, 243
Microemulsion 226
Minimal approach 246
Minimal expansion 26, 28, 34, 35
Mobility 171, 203, 206, 218, 219, 226
Molecular
-- conformation 41
-- crystal 101, 114, 119, 137, 138
-- design 20, 57-59
-- diffusion 243, 244, 265
-- fluid 243, 244, 265
-- form 1, 9, 10
-- mobility 218, 225, 226
-- motion 187
-- parameter 37
-- shape 1, 7, 10
-- structure 1, 42, 205
-- tumbling 189
Moments 10, 14
Monte-Carlo computation 246
Mosaic model 203
Motion
2D -- 227, 265
axial -- 203
diffusive -- 71, 225
electronic -- 1, 2, 13
internal -- 41
lateral -- 218
molecular -- 187
nuclear -- 1, 2, 9, 42
rotational -- 218
translational -- 260
Motional
-- averaging 211
-- narrowing 228

^{14,15}N 171, 178, 179
Naphtalene 101-104, 111-113, 120, 121
Nearest neighbour 102-105, 138
Neopentane 231, 258
Nitroxyde 171, 178, 179, 184
Non-crystalline 101
Non-rigid 24, 30, 39, 58
Nuclear 4, 7, 10, 41, 49, 51
-- configuration space 41-46, 49-52, 57, 62, 64
-- equation 2
-- motion 1, 2, 9, 42
-- probability density 23, 24, 29-37
NMR 23, 37, 88, 187, 203-205, 218, 225-227
-- pulsed field gradient method 226, 229, 260
^{13}C -- 203, 204, 216, 219
^{1}H -- 192, 203, 213, 219, 225
^{2}H -- 192, 193, 203, 204, 215, 219
^{17}O -- 192, 193
^{31}P -- 199, 203, 216
Numerical
-- calculations 88, 102, 104, 114, 137, 138, 144, 146
-- experiments 144
-- simulations 93
-- solutions 78, 119, 131

^{17}O 187, 192, 193
Octupolar interaction 104-106
Open set 42, 43, 46-49, 57, 62, 66
Order
---ing tensor 189
-- parameter 187, 189, 203, 213
local -- 172
Oriented gas model 119

^{31}P 188, 199, 203, 216
Path 60-68
reaction -- (way) 1, 5-8, 10, 13-17, 20, 57-60, 63-66
steepest descent -- 45, 46
Phase
-- diagram 187, 188
-- transition 205, 243
Phonon 131-135
Phosphate 187
Phosphocholine 204
Phospholipid 171, 172, 184, 204, 205, 219, 226
Phosphorescence 102, 104, 121-124, 128, 129, 133
Physisorbed 225, 227, 229, 243, 244, 265
Polar 187, 204
Polymer 88, 243
Potential
-- defying conformer 41, 51, 52
-- energy 1-10, 20, 24-28, 33,

34, 41-47, 51, 58-64
Probe 171-172
Process
 dynamical -- 116, 137
 interconversion -- 41, 42, 49
 molecular -- 57, 60
 phonon -- 131
 reaction -- 41, 47
 recombination -- 137, 141, 144
 stochastic -- 87
Projector 30
Protein 171, 172, 203

Quadrupol (ar)
 -- interaction 211
 -- splitting 187, 193, 196, 197, 218, 219

Radical 16, 137, 138, 243
Raman 206
Random walk 78, 80, 84, 87-92, 97-99, 131, 132
Reaction
 -- channel 57
 -- diffusion 77, 87
 -- kinetics 71, 84
 -- mechanism 16, 42, 47, 57-66
 -- path (way) 1, 5-8, 10, 13-17, 20, 57-60, 63-66
 -- polyhedra 49-51
 -- process 41, 47
 -- rate 72, 79, 243
 bimolecular -- 72, 73, 79-81
 chemical -- 20, 41, 42, 47, 57-59, 82
 monomolecular -- 42
 narcissistic -- 58
 pseudounimolecular -- 72, 73, 79, 84, 87, 88, 99
Reactive
 -- transformation 1
 -- domain 42
Rearrangement 41-43, 57
Recombination 137, 138, 141, 144, 145, 243
Relaxation
 -- rate 137, 138, 141, 144, 187, 244, 248, 255
 -- time 145, 225, 226
 cross -- 137, 144, 145
 dielectric -- 88
 dipolar -- 243, 264, 265
 intermolecular -- 260
 nuclear -- 191, 226
 spin-lattice -- 225, 244, 264
 theory of -- 225, 243-245
Rhodopsin 171, 172
Rotation (al)
 -- diffusion 187, 191, 217, 224
 -- function 29-32
 internal -- 23, 24
Ruby 149, 158, 161

Saddle point 5, 33, 34, 47, 50, 51
Saturation 149, 150
 -- recovery 159, 168
 -- steady state 153
 -- transient 156
Scattering 120
 neutron -- 225-227, 265
 x-ray -- 248
Selective excitation 120, 135
Silicagel 243
Size effect 243, 244, 264
Soap 234
Spectral
 -- density 228, 243-246, 260
 -- diffusion 149, 152
 -- dimension 79, 81-83, 91
Spin label 171, 172
Spin packet 149, 151
Stationary point 3-10, 20, 64
Superexchange 113, 131, 133, 135
Supertrap 119, 127
Surface 225-229, 243, 244, 265
Surfactant 187, 194
Survival probability 90-94, 99
Symmetry 15, 26, 30-34, 38, 58

T_1 150, 191-194, 225, 226, 229, 237, 240, 243, 244, 255, 258, 260, 265
T_2 150, 226, 229, 237, 243, 244, 255, 258, 265
Target problem 87, 88, 93-99
Thermotropic phase 203, 205
Topological 10-12, 41, 42, 46-49, 51, 57-62, 65
 -- equivalence 49
 -- invariant 42, 51, 52, 64
 -- relation 42, 52

-- space 42, 46-49, 60
-- transformation 41
Topology 42, 45-49, 58, 59, 64-66, 120-123, 226
Transfer 119-121, 130
-- rate 140
-- time 139, 145
-- yield 119, 126, 129, 135
energy -- 73, 80, 84, 88, 101, 119, 127, 129, 133, 135, 137, 141, 144, 150
exciton -- 124
Transition state/structure 17-20, 47, 49, 51, 58
Transport phenomena 243
Trapping 87-89, 93-99, 120, 129, 131
Two-dimensional 229, 243, 244
-- diffusion 225, 237, 244
-- fluid 206, 226, 244, 265
-- lattice 88, 122, 137-146, 218
-- relaxation 226, 243-245
Two-rotor molecule 23, 24, 28-33, 39

Valence 15
Vesicle 171
Vibrational energy 51, 52
Vibration-rotation states 33-35
Viscosity 171

Water 203, 204, 234, 243
-- adsorbed 227
-- biological 226, 235
-- interstitial 187, 192
Well stirred reactor 71-73

X-ray diffraction 187, 207, 246
Xylene 23, 28, 33-39

Zeolithe 226

INDEX OF NAMES

Abragam, A. 222, 266
Agranovich, V.M. 136
Alexander, S. 100
Allinger, K. 85
Allinger, N.L. 55
Altmann, S.L. 40, 70
Anacker, L.W. 86
Anderson, P.W. 117, 169
Anderson (transition) 101, 103
Anderson, S.G. 21
Angyal, S.L. 55
Antoniou, P.D. 117
Antonov-Romanovskii, V.V. 86
Arnold, K. 222, 224
Assink, R.A. 266
Atkins, P.W. 168
Atsarkin, V.A. 169
Auger (effect) 125
Avakian, P. 136
Avgul, N.N. 242
Avnir, D. 100
Avogadro, A. 241, 266

Bader, R.F.W. 21
Balagurov, Ya.B. 85, 100
Balard, H. 266
Balasubramanian, K. 70
Balian, R. 117
Barber, M.N. 100
Baroin, A. 185
Bauder, A. 70
Bearman, R.J. 267
Beckert, D. 241, 266
Beguery, M. 117, 136
Bellido, M.N. 40
BenAvraham, D. 100
Bergman, E.D. 40
Berliner, L.J. 185, 201
Berlinsky, J.A. 266
Beth, S.D. 185
Bienfait, M. 241, 267
Bienvenüe, A. 185, 223
Birgeneau, R.J. 242
Bjorkstam, J.L. 266
Blanchard, C. 169
Bloch (equations) 171, 172, 176, 179, 184
Bloch (waves) 119
Bloom, M. 222, 223, 267
Blume, A. 220-223
Blumen, A. 71, 85, 87, 100, 101, 117
Boden, N. 201
Boltzmann (distribution) 31
Bomchil, G. 241, 242, 267
Borle, F. 223
Born-Oppenheimer (approximation) 1, 2, 43, 51
Boroske, E. 222, 223, 267
Boscaino, R. 149, 168, 169
Boulanger, Y. 221
Bouman, T.D. 70
Bowman, M.K. 147
Boys, S.J. 21
Brai, M. 169
Bratoz, S. 21
Braun, S. 223
Bretz, M. 267
Brewer, R.G. 169
Broude, V.L. 136
Brown, M.F. 201, 223, 267
Brown, R. 117, 119, 136, 147
Browning, J.L. 220
Brulet, P. 241, 266
Brun, E. 169
Buishvili, L.L. 168
Büldt, G. 221
Bunker, P.R. 40
Burke, L.A. 21
Burnell, E.E. 222
Burshtein, A.I. 169
Busby, R.J. 201
Butler, K.W. 222

Cairns, S.S. 56, 69
Calef, D.F. 85
Cameron, D.G. 221, 222
Campbell, R.F. 223
Carbó, R. 40
Carslaw, H.S. 267
Casal, H.L. 221, 222
Castner, T.G. 168
Cebula, D.G. 242
Chachaty, C. 40, 187, 201
Chae, D.G. 267
Chalvet, O. 21
Chan, S.I. 222
Chan, W.K. 221
Chandrasekhar, S. 85, 100, 147
Chang, H. 222
Chao, K.A. 117
Chapman, D. 220, 221
Chapman, N.R. 222
Charvolin, J. 220, 242
Chen, S.C. 222
Cheng, D.F. 147
Cheung, T.T.P. 266
Ching, W.Y. 117
Christoffersen, R.E. 55
Ciccarello, I. 169
Clark, L.D. 201
Clary, D.C. 56
Clough, S. 168
Colson, S.D. 136
Connard, J. 266
Connor, J.N.L. 56
Cooke, R. 241
Cornell, B.A. 223
Cosgrove, T. 241
Coulomb, J.P. 241, 266
Cowan, B. 266
Cross, P.C. 21
Csizmadia, I.G. 21
Culis, P.R. 168

Danford, M.D. 267
Darke, A. 220
Dasgupta, S.K.D. 220, 223
Dash, J.G. 242
Daudel, R. 21
Davis, E.A. 117
Davis, J.H. 220, 223
Davoust, J. 171, 185
Daycock, J.T. 220
Dean, P. 117
de Boer, E. 266
Decius, J.C. 21
de Gennes, P.G. 100, 222
Delbruck, M. 265
Deutch, J.M. 85, 265
Devaux, P.F. 171, 185, 223, 266
de Voe, R.G. 169
Dezwaan, 266
Dianoux, A.J. 241, 242
Dietrich, R. 223
Dignam, K.J. 21
Dive, G. 22
Doane, J.W. 223
Doktorov, A.B. 85
Doniach, S. 221
Donsker, M.D. 85, 100
Duchamp, D.J. 55
Duke, A.J. 21
Duncan, C.D. 70
Dupuy, F. 117, 119, 136, 147
Dupuy, J. 241
Durig, J.R. 40
Duval, X. 242

Ebelhäuser, R. 222
Economou, E.N. 117
Eibel, H.J. 223
Elbers, P.F. 221
Eliel, E.L. 55
Emsley, J.W. 201
Enderby, J.E. 201
Endo, T. 149, 169
Epand, R.M. 222
Ern, V. 136
Estrade-Szwarckopf, H. 266
Evangelou, S.N. 117
Even, U. 85, 100
Evesque, P. 85

Facchini, L. 266
Fahmy, T. 222
Falcovitz, M.S. 221
Farin, D. 100
Farnell, L. 40
Favre, E. 185
Feigenson, G.W. 222
Feix, J.R. 185
Feller, W. 100
Feuer, B. 221
Fick (law) 231
Fontell, K. 220

Fookson, J. 220
Forrest, B.J. 201
Fourier (expansion) 26, 27
Fourquet, J.L. 266
Francis, A.H. 117, 136
Freed, J.H. 201, 223, 267
Frei, H. 70
Fries, P.H. 241, 266, 267
Fripiat, J.J. 241, 242, 266
Frisch, H.L. 221
Fujimoto, H. 21
Fukui, K. 21, 69
Fukuyama, H. 117
Füldner, H.H. 222, 223
Funabashi, K. 101
Fung, B. 242

Gaber, B.P. 222
Gaffney, B.J. 185, 222
Galla, H.J. 221
Gally, H.U. 221, 223
Gallot, B. 201
Gamelin, P. 241, 242, 267
Ganellin, C.R. 40
Gardiner, W.C. 86
Gayda, J.P. 169
Gawrisch, K. 222, 224
Gebhardt, C. 221
Gelardi, F.M. 149, 168, 169
Gerratt, A. 21
Gibby, M.G. 220
Giradet, J. 222
Goddard, J.D. 21
Gomer, R. 265
Grabriel-Madelmont, C. 221
Gray, P. 257
Green (function) 248
Green, T.J. 221
Greenberg, M. 69
Griffin, R.G. 220, 222, 223
Griffith, O.H. 201
Gruler, H. 221
Gudgin, E.F. 222
Guggenheim, H.J. 221
Günthard, Hs.H. 70
Gutowsky, H.S. 223

Habberkorn, R.A. 223
Habberlen, U. 220
Hall, P. 241
Halle, B. 201
Halperin, B.I. 265
Hamill, W.H. 101
Hammonds, E.M. 242
Hardy, W. 266
Harris, N. 241, 242
Harris, U.M. 267
Hashi, T. 169
Hausdorff (dimension) 76, 92
Havlin, S. 100
Hegarty, A.F. 21
Heimey, P. 242
Heinzinger, K. 201
Helfrich, W. 221
Hentschel, M. 221, 223
Hermann, G. 266
Hernández-Laguna, A. 23, 40
Hervé, J. 185
Herzfeld, J. 220, 223
Higgs, T.P. 222
Hilfer, R. 100
Hiller, R.G. 223
Hinz, H.J. 221
Hitchcock, P.B. 224
Hochstrasser, R.M. 117
Horn, G.N. 201
Horn, P.M. 242
Hosemann, R. 221
Hsi, S.C. 222
Hsung, J.C. 223
Huang, C.H. 224
Huang, T.H. 220, 223
Hubbard, P.S. 266
Huber, D.L. 100, 117
Huber, L.M. 220
Huddleston, R.K. 85
Huertas-Cabrera, A. 40
Huller, A. 242
Hundt, E.E. 169
Hutton, W.C. 224
Hwang, J.S. 201
Hwang, L.P. 267
Hyde, J.S. 185

Ishida, Z. 21
Ishioka, S. 100
Israelachvili, J.N. 201
Izmest'ev, A.A. 241, 265

Jacobs, R. 220, 223
Jacobson, K. 220
Janiak, M.J. 221

Jaeger, J.C. 267
Jeffrey, K.R. 222
Jenkin, G.T. 242
Jochensen, R. 266
Jonas, J. 266
Jortner, J. 85, 100, 117, 136
Jost, P.C. 201

Kainoshe, M. 222
Kang, K. 86, 100
Kang, S.Y. 223
Kapitza, H.G. 221
Kato, S. 21
Keniri, M. 223
Kenkre, V.M. 136
Kevan, L. 147
Khairutdinov, R.F. 85
Khutsishvili, G.R. 168
Kilpatrick, J.E. 40
Kimmick, R. 220, 223, 266
King, R.B. 55, 69
King, T.E. 223
Kishimoto, H. 266
Kiselev, A.V. 242
Klabe, W.D. 222, 224
Klafter, J. 71, 85, 87, 117
Klein, M.P. 223
Klymko, P.W. 85, 86, 136
Knopp, J.V. 40
Koch, K. 241, 266
Kofman, A.G. 169
Kohler, S.J. 223
Koiwa, M. 100
Kokin, A.A. 241, 265
Komornicki, A. 21
Kopelman, R. 85, 86, 117, 136
Korb, J.P. 137, 147, 168, 169, 241, 242, 243, 266, 267
Kothe, G. 222, 223
Kottis, Ph. 117, 119, 136, 147
Kroon, P.A. 222
Kruizinga, B. 147
Kubik, P.R. 266
Kubo, R. 266
Kunz, I.D. 241
Kyogoky, Y. 266

Laermans, C. 100
Laider, K.J. 21, 70
Lalanne, R. 136
Lambert, L.Q. 169
Landesman, A. 266
Lapersonne-Meyer, C. 147
Larsson, K. 221
Lathouwers, L. 70
Lauginie, P. 266
Lee, A.G. 185, 220
Legrand, A.P. 265
Leidermann, A.V. 136
Lemaistre, J.P. 101, 117, 119, 136, 137, 147
Leroy, G. 21, 22
Leslie, M. 241, 242, 267
Levin, I.W. 222
Levy, H.A. 267
Lewis (structure) 13, 15
Lichtenberg, P. 222
Lin, C.C. 40
Lin, D.P. 147
Linnett (structure) 13, 15
Loebel, E. 224
Look, D.C. 266
Lowe, I.J. 266
Luckhurst, G.R. 201
Luitingh, A.J. 221
Luna, E.H. 221
Luzzati, V. 221

Mackay, A.L. 222
Maggiora, G.M. 55
Man, P.P. 266
Mandelbrot, B.B. 85, 100
Manor, N. 85, 100
Mantegna, R.N. 149
Mantsch, H.H. 220, 222
Manz, J. 56
Marcelja, S. 201
Marlow, I. 242
Marsolais, R.M. 265
Marsh, D. 185, 222
Martin, R.B. 224
Maruani, J. 40, 55, 70, 147, 168
Mason, R. 224
Maynard, R. 117
McAmis, L. 242
McConnell, H.M. 86, 117, 185, 221, 241, 242, 265, 266
McCullough, E.A. 21
McIver, J.V. 21, 70
McLauchlan, K.A. 168
Meadows, M. 220
Meakin, P. 86

Megel, J. 117, 136, 147
Mehring, M. 220, 222
Meir, P. 222, 223
Meirovitch, E. 223
Mendeleev (table) 15
Merrifield, R.E. 136
Meyer, R. 56
Mezey, P.G. 41. 55-57, 69
Miller, A. 242
Miller, J.R. 85
Mills, M. 21
Milnor, J. 56
Mims, W.B. 169
Monberg, E.M. 136
Montroll, E.W. 100, 101
Morokuma, K. 21
Morrison, G.A. 55
Morrow, M. 266
Morse, M. 56, 69
Morsink, J.B.W. 147
Mott, N.F. 117
Muramoto, T. 147, 169
Murrell, J.N. 21, 70
Murugesan, R. 266

Nagaoka, Y. 117
Nagle, J.F. 221, 222
Neugebauer, F.A. 222
Newhouse, J.S. 86
Ngo, F.Q.H. 147
Nguyen, M.T. 21, 22
Nguyen-Dang, T.T. 21
Nichols, C.P. 222
Nicolson, G.L. 220
Niebuhr, H.H. 169
Niederberger, W. 223
Nieman, G.C. 136
Ninham, B.W. 100
Norris, J.R. 147
Norten, A.H. 267
Noyes, R.M. 86

Ochs, F.W. 136
Ohmes, E. 222, 223
Oie, T. 55
Olejniczak, E.T. 223
Oldfield, E. 220, 223, 224
Olivier, J.P. 242
Oppenheim, I. 267
Orbach, R. 100
Ostlund, S. 255
Ovchinnikov, A.A. 85
Overath, P. 223

Packer, K.J. 241
Papamadjopoulos, D. 220
Papon, P. 266
Park, A.H. 185
Parsegian, V.A. 221
Pasher, I. 224
Pearce, P.A. 221
Pearson, R.H. 224
Pee, Ph. 117, 119, 136, 147
Peeters, D. 21
Percival, P.W. 168
Percus (equation) 246
Percus, J.K. 267
Perly, B. 40, 187, 201
Perron, R. 221
Pershan, P.S. 223
Peters, A. 223
Petersen, N.O. 222
Peticolas, W.L. 222
Pfeifer, H. 241
Pfeifer, P. 100
Pines, A. 220
Pink, D.A. 221
Pinto da Silva, P. 221
Pitzer, K.S. 40
Pluschke, G. 223
Poisson (distribution) 94, 96, 97, 99
Pollak, E. 56
Papp, C. 185
Port, H. 117, 136
Portis, A.M. 168
Powell, R.C. 85
Powers, L. 223
Prasad, P.N. 136
Pratt, G.L. 70
Provotorov, B.N. 168
Pulay, P. 21
Pullman, A. 21

Quade, C.M. 40
Quaegebeur, J.P. 187, 201
Quateman, J.H. 267

Rademann. K. 85. 100
Raghavan, R. 117
Rammal, R. 100

Rand, R.P. 231
Rand, S.C. 169
Rao, K.V.S. 201
Rayment, T. 242
Rebiere, Y. 119, 136, 147
Reckinger, G. 21
Redner, S. 86, 100
Ree, F.H. 267
Ree, T. 267
Reeves, L.W. 201
Reisfeld, R. 85, 100
Reman, F.C. 221
Resing, M.A. 241
Reuben, J. 185
Rice, D. 220, 223
Rice, S.A. 265
Richards, P.M. 100, 266
Richards, W.G. 40
Ricklund, R. 117
Riehl, J.W. 241, 266
Rigny, P. 242
Robinson, G.W. 136
Röder, U. 266
Roeder, S.B.W. 222
Romelt, J. 56
Rosen, S. 242
Rosenberg, H.M. 100
Rosenstock, H.B. 85, 100
Ross, D.K. 241
Ross, S. 242
Roux, M. 21
Rubin, R.J. 100
Rund, D. 117
Ruocco, M.J. 222, 223
Rüjpel, D.A. 221
Ryabushkin, O.A. 169

Sackmann, E. 185, 221
Saffman, P.G. 265
Saito, H. 220
Salamon, M.B. 265, 266
Salem, L. 21, 70
Salsbury, N.J. 220
Sana, M. 1, 21, 22
Sander, L.M. 100
Saype, A. 222
Scandella, C.J. 185, 266
Scheuermann, A. 220, 265
Schindler, H. 201
Schlegel, H.B. 21
Schmidt, C.F. 223
Schneider, M.B. 221
Schnur, G. 220, 266
Schöllhorn, R. 266
Schrödinger (equation) 1, 21, 24, 25, 29
Schulz du Bois, E.O. 169
Scott, C.A. 168
Scott, M.L. 221
Seelig, A. 220, 221
Seelig, J. 201, 220, 221, 223
Seigneuret, M. 171, 185
Seiter, C.H.A. 222
Selzer, P.M. 100, 117
Serre, J. 55, 70
Seul, M. 221
Shipley, C.G. 221, 222, 224
Shlesinger, M.F. 85, 100
Sholl, C.A. 266
Sierpinski (gasket) 80-83, 87, 88, 91-99
Silbey, R. 85, 100, 136
Silver, D.M. 21
Siminovitch, D.J. 223
Singer, S.J. 220
Skarjune, R.P. 223, 224
Skidanov, V.A. 169
Small, D.M. 221
Smalley, M. 242
Smeyers, Y.G. 23, 40
Smith, B.A. 221
Smith, I.C.P. 220, 224
Smith, J.M. 85
Smith, R. 223
Smoluchowski (form) 71, 79, 81-83
Sommer, B.S. 136
Soos, Z.G. 85, 265
Spanier, E.H. 69
Spiess, H.W. 221, 222
Springer, T. 241
Srinavasan, R. 266
Stafström, S. 117
Stamatoff, J. 221
Stanley, H.E. 86
Stanton, R.E. 21, 70
Stejskal, E.O. 242
Stephens, M.E. 21
Sternlicht, H. 117, 136
Stockley, C.P. 201
Strenk, L.M. 223
Sttat, B.W. 266
Sturtevant, J.M. 221, 222

Suhai, S. 40
Suna, A. 136
Sundaralingan, M. 223
Swanström, P. 21
Swenberg, C.E. 136

Tabony, J. 225, 241, 242, 266
Tachiba, A. 21, 69
Tajaori, T. 266
Takizawa, O. 266
Tanford, C. 220
Tanner, J.E. 242
Tardieu, A. 220, 221
Tellez, G. 221
Theveneau, M. 266
Thom, R. 21
Thomas, K.M. 224
Thomas, R.K. 241, 242, 267
Thompson, M.J. 222
Thomsen, K.T. 21
Thomy, A. 242
Thorel, P. 241, 267
Thouless, D.J. 117
Throop, G.J. 267
Tol, Y. 21
Tomita, K. 266
Torney, D.C. 86, 241, 265, 266
Tougne, P. 264
Toulouse, G. 100, 117
Toussaint, D. 85
Trahms, L. 203, 222-224, 267
Tratas, T.G. 136
Traüble, H. 185
Trewern, T.D. 241, 242, 267
Trindle, C. 70
Tsien, F. 267
Tudron, F.B. 136

Ueno, M. 266
Urbina, J. 223

Vaks, V.G. 85, 100
Valic, M.I. 222
Valleau, J.P. 267
van Deenen, L.L.M. 221
van der Berg, H.J. 221
van Leuven, P. 70
Varadhan, S.R.S. 85, 100
Vaucamps, C. 136
Vaz, M.J. 223
Veillard, A. 21
Veksli, Z. 220
Venkataramu, S.D. 185
Verkleij, A.J. 221
Verlet, L. 267
Verma, S.P. 220
Ververgaert, P.M.J. 221
Vilfan, M. 266
Villa, M. 241, 266
Villaveces, J.L. 21
Vogel, P.C. 201
Volino, F. 241
Volke, F. 222, 224
von Smoluchowski, M. 85

Waite, T.R. 85
Wallack, D.F.W. 220
Wang, Y.M. 136
Waugh, J.S. 220, 222, 223
Webb, W.W. 221
Webber, S.E. 119
Weeks, G. 222
Weidner, J. 223
Weiss, G.H. 100
Weiss, J.J. 265
Weiss, P.R. 169
Wennerström, H. 201, 222
West, B.J. 85, 100
Westermann, P.W. 223
White, J.W. 241, 242, 267
Wiersma, D.A. 147
Wilczek, F. 85
Wilkinson, A. 222
Williams-Watts (form) 88
Willig, F. 85
Wilson, E.B. 21
Winterhalter, M. 242, 266
Witschel, J. 242
Wittebort, R.J. 220, 223
Witten, T. 100
Wohlgemuth, R. 223
Wokaun, A. 169
Wolf, E.L. 168
Wolf, H.C. 117
Wong, T.C. 222
Wu, E.S. 220

Yamane, T. 221
Yamashita, K. 69
Yeagle, P.L. 224
Yellin, N. 222
Yen, W.M. 100, 117

Yevick (equation) 248
Yevick, G.J. 267
Yonezawa, F. 117

Zaccai, G. 221
Zamaraev, K.I. 85
Zeldovich, Ya.B. 85
Zemb, T. 40
Zewail, A.H. 117
Zumer, S. 266
Zumofen, G. 71, 85, 87
Zviadadze, M.D. 168

Batch number: 10584469

Printed by Printforce, the Netherlands